Chapman & Hall/CRC
Studies in Informatics Series

STOCHASTIC RELATIONS

Foundations for Markov Transition Systems

Chapman & Hall/CRC
Studies in Informatics Series

SERIES EDITOR

G. Q. Zhang
Case Western Reserve University
Department of EECS
Cleveland, Ohio, U.S.A

PUBLISHED TITLES

Stochastic Relations: Foundations for Markov Transition Systems
Ernst-Erich Doberkat

FORTHCOMING TITLES

Conceptual Structures in Practice
Pascal Hitzler and Henrik Scharfe

Chapman & Hall/CRC
Studies in Informatics Series

STOCHASTIC RELATIONS

Foundations for Markov Transition Systems

Ernst-Erich Doberkat

CRC Press
Taylor & Francis Group
Boca Raton London New York

CRC Press is an imprint of the
Taylor & Francis Group, an **informa** business

A CHAPMAN & HALL BOOK

CRC Press
Taylor & Francis Group
6000 Broken Sound Parkway NW, Suite 300
Boca Raton, FL 33487-2742

First issued in paperback 2019

ISBN-13: 978-1-58488-941-0 (hbk)
ISBN-13: 978-0-367-38911-6 (pbk)

Library of Congress Cataloging-in-Publication Data

Doberkat, Ernst-Erich.
    Stochastic relations : foundations for Markov transition systems / Ernst-Erich Doberkat.
        p. cm. -- (Chapman & hall/crc studies in informatics series)
    Includes bibliographical references and index.
    ISBN-13: 978-1-58488-941-0 (alk. paper)
    ISBN-10: 1-58488-941-1 (alk. paper)
    1. Computer science--Mathematics. 2. Stochastic processes. 3. Markov processes. I. Title. II. Series.

QA76.9.M35D578 2007
004.01'51--dc22   2007004444

Für Gudrun, Julia und Thomas.

Contents

Preface

This book develops the theory of stochastic relations as a foundation for Markov transition systems. Central topics such as congruences and morphisms are investigated and applied to the monoidal structure. Bisimilarity and behavioral equivalence are defined and investigated within this framework; developments from the general theory of coalgebras are viewed from the context provided by the subprobability functor. It is shown with these tools that bisimilarity, behavioral and logical equivalence are the same for general modal logics and for continuous time stochastic logic with and without fixed point operator.

The book starts with an extensive and gentle introduction to the basic mathematical tools from topology, measure theory and categories.

Motivation

A coalgebra for the endofunctor $\mathfrak{F} : \mathfrak{C} \to \mathfrak{C}$ is a pair $\langle x, t \rangle$, where $l : x \to \mathfrak{F}(x)$ is a morphism in category $\mathfrak{C}$. The study of coalgebras provides many interesting vistas into the landscape of (theoretical) computer science and algebra, as can be witnessed for example from the survey paper (Rutten, 2000). Particularly interesting are the connections to modal logics (Blackburn et al., 2001), in which the power set functor $\mathfrak{Pow}$ in the category of sets plays a prominent rôle. For example, a coalgebra $\langle x, t \rangle$ for $\mathfrak{Pow}$ may be identified with a relation on set x.

Consider this example. A formula in a simple modal logic with A and AP as set of actions resp. atomic propositions is defined recursively through

$$\phi ::= \top \mid p \mid \phi' \wedge \phi'' \mid \langle a \rangle \phi'$$

Thus $\top$ is a formula indicating *truth*, each atomic proposition $p \in AP$ is a formula, the conjunctions of two formulas is one, and whenever we have a formula, then prefixing it with $\langle a \rangle$ for an action $a \in A$ will yield a formula again. The intuitive meaning of $\langle a \rangle \phi$ is "it is *possible* that ϕ holds after action a" (just like the diamond is interpreted in ordinary modal logic as indicating possibility). An interpretation will take a set S of states, assign to each action $a \in A$ a relation $R_a \subseteq S \times S$, and to each atomic proposition $p \in AP$ a subset $L(p) \subseteq S$ of states. $L(p)$ indicates the set of states in which p is valid. The underlying Kripke model $\mathcal{K} = (S, AP, (R_a)_{a \in A})$ holds these data as a container and is used for defining the semantics, which is done recursively through ($p \in AP, a \in A$, writing R_a as $\rightarrow_a$):

$$\mathcal{K}, s \models \top \Leftrightarrow s \in S$$
$$\mathcal{K}, s \models p \Leftrightarrow s \in L(p)$$
$$\mathcal{K}, s \models \phi' \wedge \phi'' \Leftrightarrow \mathcal{K}, s \models \phi' \text{ and } \mathcal{K}, s \models \phi''$$
$$\mathcal{K}, s \models \langle a \rangle \phi \Leftrightarrow \mathcal{K}, s' \models \phi \text{ for some } s' \text{ with } s \rightarrow_a s'.$$

Thus we have $\mathcal{K}, s \models \langle a \rangle \phi$ iff we can find a $\rightarrow_a$-successor s' to s so that $\mathcal{K}, s' \models \phi$.

In terms of coalgebras, this Kripke structure can be seen essentially as a coalgebra $\langle S, t \rangle$ for the functor that maps each set X to $\mathfrak{Pow}\,(A \times X)$.

Now look at this: We have a system composed of two processors which may fail, but which may be repaired; the processors work independently and in parallel.

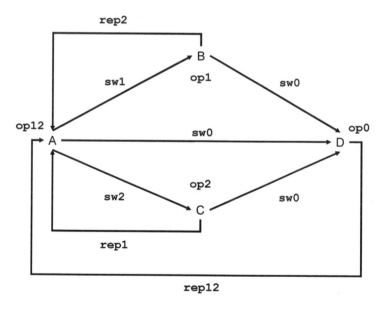

The atomic propositions tell us which processor works, so that

$$AP = \{\mathsf{op0}, \mathsf{op1}, \mathsf{op2}, \mathsf{op12}\},$$

the actions either indicate the failure of a processor, saying which processor survives (e.g., if both processors are operational, action $\mathsf{sw1}$ indicates that processor 2 fails and processor 1 survives; if at least one processor operates, action $\mathsf{sw0}$ has the effect that all processors fail), or indicate a repair action, telling us which processor has to be repaired, hence

$$A = \{\mathsf{sw2}, \mathsf{sw1}, \mathsf{sw0}, \mathsf{rep1}, \mathsf{rep2}, \mathsf{rep12}\}.$$

The figure gives a Kripke structure; each arrow has an action as a label, the label indicates, too, to which relation the pair belongs.

In state A, both processors are operational; in state D, none is. State B has system 2 fail, etc. The map L is evident from the figure as well.

Put

$$\phi_1 := \langle \mathtt{rep12} \rangle \langle \mathtt{sw2} \rangle \langle \mathtt{sw0} \rangle \mathtt{op0},$$
$$\phi_2 := \langle \mathtt{rep12} \rangle \langle \mathtt{sw1} \rangle \langle \mathtt{sw0} \rangle \mathtt{op0},$$

then it is easy to see that both, e.g., $D \models \phi_1$ and $D \models \phi_2$, hold; so state D cannot distinguish between these formulas. So it is not really important whether processor 1 fails, or processor 2. But probably it is.

Assume that probabilities are attached to the state transitions, as indicated in the figure below (to be precise, sub-probabilities, because the values do not add up to unity). Computing the respective probabilities, it is clear that the probability for D accepting ϕ_1 is $4, 5 \cdot 10^{-2}$, whereas for ϕ_2 it is $18 \cdot 10^{-2}$, thus D can distinguish ϕ_1 and ϕ_2 now on the, say, 10% level. Thus probabilities add to a more precise understanding of this system.

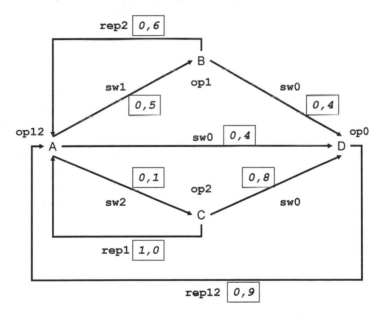

Addressing the question what kind of model might be adequate for probabilistic modeling, one wants to pursue a development that runs in parallel to the coalgebraic one above and considers coalgebras for the probability functor $\mathfrak{P}$. Thus a coalgebra for the functor $S \mapsto \mathfrak{Pow}(A \times S)$ would be replaced by a coalgebra for the functor $S \mapsto \mathfrak{P}(A \times S)$ where $\mathfrak{P}(A \times S)$ is the set of all probabilities over the set $A \times S$. Though straightforward, this approach raises some questions and needs to be refined. First, we have assumed implicitly that the sets we are dealing with are finite. If A and S are finite, so are $\mathfrak{Pow}\,(A \times S)$, but $\mathfrak{P}\,(A \times S)$ is an uncountable set as soon as $A \times S$ contains more than one element; so we leave the realm of finite sets with this construction. Thus it is conceptually difficult to iterate this construction, even for finite base sets; this, however, may be necessary when discussing monads. Moreover, some applications are intrinsically based on nonfinite sets: consider, e.g., a continuous time logic, where the residence times for the states may be nonnegative real numbers. Second, when the universe may no longer assumed to be finite or countable, it may be difficult assigning positive probabilities to single elements; subsets are a more adequate domain.

But we know that we end up in considerable foundational difficulties when assuming that we assign a probability to each subset of an arbitrary set (El-strodt, 1999; Wagon, 1981). Thus we need a structured subset of the power set as the domain for the probabilities, the structure being, as a student of measure theory knows, a σ-algebra. So we are poised to consider coalgebras for the functor that no longer takes an arbitrary set X but rather a measurable space $(X, \mathcal{A})$ and assigns it all probabilities $\mathfrak{P}\,(A \times X, \mathcal{H} \otimes \mathcal{A})$ on the measurable space $(A \times X, \mathcal{H} \otimes \mathcal{A})$ (since we assume that the actions are coming from a measurable space $(A, \mathcal{H})$ as well). Note that this switch entails changing the base category from the category of all sets with maps to the category of measurable spaces with measurable maps as morphisms.

But this is not yet enough. We want to model properties for these coalgebras that permit sensible applications like the study of bisimulations of various sorts or modeling probabilistically infinite paths in a logic for reactive systems. This is nearly hopeless to do in general measurable spaces, because these spaces are not rich enough for supporting an interesting probabilistic structure. It is well known that Polish spaces, i.e., topological spaces that have a countable dense subset and for which a complete metric exists, provide enough support for the probability measures defined on their Borel sets (the smallest σ-algebra containing the open subsets) to permit the kind of constructions that we need. Examples of Polish spaces are countable discrete spaces (with the discrete topology), the real numbers $\mathbb{R}$, open or closed subsets of some Euclidean space, and even the measures on a Polish space under the weak topology; Polish spaces are closed under countable sums and products, and the general topological structure of Polish spaces has long been known very well. Thus if we have a Polish space S, e.g., the space $\prod_{i \in \mathbb{N}}(\mathbb{R}_+ \times S)$ of all infinite paths with timing information, is Polish. Consequently we will work in the base category of all Polish spaces with Borel maps as morphisms, occasionally assuming continuity for the morphisms. Sometimes we will be able to extend the results to analytic spaces, so we can even go a step further and consider for most applications analytic spaces, hence measurable spaces that are the Borel images of Polish spaces.

Let us briefly return to the basic configuration of a coalgebra $\langle x, t \rangle$ with $t : x \to \mathfrak{F}\,(x)$. Here the domain x for the dynamic t coincides with the domain for the functor $\mathfrak{F}$. This is sometimes an uneasy restriction, so it is sometimes more adequate to work in a scenario that would look like $t : x \to \mathfrak{F}(y)$. Separating the domain of the morphism t from the functor's domain gives rise to a finer mode of description. For example a morphism $g : \langle x, t \rangle \to \langle y, s \rangle$ in the coalgebraic case is a morphism $g : x \to y$ with $s \circ g = \mathfrak{F}(g) \circ t$, making this diagram commutative.

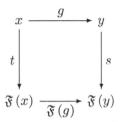

The extended case requires a morphism $\langle x, y, t \rangle \to \langle a, b, s \rangle$ to be a pair $\langle g, h \rangle$ of morphisms $g : x \to a$ and $h : y \to b$ with $g \circ t = \mathfrak{F}(h) \circ s$. This makes the diagram

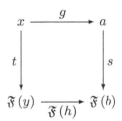

commutative. This observation permits separating the concerns of the domain from the codomain, which will be helpful in understanding some phenomena, as we will see.

Another instance where this separation of concerns pays is the description of congruences. In the coalgebraic case a congruence on $\langle x, t \rangle$ is essentially an equivalence relation on the carrier x of the coalgebra that is compatible both with the dynamics t and the functor $\mathfrak{F}$. In the extended case we deal with a pair of equivalence relations (α, β) that has to satisfy compatibility conditions with respect to both the dynamics and the functor. This separation will be of advantage in many places, for example when discussing Kripke models.

To summarize, we will discuss here morphisms of the kind $x \to \mathfrak{F}(y)$ with $\mathfrak{F}$ as the probability functor or one of its close relatives, defined usually on the category of Polish spaces. Again using the analogy to the power set functor we will see the objects we are dealing with as relations, albeit as stochastic ones. What we discuss will be outlined next.

Overview

We give a brief overview of the chapters' contents, the graph indicating the dependencies between the sections[1].

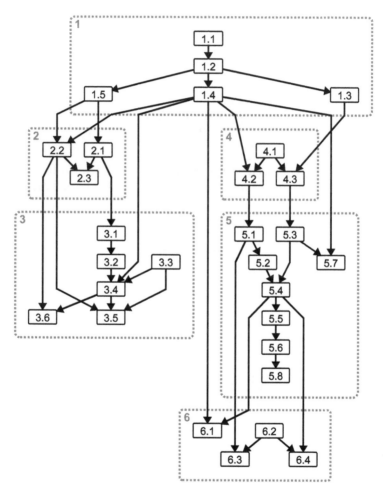

Chapter 1   The reader is assumed to be familiar with the basic properties of topological spaces, and we assume that the very fundamental constructions in probability theory are known up to and including the Radon-Nikodym Theorem. This is not sufficient, however, for the purposes of exploring Markov transition systems in depth. It becomes clear rather rapidly that Polish and

---

[1]The   book's   web   site   can   be   found   at   the   address http://ls10-www.cs.uni-dortmund.de/momts.

analytic spaces are needed for meaningful constructions: the former, because they balance generality with applicability, the latter because they permit factorizations (which Polish spaces do not). The reader finds an introduction to these spaces here, together with a careful study of Borel sets, culminating in Souslin's Theorem, a classic that will turn out to be most helpful. From a measure theoretic point of view it is necessary to know a wee bit about the topology of weak convergence in the space of all subprobabilities over a Polish space, and to know about disintegration of measures on products. And, yes, sometimes one can establish the existence of a badly needed object through a selection argument, which then yields a completely new and unexpected insight for Markov transition systems as well. This is why a brief introduction to measurable selectors is provided as well.

Algebraically, the reader should have some basic understanding of category theory; adjunctions, monads and Eilenberg-Moore algebras will be discussed here.

We have made an attempt to make this book self contained. Hence material that is deemed necessary for the mastery of the material outlined here but that is scattered in the vast literature and somewhat difficult to find in one place has been collected and compiled.

Chapter 2 This chapter will investigate categorial and probabilistic properties of stochastic relations. We take the properties of set-theoretic relations as a firm guide by looking at the power set functor on the category of sets with maps as morphisms. This functor assigns each set its power set, and it is the functorial part of the Manes monad (Section 2.2). In parallel, we define the functor on measurable spaces that assigns each measurable space all its sub-probabilities. Under a suitable chosen σ-algebra, this is a measurable space again, so the functor is an endofunctor, and it can be seen as well as the functorial part of a monad, the Giry monad. It is shown that this is actually a monad; all this is done in Section 2.3. These Sections define and investigate the Kleisli construction for each monad, fairly basic for the discussions to follow. Since both Kleisli constructions have similar properties, we investigate in the case study in Section 2.4 how these similarities can be exploited. A simple software architecture is modeled through a monad that enjoys a tensorial strength operator (both the Giry monad and the Manes monad belong to this family).

Chapter 3 The investigation of the subprobability functor and the Giry monad would not be complete without having a look at the algebras that are associated with the monad. Mac Lane argues that the Kleisli construction and the Eilenberg-Moore algebras live at opposite ends of the spectrum of adjoint pairs of those functors which define a given monad; thus, having investigated the Kleisli construction, it is challenging to explore the other end. We identify the algebras for the Giry monad through positive convex structures in Section 3.4; some illustrating examples are given in Section 3.5.

As always, the free algebra $\langle \mathfrak{S}(X), \mathfrak{m}_X \rangle$ is an Eilenberg-Moore algebra for each Polish space X, where $\mathfrak{S}$ is the subprobability functor and $\mathfrak{m}$ is the multiplication for the Giry monad. This example helps to identify in Section 3.6 the left adjoint to the forgetful functor that associates with each algebra the associated Polish space.

This chapter lives in the category of Polish spaces with continuous maps. Transferring the results to the more general category of Polish spaces with Borel measurable maps (or even to the category of analytic spaces with Borel maps) is an open question.

Chapter 4 Many constructions in the theory of coalgebras depend on the assumption that the functor involved has at least weak pullbacks, see, e.g., Rutten's survey (Rutten, 2000) for an overview. For practical purposes like the study of bisimulations it would be good to know whether or not the subprobability functor has a weak pullback as well. These hopes are shattered rather quickly, so what one wants to look for is the existence of semi-pullbacks (thus each co-span of morphisms has a span, so that a commutative diagram is created). By formulating this problem as a selection problem and using the theory of measurable selectors, it is shown that in Polish spaces semi-pullbacks do exist. This solution can even be carried over to analytic spaces. Most work for a solution to this problem in Section 4.4 concerns actually a measure extension problem — in probabilistic terms, a distribution on a sub-σ-algebra of the Borel sets has to be extended to the full Borel sets; see Section 4.3. The problem is tackled using tools from classical analysis, based partially on the axiom of choice.

Chapter 5 Stochastic interpretations of modal logic show that equivalence relations are of interest that are countably generated, or, what amounts to the same, that are represented as the kernel of a Borel map. We study these relations that are called smooth in Section 5.2. The interest in these relations stems in part from the fact that factoring an analytic space with a smooth relation yields an analytic space again; this is well known (Srivastava, 1998; Kechris, 1994; Arveson, 1976). We stress in particular the rôle of invariant Borel sets. They are important for two reasons: first, the inverse images of the Borel sets on the factor space under the factor map are just the invariant Borel sets. Second, this σ-algebra determines the equivalence relation uniquely (an observation that will be capitalized upon later, in Section 6.4). These properties are investigated in greater detail in Section 5.2; they are being made use of heavily for understanding congruences for stochastic relations. A pair (α, β) of smooth equivalence relations is called a congruence for a stochastic relation iff objects that cannot be separated through α and β cannot be separated through the relation; see Section 5.3. This section studies algebraic properties of congruences as well. An Isomorphism Theorem quite akin to the one in classical algebra can be established, providing the proper algebraic context.

Congruences are right at the heart of bisimulations. This is not really evident at first: bisimulations are introduced in Section 5.4 essentially through spans of morphisms. This is a rather fruitful notion for investigating modal logics in Chapter 6, but we have a look at immediate properties first, deriving a criterion for two stochastic relations to be bisimilar. Here we employ smooth equivalence relations and simulation equivalent congruences — roughly, the relation may simulate each other. This leads to an intrinsic condition for bisimilarity: one looks only at the relations and does not have another, external, institution like a logic that assists in the decision. It has the interesting consequence that two relations are bisimilar provided they have isomorphic factor spaces. This criterion is also sufficient for compact metric spaces. The question remains open, however, whether or not this condition is sufficient in the general Polish or analytic case. While bisimilarity requires a span, behavioral equivalence of stochastic relations is defined through the existence of a cospan. It is shown that simulation equivalent congruences permit the construction of such a cospan through factoring.

Thus, in order to investigate bisimilarity of stochastic relations or for finding out about their behavioral equivalence, it is helpful to find congruences and to show that they are simulation equivalent. This will then permit constructing a span or a cospan of morphisms. The span will be constructed through a semi-pullback as in Section 4.4; the cospan will be constructed through factoring as in Section 5.4.

A case study shows that bisimilarity does not break easily: we show that forming the converse of a stochastic relation — an interesting problem in itself — respects bisimilarity. If two relations are bisimilar, then their converses are as well; the properties of the converse relation are also explored, see Section 5.8.

Bisimilarity may be specialized by taking projections as the morphisms involved; this is introduced in Section 5.6 and leads to the notion of 2-bisimulation. Here the connections between bisimulations and congruences becomes fully visible: we show that each congruence can be used as the basis for a 2-bisimulation on a stochastic relation. This fairly deep property is reformulated as a selection problem and then solved through a measurable selector. Once we know that, we are in a good position to tackle simple stochastic relations. These are those relations that have no nontrivial subsystems. It can be shown that they are completely characterized through injective Borel maps, providing a rather easy criterion for recognizing them (when encountering them in the street, say). As a consequence one derives that the subprobability functor does not have a final system save for the case of proper probabilistic relations.

The interplay between bisimulations and simple systems is used in the theory of coalgebras for a calculus of coinduction, see (Rutten, 2000; Rutten,

2002; Arbab and Rutten, 2002). Given the very simple structure of simple systems for the functor considered here, such an application does not seem to be realistic. Section 5.9 shows, however, that not all is lost. We indicate in this case study that the knowledge of simple systems permits at least some transfer results between discrete and continuous probability spaces when analyzing algorithms.

Chapter 6 This final chapter is devoted to stochastic interpretations of modal and continuous time logics. We extend the usual notion of modal logics for probabilistic purposes (incorporating into the language the notion of a probability with which a formula should be satisfied). Stochastic relations are required here in full generality. The interpretation needs not be confined to labeled Markov transition systems; rather, a treatment of general Kripke models becomes feasible. This is proposed in Section 6.2. The relationship between stochastic Kripke models and those based on set theoretic relations is investigated in this section as well, where we capitalize on the support function for rendering a nondeterministic Kripke model from a stochastic one. The relationship between bisimulations and the Hennessy-Milner equivalence relation is scrutinized. It is shown how the characterization of bisimilar stochastic relations can be used as a basis for establishing that bisimilarity behavioral and logical equivalence are really the same.

Whereas modal logic deals with finite paths, continuous stochastic logic is used to model reactive systems; hence infinite paths have to be taken into account. These paths are usually written down as sequences alternating between states the system is in and residence times for indicating how long the system remains in this state before a state change occurs. Probabilistically, this is modeled through the projective limit of a process in which state changes and residence times are stochastically independent (this refers to one step in the system); see Section 6.3. The logic, dubbed **CSL**, distinguishes state formulas, which display some sort of local behavior, from path formulas, which entertain properties that are manifested in the long run. We discuss in Section 6.4 another kind of bisimilarity: call two states F-bisimilar iff they satisfy exactly the same formulas from a given set F of state formulas. Because F is at most countable, this relation is smooth; hence it gives rise to a congruence on the interpreting relation. And here we are again: we can use the invariant sets of this smooth equivalence to determine sets G for which F-bisimilarity and G-bisimilarity are identical. One wants these sets G of course to be as large as possible for maximizing the effect with minimal resources. It will help solving the problem of deciding whether for the set AP of atomic propositions AP-bisimilarity is equal to $\mathfrak{L}_{AP}$-bisimilarity, where $\mathfrak{L}_{AP}$ is the set of all formulas.

It is clear that this question is motivated through practical considerations: If the answer would be in the positive, one would have only to test the atomic propositions in order to make statements regarding the entire set of formulas.

These questions are investigated in Section 6.4; unfortunately, there is no clear-cut, simple answer: as usual, it depends, in this case on the invariant sets. We define in this section the extension of a set F of formulas as the set of all formulas that have the same invariant sets as F and investigate an equivalence result involving 2-bisimulations.

The problem of bisimilarity, logical and behavioral equivalence is discussed again for the logic μ**CSL**, a variant to **CSL** that also has the mu-operator. Bisimilarity is related to behavioral and logical equivalence essentially through properties that are determined by theories for states and paths. It turns out that the functor which assigns to each state space the space of infinite paths of residence times and states plays a rather decisive rôle, albeit indirectly, hinting at a more general picture that may be painted through coalgebraic logic.

Acknowledgments

Most of the work is based on the author's research over the last couple of years, and it has been shaped by comments of many people, some of whom I do not even know, because they were the anonymous referees for conferences and journals. Prakash Panangaden and Josèe Desharnais provided some helpful and constructive insights. Michail Jerschov and Dieter Pumplün kindly offered comments and suggestions at crucial points in the development of all this. The DEUTSCHE FORSCHUNGSGEMEINSCHAFT funded part of the research leading to this work through grants for the project *Algebraische Eigenschaften stochastischer Relationen*. Writing the first draft of this book was done mostly during a sabbatical, part of which I was fortunate enough to spend in the *colline pisane* in the congenial atmosphere of Fritz Hans Kaltenbach's FATTORIA PUGNANO. I was permitted access to the impressive Computer Science library of the University of Pisa, which was made possible by Professor Carlo Montangero. Alla Stankjawitschene, my secretary, provided help and support in many things. Helmut Henning has patiently given me advice whenever I ran into problems with typesetting. Georgios Lajios, Ignacio Viglizzo and Sibylle Hess offered comments that did improve the representation, and I profited from discussions with Robin Cockett, José Meseguer, Larry Moss, Peter Padawitz and Ana Sokolova. Petra Mutzel helped drawing the dependency graph. Stefan Dißmann's superb organizational skills made many things easier for me. The cooperation with Randi Cohen from Chapman & Hall was excellent, helpful and efficient. I want to thank them all.

Ernst-Erich Doberkat
Bochum and Dortmund

Chapter 1

A Gentle Tutorial to All Things Considered

1.1 Introduction

The study of Markov transition systems and stochastic relations as their mathematical foundation requires some familiarity with concepts from topology and measure theory when one wants to investigate phenomena that go beyond elementary observations. For example, when discussing behavioral equivalence, one wants to factor the state space of the transition system, and encounters then the problem that one has to determine the structure of the factor space. Thus one is led to an analysis of an analytic space. Similarly, when looking for the converse of a stochastic relation one is all of a sudden confronted with problems of disintegration of a measure on a product space.

The mathematical tools for this enterprise are coming from topology and measure theory. They are somewhat scattered in the literature, and, as far as I could see, not available in a single place. Hence one has to hunt for properties of, say, analytic spaces in one place while properties of measures can be found in another source, and where do I learn about the existence or the nonexistence of the right inverse for a measurable map?

So I tried to provide a single source for all the crucial topological and measure theoretic properties that occurred to me as being important for working in this area. Of course, one has to start somewhere. I assume that the reader is familiar with the basic notions from topology (like open sets, continuity and the like, most likely from discussing the real line) and from measure theory. Here I assume some familiarity with the concept of Lebesgue integration which is usually given in an advanced course on Calculus or an introductory course to Probability Theory. Between the continents of topology and mea-

1

sure theory lies the ocean of Borel sets; I will make also some results from the theory of Borel sets available which are used here but which are to be found only in more specialized treatises. This includes in particular some glances at the theory of measurable selectors which permit to establish some existential statements that are difficult to obtain otherwise.

The final part of this overview deals with categories, in particular with constructions related to monads and to Eilenberg-Moore algebras through adjunctions. Since categories are somewhat more familiar in Theoretical Computer Science than, say, Polish spaces, the representation is a little bit more sketchy for categories than for the topics related to measures and all that. I did make an attempt, however, to keep things self contained (for example, only Yoneda's Lemma is really required).

1.2   Measurable Spaces

A measurable space $(M, \mathcal{M})$ consists of a set M with a σ-algebra $\mathcal{M}$, which is an algebra of subsets of M that is closed under countable unions (hence countable intersections or countable disjoint unions). If $\mathcal{M}_0$ is a family of subsets of M, then

$$\sigma\left(\mathcal{M}_0\right) = \bigcap\{\mathcal{M} \mid \mathcal{M} \text{ is a } \sigma\text{-algebra on } M \text{ with } \mathcal{M}_0 \subseteq \mathcal{M}\}$$

is the smallest σ-algebra on M which contains $\mathcal{M}_0$. This construction works since the power set $\mathcal{P}(M)$ is a σ-algebra on M. Take for example as a generator $\mathcal{I}$ all open intervals in the real numbers $\mathbb{R}$, then $\sigma(\mathcal{I}) =: \mathcal{B}(\mathbb{R})$ is the σ-algebra of real *Borel sets*. We will encounter the Borel sets again in Section 1.3.

An important tool is the π-λ-Theorem which makes it sometimes simpler to identify the σ-algebra generated from some family of sets.

THEOREM 1.1

(π-λ-**Theorem**) *Let $\mathcal{P}$ be a family of subsets of a set X that is closed under finite intersections (a π-class). Then $\sigma(\mathcal{P})$ is the smallest λ-class containing $\mathcal{P}$, where a family $\mathcal{L}$ of subsets of X is called a λ-class iff it is closed under complements and countable disjoint unions.*

PROOF   1. Let $\mathcal{L}$ be the smallest λ-class containing P, then we show that $\mathcal{L}$ is a σ-algebra.

2. We show first that it is an algebra. Being a λ-class, $\mathcal{L}$ is closed under complementation. Let $A \subseteq X$, then $\mathcal{L}_A := \{B \subseteq X \mid A \cap B \in \mathcal{L}\}$ is a λ-class again: if $A \cap B \in \mathcal{L}$, then

$$A \cap (X \setminus B) = A \setminus B = X \setminus ((A \cap B) \cup (X \setminus A)),$$

which is in $\mathcal{L}$, since $(A \cap B) \cap X \setminus A = \emptyset$, and since $\mathcal{L}$ is closed under disjoint unions.

If $A \in \mathcal{P}$, then $\mathcal{P} \subseteq \mathcal{L}_A$, because $\mathcal{P}$ is closed under intersections. Because $\mathcal{L}_A$ is a λ-system, this implies $\mathcal{L} \subseteq \mathcal{L}_A$ for all $A \in \mathcal{P}$. Now take $B \in \mathcal{L}$, then the preceding argument shows that $\mathcal{P} \subseteq \mathcal{L}_B$, and again we may conclude that $\mathcal{L} \subseteq \mathcal{L}_B$. Thus we have shown that $A \cap B \in \mathcal{L}$, provided $A, B \in \mathcal{L}$, so that $\mathcal{L}$ is closed under finite intersections. Thus $\mathcal{L}$ is a Boolean algebra.

3. $\mathcal{L}$ is a σ-algebra as well. It is enough to show that $\mathcal{L}$ is closed under countable unions. But since

$$\bigcup_{n \in \mathbb{N}} A_n = \bigcup_{n \in \mathbb{N}} \left(A_n \setminus \bigcup_{i=1}^{n-1} A_i \right),$$

this follows immediately.   □

If $(N, \mathcal{N})$ is another measurable space, then a map $f : M \to N$ is called *$\mathcal{M}$-$\mathcal{N}$-measurable* iff the inverse image under f of each set in $\mathcal{N}$ is a member of $\mathcal{M}$, hence iff $f^{-1}[G] \in \mathcal{M}$ holds for all $G \in \mathcal{N}$.

Checking measurability is made easier by the observation that it suffices for the inverse images of a generator to be measurable sets.

LEMMA 1.2
Let $(M, \mathcal{M})$ and $(N, \mathcal{N})$ be measurable spaces, and assume that $\mathcal{N} = \sigma(\mathcal{N}_0)$ is generated by a family $\mathcal{N}_0$ of subsets of N. Then $f : M \to N$ is $\mathcal{M}$-$\mathcal{N}$-measurable iff $f^{-1}[G] \in \mathcal{M}$ holds for all $G \in \mathcal{N}_0$.

PROOF   Clearly, if f is $\mathcal{M}$-$\mathcal{N}$-measurable, then $f^{-1}[G] \in \mathcal{M}$ holds for all $G \in \mathcal{N}_0$.

Conversely, suppose $f^{-1}[G] \in \mathcal{M}$ holds for all $G \in \mathcal{N}_0$, then we need to show that $f^{-1}[G] \in \mathcal{M}$ for all $G \in \mathcal{N}$. In fact, consider the set $\mathcal{G}$ for which the assertion is true,

$$\mathcal{G} := \{G \in \mathcal{N} \mid f^{-1}[G] \in \mathcal{M}\}.$$

An elementary calculation shows that the empty set and N are both members of $\mathcal{G}$, and since $f^{-1}[N \setminus G] = M \setminus f^{-1}[G]$, $\mathcal{G}$ is closed under complementation. Because

$$f^{-1}\left[\bigcup_{i \in I} G_i \right] = \bigcup_{i \in I} f^{-1}[G_i]$$

holds for any index set I, $\mathcal{G}$ is closed under finite and countable unions. Thus $\mathcal{G}$ is a σ-algebra, so that $\sigma(\mathcal{G}) = \mathcal{G}$ holds. By assumption, $\mathcal{N}_0 \subseteq \mathcal{G}$, so that

$$\mathcal{M} = \sigma(\mathcal{N}_0) \subseteq \sigma(\mathcal{G}) = \mathcal{G} \subseteq \mathcal{M}$$

is inferred. Thus all elements of $\mathcal{N}$ have their inverse image in $\mathcal{M}$. ▯

An example is furnished by a real valued function $f : M \to \mathbb{R}$ on M which is $\mathcal{M}$-$\mathcal{B}(\mathbb{R})$-measurable iff $\{m \in M \mid f(m) \bowtie t\} \in \mathcal{M}$ holds for each $t \in \mathbb{R}$; the relation $\bowtie$ may be taken from $<, \leq, \geq, >$. This observation will be used frequently.

The proof's technique deserves some attention as well. The strategy is that we have a look at all objects that have the desired property, and that we show that this set of good guys is a σ-algebra. This is similar to showing in a proof by induction that the set of all natural numbers having a certain property is closed under constructing the successor. Then we show that the generator of the σ-algebra is contained in the good guys, which is rather similar to establishing the start of the induction. Taking both steps together then yields the desired properties for the induction case as well as for the case of σ-algebras. We will encounter this pattern of proof over and over again.

If $(M, \mathcal{M})$ is a measurable space and $f : M \to N$ is a map, then

$$\mathcal{N} := \{D \subseteq N \mid f^{-1}[D] \in \mathcal{M}\}$$

is the largest σ-algebra $\mathcal{N}_0$ on N that renders f $\mathcal{M}$-$\mathcal{N}_0$-measurable ($\mathcal{N}$ is the *final* σ-algebra w.r.t. f). In fact, because the inverse set operator f^{-1} is compatible with the Boolean operations, it is immediate that $\mathcal{N}$ is closed under the operations for a σ-algebra, and a little moment's reflection shows that this is also the largest σ-algebra with this property.

Symmetrically, if $g : P \to M$ is a map, then

$$g^{-1}[\mathcal{M}] := \{g^{-1}[E] \mid E \in \mathcal{M}\}$$

is the smallest σ-algebra $\mathcal{P}_0$ on P that renders $g : \mathcal{P}_0 \to \mathcal{M}$ measurable (accordingly, $g^{-1}[\mathcal{M}]$ is called *initial* w.r.t. f). Similarly, $g^{-1}[\mathcal{M}]$ is a σ-algebra, and it is fairly clear that this is the smallest one with the desired property. In particular, the inclusion $i_Q : Q \to M$ becomes measurable for a subset $Q \subseteq M$ when Q is endowed with the σ-algebra $\{Q \cap B \mid B \in \mathcal{M}\}$. It is called the *trace of $\mathcal{M}$ on Q* and is denoted — in a slight abuse of notation — by $\mathcal{M} \cap Q$.

Initial and final σ-algebras generalize in an obvious way to families of maps. For example, $\sigma\left(\bigcup_{i \in I} g_i^{-1}[\mathcal{M}_i]\right)$ is the smallest σ-algebra $\mathcal{P}_0$ on P which makes all the maps $g_i : P \to M_i$ $\mathcal{P}_0$-$\mathcal{M}_i$-measurable for a family $((M_i, \mathcal{M}_i))_{i \in I}$ of measurable spaces.

This is an intrinsic, universal characterization of the initial σ-algebra for a single map.

LEMMA 1.3

Let $(M, \mathcal{M})$ be a measurable space and $f : M \to N$ be a map. The following conditions are equivalent:

a. *The σ-algebra $\mathcal{N}$ on N is final with respect to f.*

b. *If $(P, \mathcal{P})$ is a measurable space, and $g : N \to P$ is a map, then the $\mathcal{M}$-$\mathcal{P}$-measurability of $g \circ f$ implies the $\mathcal{N}$-$\mathcal{P}$-measurability of g.*

PROOF   1. Taking care of $a \Rightarrow b$, we note that

$$(g \circ f)^{-1} [\mathcal{P}] = f^{-1} \left[g^{-1} [\mathcal{P}] \right] \subseteq \mathcal{M}.$$

Consequently, $g^{-1} [\mathcal{P}]$ is one of the σ-algebras $\mathcal{N}_0$ with $f^{-1} [\mathcal{N}_0] \subseteq \mathcal{M}$. Since $\mathcal{N}$ is the largest of them, we have $g^{-1} [\mathcal{P}] \subseteq \mathcal{N}$. Hence g is $\mathcal{N}$-$\mathcal{P}$-measurable.

     2. In order to establish $b \Rightarrow a$, we have to show that $\mathcal{N}_0 \subseteq \mathcal{N}$ whenever $\mathcal{N}_0$ is a σ-algebra on $\mathcal{N}$ with $f^{-1} [\mathcal{N}_0] \subseteq \mathcal{M}$. Put $(P, \mathcal{P}) := (N, \mathcal{N}_0)$, and let g be the identity id_N. Because $f^{-1} [\mathcal{N}_0] \subseteq \mathcal{M}$, we see that $id_N \circ f$ is $\mathcal{N}_0$-$\mathcal{M}$-measurable. Thus id_N is $\mathcal{N}$-$\mathcal{N}_0$-measurable. But this means $\mathcal{N}_0 \subseteq \mathcal{N}$.
$\square$

We will use the final σ-algebra mainly for factoring through an equivalence relation. In fact, let α be an equivalence relation on a set X, where $(X, \mathcal{M})$ is a measurable space. Then the factor map

$$\eta_\alpha : \begin{cases} X & \to X/\alpha \\ x & \mapsto [x]_\alpha \end{cases}$$

that maps each element to its class can be made measurable by taking the final σ-algebra $\mathcal{M}/\alpha$ with respect to η_α and $\mathcal{M}$ as the σ-algebra on X/α.

     Dual to Lemma 1.3, the initial σ-algebra is characterized.

LEMMA 1.4

Let $(N, \mathcal{N})$ be a measurable space and $f : M \to N$ be a map. The following conditions are equivalent:

a. *The σ-algebra $\mathcal{M}$ on M is initial with respect to f.*

b. *If $(P, \mathcal{P})$ is a measurable space, and $g : P \to M$ is a map, then the $\mathcal{P}$-$\mathcal{N}$-measurability of $f \circ g$ implies the $\mathcal{P}$-$\mathcal{M}$-measurability of g.*

Let $((M_i, \mathcal{M}_i))_{i \in I}$ be a family of measurable spaces, then the product-σ-algebra $\bigotimes_{i \in I} \mathcal{M}_i$ denotes that initial σ-algebra on $\prod_{i \in I} M_i$ for the projections

$$\pi_j : \langle m_i \mid i \in I \rangle \mapsto m_j.$$

It is not difficult to see that $\bigotimes_{i \in I} \mathcal{M}_i = \sigma(\mathcal{Z})$ with

$$\mathcal{Z} := \{ \prod_{i \in I} E_i \mid \forall i \in I : E_i \in \mathcal{M}_i, E_i = M_i \text{ for almost all indices} \}$$

as the collection of *cylinder sets* (use Theorem 1.1 and the observation that $\mathcal{Z}$ is closed under intersection); we will make frequent use of cylinders when dealing with infinite products for interpreting continuous time stochastic logics in Chapter 6.

For $I = \{1, 2\}$, the σ-algebra $\mathcal{M}_1 \otimes \mathcal{M}_2$ is generated from the set of *measurable rectangles*

$$\{E_1 \times E_2 \mid E_1 \in \mathcal{M}_1, E_2 \in \mathcal{M}_2\}.$$

Dually, the sum $(M_1 + M_2, \mathcal{M}_1 + \mathcal{M}_2)$ of the measurable spaces $(M_1, \mathcal{M}_1)$ and $(M_2, \mathcal{M}_2)$ is defined through the final σ-algebra on the sum $M_1 + M_2$ for the injections $M_i \to M_1 + M_2$. This is the special case of the coproduct $\bigoplus_{i \in I}(M_i, \mathcal{M}_i)$, where the σ-algebra $\coprod_{i \in I} \mathcal{M}_i$ is initial with respect to the injections.

We need occasionally the representation of sets through indicator functions. Define for $A \subseteq N$ the *indicator function*

$$\chi_A(x) := \begin{cases} 1, & \text{if } x \in A \\ 0, & \text{if } x \notin A. \end{cases}$$

Clearly, if $\mathcal{N}$ is a σ-algebra on N, then $A \in \mathcal{N}$ iff χ_A is a $\mathcal{N}$-$\mathcal{B}(\mathbb{R})$-measurable function. This is so since we have for the inverse image of an interval under χ_A

$$\chi_A^{-1}[[0, q]] = \begin{cases} \emptyset, & \text{if } q < 0, \\ X \setminus A, & \text{if } 0 \leq q < 1, \\ X, & \text{if } q \geq 1. \end{cases}$$

A measurable *step function*

$$f = \sum_{i=1}^{n} \alpha_i \cdot \chi_{A_i}$$

is a linear combination of indicator functions with $A_i \in \mathcal{N}$. The following statement is folklore in measure theory (Halmos, 1950, Chapter IV), where it is used among others for the construction of the Lebesgue integral. It will come in quite handy in many situations when we have information about the behavior of a construction for measurable sets (i.e., for indicator functions), when the construction is linear, and when we can guarantee closedness under monotone convergence.

PROPOSITION 1.5
Denote for a measurable space $(N, \mathcal{N})$ by

$$\mathcal{F}(N, \mathcal{N}) := \{f : N \to \mathbb{R} \mid f \text{ is } \mathcal{N} - \mathcal{B}(\mathbb{R}) \text{ measurable and bounded}\}$$

the linear space of all bounded measurable real functions on N. Then

a. *For $f \in \mathcal{F}(N,\mathcal{N})$ with $f \geq 0$ there exists an increasing sequence $(f_n)_{n \in \mathbb{N}}$ of step functions $f_n \in \mathcal{F}(N,\mathcal{N})$ with*

$$f(x) = \sup_{n \in \mathbb{N}} f_n(x)$$

for all $x \in X$.

b. *For $f \in \mathcal{F}(N,\mathcal{N})$ there exists a sequence $(f_n)_{n \in \mathbb{N}}$ of step functions $f_n \in \mathcal{F}(N,\mathcal{N})$ with*

$$f(x) = \lim_{n \to \infty} f_n(x)$$

for all $x \in X$.

Convention. Measurability of real-valued functions always means measurability with respect to the Borel sets $\mathcal{B}(\mathbb{R})$ of the real numbers, unless otherwise stated.

1.3 Polish and Analytic Spaces

General measurable spaces are sometimes too general for supporting specific structures. We deal with Polish and analytic spaces which are general enough to support interesting applications but have specific properties which help establishing vital properties. We remind the reader first of some basic facts and provide then some helpful tools for working with Polish spaces, and their more general cousins, analytic spaces.

A *topology* $\mathcal{T}$ on a set X is a family of subsets that is closed under finite intersections and arbitrary unions, and that contains the empty set and the entire set X; the pair $(X,\mathcal{T})$ is called a *topological space*. The elements of $\mathcal{T}$ are called the *open sets*; their complements are called *closed sets*. The space $(X,\mathcal{T})$ is called a *Hausdorff space* iff two distinct points can be separated through disjoint open sets. Thus, given $x \neq y$, there exist disjoint open sets U,V with $x \in U, y \in V$.

> *All topological spaces considered here will be Hausdorff spaces, ever.*

A family $\mathcal{B}$ of open subsets of X is called a *base* for topology $\mathcal{T}$ iff each element of $\mathcal{T}$ can be represented as the union of elements of $\mathcal{B}$. This is equivalent to saying that $\bigcup\{B \mid B \in \mathcal{B}\} = X$, and that we can find for each $x \in B_1 \cap B_2$ with $B_1, B_2 \in \mathcal{B}$ an element $B_3 \in \mathcal{B}$ with $x \in B_3 \subseteq B_1 \cap B_2$. A *subbase* $\mathcal{S}$ for $\mathcal{T}$ has the property that the set $\{\bigcap \mathcal{F} \mid \mathcal{F} \subseteq \mathcal{S} \text{ finite}\}$ of finite intersections of elements of $\mathcal{S}$ forms a base for $\mathcal{T}$.

Given another topological space $(Y,\mathcal{S})$, a map $f : X \to Y$ is called $\mathcal{T}$-$\mathcal{S}$-*continuous* iff the inverse image of an open set from Y is open in X again, i.e., iff $f^{-1}[S] \subseteq T$. The topological spaces (X, T) and (Y, S) are called *homeomorphic* iff there exists a T-S-continuous bijection $f : X \to Y$ the inverse of which is S-T-continuous.

Proceeding in analogy to measurable spaces, a topology T on a set X is called *initial* for a map $f : T \to S$ with a topological space (Y, S) iff T is the smallest topology T_0 on X rendering f a T_0-S-continuous map. For example, if $Y \subseteq X$ is a subset, then the topological subspace $(Y, \{Y \cap G \mid G \in T\})$ is just the initial topology with respect to the inclusion map $i_Y : Y \to X$.

Dually, if (X, T) is a topological space and $f : X \to Y$ is a map, then the *final topology* S on Y is the largest topology S_0 on Y making f T-S_0-continuous. Both initial and final topologies generalize to families of spaces and maps.

The *topological product* $\prod_{i \in I}(X_i, T_i)$ of the topological spaces $((X_i, T_i))_{i \in I}$ is the Cartesian product $\prod_{i \in I} X_i$ endowed with the initial topology with respect to the projections, and the *topological sum* $\coprod_{i \in I}(X_i, T_i)$ of the topological spaces $((X_i, T_i))_{i \in I}$ is the direct $\coprod_{i \in I} X_i$ endowed with the final topology with respect to the injections.

Given a topological space (X, T), a measurable structure comes for free: denote by $\mathcal{B}(X, T)$ the smallest σ-algebra on X that contains the open sets, so that $\mathcal{B}(X, T) = \sigma(T)$. These sets are called the *Borel sets* of (X, T); measurability of maps with respect to the Borel sets is referred to as *Borel measurability*.

An immediate consequence of Lemma 1.2 is that continuity implies Borel measurability.

LEMMA 1.6

Let (X_1, T_1) and (X_2, T_2) be topological spaces. Then $f : X_1 \to X_2$ is $\mathcal{B}(X_1, T_1)$-$\mathcal{B}(X_2, T_2)$ measurable, provided f is T_1-T_2-continuous.

1.3.1   Metric Spaces

Metric spaces are particularly important topological spaces.

DEFINITION 1.7   *A metric d on a set X is a map $d : X \times X \to \mathbb{R}_+$ such that for all $x, y, z \in X$*

a. $d(x, y) = 0 \Leftrightarrow x = y$,

b. $d(x, y) = d(y, x)$ (symmetry),

c. $d(x, z) \leq d(x, y) + d(y, z)$ (triangle inequality)

holds. The pair (X, d) is called a metric space.

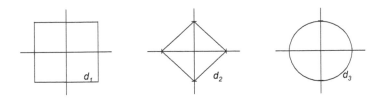

Figure 1.1:   Unit Balls

EXAMPLE 1.8

Let $X := \mathbb{R}^2$ be the Euclidean plane, and define

a. $d_1(\langle x_1, x_2 \rangle, \langle y_1, y_2 \rangle) := \max\{|x_1 - y_1|, |x_2 - y_2|\},$

b. $d_2(\langle x_1, x_2 \rangle, \langle y_1, y_2 \rangle) := |x_1 - y_1| + |x_2 - y_2|,$

c. $d_3(\langle x_1, x_2 \rangle, \langle y_1, y_2 \rangle) := \sqrt{|x_1 - y_1|^2 + |x_2 - y_2|^2}.$

Then these are metrics on X for the usual topology.

Each set M can be made into a metric space (M, d) upon setting

$$d(x, y) := \begin{cases} 1, & \text{if } x \neq y \\ 0, & \text{if } x = y \end{cases}$$

(d is called the *discrete metric*).

Given a metric space (X, d), define for $x \in X, r > 0$ the *open ball around x with radius r*

$$B_{r,d}(x) := B_r(x) := \{x' \in X \mid d(x, x') < r\}.$$

EXAMPLE 1.9

If (M, d) is discrete, then

$$B_r(x) = \begin{cases} \{x\}, & \text{if } r \leq 1 \\ M, & \text{if } r > 1. \end{cases}$$

The unit ball around the origin for the metrics d_1, d_2, d_3 on the plane from Example 1.8 are shown in Figure 1.1.

Call $G \subseteq X$ *open* iff given $x \in G$ there exists $r > 0$ such that the open ball around x with radius r is entirely contained in G, so that $B_r(x) \subseteq G$.

Open balls are open sets: let $G := B_r(x)$ be the open ball around x, and take $y \in B_r(x)$, so that $s := r - d(x,y) > 0$. We claim that $B_s(y) \subseteq B_r(x)$. In fact, if $z \in B_s(y)$, then

$$d(x,z) \le d(x,y) + d(y,z) < d(x,y) + r - d(x,y) = r$$

by the triangle inequality.

The open sets for a metric space form a topology (the *metric topology*); the topology is said to be metrized through the metric. Vice versa, a topological space is called metrizable iff its topology comes from a metric; we sometimes talk simply about *metrizable spaces*. A metrizable topology is Hausdorff: given $x \ne y$, we know that $s := d(x,y) > 0$, so balls with radius $s/3$ are open sets that separate these points. Continuity now takes this — probably more familiar — form:

LEMMA 1.10
Let (X_1, d_1) and (X_2, d_2) be metric spaces, and denote by $\mathcal{T}_i$ the topology on X_i induced by d_i. The following conditions are equivalent for a map $f : X_1 \to X_2$.

a. f is $\mathcal{T}_1$-$\mathcal{T}_2$-continuous.

b. Given $x_1 \in X_1$ and $\varepsilon > 0$, there exists $\delta > 0$ such that

$$d_1(x_1, x_1') < \delta \Rightarrow d_2(f(x_1), f(x_1')) < \varepsilon.$$

PROOF The proof is rather immediate, because the implication in condition b may be rephrased as $B_{\delta, d_1}(x_1) \subseteq f^{-1}[B_{\varepsilon, d_2}(f(x_1))]$. ∎

Let (X, d) be metric, then define for $x \in X$ and for $A \subseteq X$ the distance $d(x, A)$ of x to A through

$$d(x, A) := \inf_{y \in A} d(x, y)$$

(with $(d(x, \emptyset) := 0$). Thus we know that a point x has a distance to $A \ne \emptyset$ of less than $r > 0$ iff we can find $y \in Y$ with $d(x, y) < r$. Because

$$|d(x, A) - d(y, A)| \le d(x, y),$$

we know that $x \mapsto d(x, A)$ is uniformly continuous. If $F \subseteq X$ is closed, then $x \in F$ iff $d(x, F) = 0$ (take a sequence $(x_n)_{n \in \mathbb{N}}$ of elements in F with $d(x_n, x) < 1/n$). F can be written as

$$F = \{x \in X \mid d(x, F) = 0\} = \bigcap_{n \in \mathbb{N}} \{x \in X \mid d(x, F) < \frac{1}{n}\},$$

the latter being a countable intersection of open sets. Thus a closed set in a metric space is a G_δ-set:

DEFINITION 1.11 *The countable intersection of open sets in a topological space is called a G_δ-set; the countable union of closed sets is called an F_σ-set.*

Whenever feasible, we will omit the notation of a topology or a metric from a space.

Recall that a subset D in a topological space is said to be *dense* iff D meets each nonempty open set. A metric space has a countable dense subset iff its topology has a countable base. This is fairly obvious: take in this case from each element of the base an arbitrary element, then this will form a countable dense subset. Conversely, if D is the countable dense set, then all balls with rational radii, $\{B_r(x) \mid 0 < r \in \mathbb{Q}, d \in D\}$, will form a countable base. Spaces with a countable base can be embedded into the cube $[0,1]^\infty := \prod_{n \in \mathbb{N}}[0,1]$, the *Hilbert cube* as we will see in Theorem 1.23 (by the way, we define

$$X^\infty := \prod_{n \in \mathbb{N}} X = X^{\mathbb{N}}$$

for any set X).

DEFINITION 1.12 *A metric space is called* separable *iff it has a countable dense subset, or, equivalently, iff its topology has a countable base.*

1.3.2 Polish Spaces: Elementary Properties

Neither general topological spaces nor metric spaces offer a structure rich enough for the study of the transition systems that we will enter into. We need to restrict the class of topological spaces to a particularly interesting class of spaces that are traditionally called *Polish* (rumor (Kellerer, 1972) has it that Bourbaki (Bourbaki, 1989, Definition IX.6.1) proposed this name for honoring the contributions of the Polish School of Topology to this field).

Remember that a metric space (X, d) is called *complete* iff each Cauchy sequence has a limit.

DEFINITION 1.13 *A Polish space X is a topological space the topology of which is metrizable through a complete metric, and which has a countable dense subset.*

Familiar spaces are Polish, as these examples show.

EXAMPLE 1.14

Denote by $\mathbb{R}$ the reals with their usual topology, which is induced by the open intervals. Then this is a Polish space.

EXAMPLE 1.15

The open unit interval $]0, 1[$ with the usual topology induced by the open intervals form a Polish space.

This comes probably as a surprise, because $]0, 1[$ is known not to be complete with the usual metric. But all we need is a dense subset (take the rationals $\mathbb{Q} \cap]0, 1[$), and a metric that generates the topology, and that is complete. Define

$$d(x, y) := \left| \ln \frac{x}{1-x} - \ln \frac{y}{1-y} \right|,$$

then this is a complete metric for $]0, 1[$. This is so since $x \mapsto \ln(x/(1-x))$ is a continuous bijection from $]0, 1[$ to $\mathbb{R}$, and the inverse $y \mapsto e^y/(1 + e^y)$ is also a continuous bijection.

LEMMA 1.16

Let X be a Polish space, and assume that $F \subseteq X$ is closed, then the subspace F is Polish as well.

PROOF   Because F is closed, each Cauchy sequence in F has its limit in F, so F is complete. The topology that F inherits from X has a countable base and is metrizable, so F has a countable dense subset, too.   ⬜

LEMMA 1.17

Let $(X_n)_{n \in \mathbb{N}}$ be a sequence of Polish spaces, then the product $\prod_{n \in \mathbb{N}} X_n$ and the coproduct $\coprod_{n \in \mathbb{N}} X_n$ are Polish spaces.

PROOF   Assume that the topology $\mathcal{T}_n$ on X_n is metrized through metric d_n, where it may be assumed that $d_n \leq 1$ holds (otherwise use for $\mathcal{T}_n$ the complete metric $d_n(x, y)/(1 + d_n(x, y))$). Then

$$d((x_n)_{n \in \mathbb{N}}, (y_n)_{n \in \mathbb{N}}) := \sum_{n \in \mathbb{N}} 2^{-n} d_n(x_n, y_n)$$

is a complete metric for the product topology $\prod_{n \in \mathbb{N}} \mathcal{T}_n$. For the coproduct, define the complete metric

$$d(x, y) := \begin{cases} 2, & \text{if } x \in X_n, y \in X_m, n \neq m \\ d_n(x, y), & \text{if } x, y \in X_n. \end{cases}$$

All this is established through standard arguments.   ⬜

EXAMPLE 1.18

The set $\mathbb{N}$ of natural numbers with the discrete topology is a Polish space on account of being the topological sum of its elements. Thus the set $\mathbb{N}^{\infty}$ of all infinite sequences is a Polish space. The sets

$$\Sigma_{\alpha} := \{\tau \in \mathbb{N}^{\infty} \mid \alpha \text{ is an initial piece of } \tau\}$$

for $\alpha \in \mathbb{N}^*$, the free monoid generated by $\mathbb{N}$, constitute a base for the product topology.

This last example will be discussed in much greater detail later on. It permits sometimes reducing the discussion of properties for general Polish spaces to an investigation of the corresponding properties of $\mathbb{N}^{\infty}$, the structure of the latter space being more easily accessible than that of a general space. We apply Example 1.18 directly to show that all open subsets of a metric space X with a countable base can be represented through a single closed set in $\mathbb{N}^{\infty} \times X$.

Define for $D \subseteq X \times Y$ the *vertical cut*

$$D_x := \{y \in Y \mid \langle x, y \rangle \in D\}$$

and the *horizontal cut*

$$D^y := \{x \in X \mid \langle x, y \rangle \in D\}.$$

Note that $((X \times Y) \setminus D)_x = Y \setminus D_x$.

PROPOSITION 1.19

Let X be a separable metric space. Then there exists an open set $U \subseteq \mathbb{N}^{\infty} \times X$ and a closed set $F \subseteq \mathbb{N}^{\infty} \times X$ with these properties:

a. For each open set $G \subseteq X$ there exists $\tau \in \mathbb{N}^{\infty}$ such that $G = U_{\tau}$.

b. For each closed set $C \subseteq X$ there exists $\tau \in \mathbb{N}^{\infty}$ such that $C = F_{\tau}$.

PROOF 0. It is enough to establish the property for open sets; taking complements will prove it for closed ones.

1. Let $(V_n)_{n \in \mathbb{N}}$ be a basis for the open sets in X with $V_n \neq \emptyset$ for all $n \in \mathbb{N}$. Define

$$U := \{\langle \tau, x \rangle \mid x \in \bigcup_{n \in \mathbb{N}} V_{\tau_n}\},$$

then $U \subseteq \mathbb{N}^{\infty} \times X$ is open. In fact, let $\langle \tau, x \rangle \in U$, then there exists $n \in \mathbb{N}$ with $x \in V_n$, thus

$$\langle \tau, x \rangle \in \Sigma_n \times V_n \subseteq U,$$

and $\Sigma_n \times V_n$ is open in the product.

2. Let $G \subseteq X$ be open. Because $(V_n)_{n \in \mathbb{N}}$ is a basis for the topology, there exists a sequence $\tau \in \mathbb{N}^\infty$ with $G = \bigcup_{n \in \mathbb{N}} V_{\tau_n} = U_\tau$.   ☐

The set U is usually called a *universal open set*, similar for F. These universal sets will be used rather heavily when we discuss analytic sets.

Let us briefly look into the order structure of $\mathbb{N}^\infty$. We can lexicographically order this set by saying that $\tau \preceq \tau'$ iff there exists $k \in \mathbb{N}$ such that $\tau_k \leq \tau'_k$, and $\tau_\ell = \tau'_\ell$ for all ℓ with $1 \leq \ell < k$. Then $\preceq$ defines a total order. This construction will be needed in Section 1.6.1.

LEMMA 1.20

Each nonempty closed set $F \subseteq \mathbb{N}^\infty$ has a minimal element in the lexicographic order.

PROOF   Let n_1 be the minimal first component of all elements of F, n_2 be the minimal second component of those elements of F that start with n_1, etc. This defines an element $\tau := \langle n_1, n_2, \ldots \rangle$ which is in F since F is closed (because τ is an accumulation point of elements in F).   ☐

We have seen that a closed subset of a Polish space is a Polish space in its own right; a similar argument shows that an open subset of a Polish space is Polish as well. Both observations turn out to be special cases of the characterization of Polish subspaces through G_δ-sets.

We need an auxiliary statement due to Kuratowski which permits the extension of a continuous map from a subspace to a G_δ-set containing it — just far enough to be interesting to us. Denote by $\mathsf{cl}\,(A)$ the topological closure of a set A.

LEMMA 1.21

Let Y be a complete metrizable space, W a metric space, then a continuous map $f : A \rightarrow Y$ can be extended to a continuous map $f_* : G \rightarrow Y$ with G a G_δ-set such that $A \subseteq G \subseteq \mathsf{cl}\,(A)$.

PROOF   (Sketch) 0. We may and do assume that the complete metric d for Y is bounded by 1. Define the *diameter* $\mathsf{diam}(Q)$ of $Q \subseteq Y$ as

$$\mathsf{diam}(Q) := \sup\{d(y_1, y_2) \mid y_1, y_2 \in Q\}.$$

The *oscillation* $\varnothing_f(x)$ of f at $x \in \mathsf{cl}\,(A)$ is defined as the smallest diameter of the image of an open neighborhood of x, formally,

$$\varnothing_f(x) := \inf\{\mathsf{diam}(f\,[A \cap V]) \mid x \in V, V \text{ open}\}.$$

Because f is continuous on A, we have $\varnothing_f(x) = 0$ for each element of A.

1. Put

$$G := \{x \in \mathsf{cl}\,(A) \mid \emptyset_f(x) = 0\},$$

then $A \subseteq G \subseteq \mathsf{cl}\,(A)$, and G is a G_δ in W. In fact, represent G as

$$G = \bigcap_{n \in \mathbb{N}} \{x \in \mathsf{cl}\,(A) \mid \emptyset_f(x) < \frac{1}{n}\},$$

where

$$\{x \in \mathsf{cl}\,(A) \mid \emptyset_f(x) < q\} = \bigcup \{V \cap \mathsf{cl}\,(A) \mid \mathsf{diam}(f\,[V \cap A]) < q\}$$

is open in $\mathsf{cl}\,(A)$, and note that $\mathsf{cl}\,(A)$ is — as a closed set — a G_δ in W.

2. Now take an element $x \in G \subseteq \mathsf{cl}\,(A)$. Then there exists a sequence $(x_n)_{n \in \mathbb{N}}$ of elements $x_n \in A$ with $x_n \to x$. Given $\epsilon > 0$, we find a neighborhood V of x with $\mathsf{diam}(f\,[A \cap V]) < \epsilon$. Since $x_n \to x$, we know that $x_m \in V \cap A$ for all $m > n_\epsilon$, so that the sequence $(f(x_n))_{n \in \mathbb{N}}$ is a Cauchy sequence in Y; it converges because Y is complete. Put

$$f_*(x) := \lim_{n \to \infty} f(x_n).$$

It is then not difficult to see that the map f_* is well defined and extends f, and that f_* is continuous. ◻

This technical Lemma is an important step in establishing a far reaching characterization of subspaces of Polish spaces that are Polish in their own right.

THEOREM 1.22
Let Y be a Polish space. Then the subspace $X \subseteq Y$ is a Polish space iff X is a G_δ-set.

PROOF   (Sketch) 1. Let X be a complete subset of Y, then by Lemma 1.21 the inclusion $id_X : X \to Y$ can be extended to a continuous map $g_* : G \to Y$ for a G_δ-set G with $X \subseteq G \subseteq \mathsf{cl}\,(X)$. Since X is dense in G, we see $ig_* = id_X$ and $X = G$.

2. Assume that $X = \bigcap_{n \in \mathbb{N}} G_n$ with open sets G_n. Let d be the complete metric for Y, then put for $x, y \in X$

$$d'(x, y) := d(x, y) + \sum_{n \in \mathbb{N}} \min \left\{ 2^{n+1}, \left| \frac{1}{d(x, Y \setminus G_n)} - \frac{1}{d(y, Y \setminus G_n)} \right| \right\}.$$

One shows that the identity $id_Y : (Y, d) \to (Y, d')$ is continuous in both directions. Thus the open sets are the same for both metrics.

The next step is to show that (X, d') is complete. In fact, let $(x_n)_{n \in \mathbb{N}} \subseteq X$ be a d'-Cauchy sequence, then it is a d-Cauchy sequence in Y, so there exists

$x \in Y$ with $x_n \to x$. But for each n the sequence $(1/d(x_j, Y \setminus G_n))_{j \in \mathbb{N}}$ converges as well, and we have $d(x_j, Y \setminus G_n) \to d(x, Y \setminus G_n) \neq 0$ by construction. This means that $x \notin Y \setminus G_n$ for each $n \in \mathbb{N}$, thus $x \in X$ is established. Hence (X, d') is complete.   ⬚

Conversely, each Polish space can be represented as a G_δ-set in the *Hilbert cube* $[0, 1]^\infty$; this is the famous characterization of Polish spaces due to Alexandrov (Kuratowski, 1966, III.33.VI), which will be used, e.g., in Section 4.3.2.

THEOREM 1.23
(Alexandrov) *Let X be a separable metric space, then X is homeomorphic to a subspace of the Hilbert cube. If X is Polish, this subspace is a G_δ.*

PROOF   1. We may and do assume that the metric d is bounded by 1. Let $(x_n)_{n \in \mathbb{N}}$ be a countable and dense subset of X, and put $f(x) := \langle d(x, x_1), d(x, x_2), \ldots \rangle$. Then f is injective and continuous. Define $g : f[X] \to X$ as f^{-1}, then g is continuous as well: assume that $f(y_m) \to f(y)$ for some y, hence $\lim_{m \to \infty} d(y_m, x_n) = d(y, x_n)$ for each $n \in \mathbb{N}$. Since $(x_n)_{n \in \mathbb{N}}$ is dense, we find for a given $\epsilon > 0$ an index n with $d(y, x_n) < \epsilon$; by construction we find for n an index m_0 with $d(y_m, x_n) < \epsilon$ whenever $m > m_0$. Thus $d(y_m, y) < 2 \cdot \epsilon$ for $m > m_0$, so that $y_m \to y$. This demonstrates that g is continuous, thus f is a homeomorphism.

2. If X is Polish, $f[X] \subseteq [0, 1]^\infty$ is Polish as well. Thus the second assertion follows from Theorem 1.22.   ⬚

Recall that a topological Hausdorff space X is *compact* iff each open cover of X contains a finite cover of X. These properties of compact spaces will be used from time to time.

THEOREM 1.24
Let X be a topological Hausdorff space. Then

a. *If Y is a compact subspace of X, then Y is closed.*

b. *If X is compact and Y is closed, then Y is compact.*

c. *The union of a finite number of compact sets is compact.*

d. *The continuous image of a compact set into a Hausdorff space is compact.*

e. *The product of compact spaces is compact* (Tihonov's Theorem).

f. *If X is metrizable, then X is compact iff every sequence has a convergent subsequence.*

g. *If X is metrizable, then X is compact iff it is complete and totally bounded (thus given $\epsilon > 0$, there exists a finite subset $\{x_1, \ldots, x_n\} \subseteq X$ with $X \subseteq B_\epsilon(x_1) \cup \cdots \cup B_\epsilon(x_n)$) (Bolzano-Weierstraß Theorem).*

The Bolzano-Weierstraß Theorem implies that compact metrizable spaces are Polish. It is inferred from Tihonov's Theorem that the Hilbert cube $[0,1]^\infty$ is compact, because the unit interval $[0,1]$ is compact, again by the Bolzano-Weierstraß Theorem. Thus Alexandrov's Theorem 1.23 embeds a Polish space as a G_δ into a compact metric space, the closure of which will be compact by Proposition 1.24, part *b*.

1.3.3 Manipulating Polish Topologies

We will show now that Borel maps between Polish spaces can be turned into continuous maps. Specifically, we will show that, given a measurable map between Polish spaces, we can find on the domain a finer Polish topology with the same Borel sets which renders the map continuous. This will be established through a sequence of auxiliary statements, each of which will be of interest and of use in its own right.

We fix for the discussion to follow a Polish space X with topology $\mathcal{T}$.

LEMMA 1.25
Let F be a closed set in X. Then there exists a Polish topology $\mathcal{T}'$ such that $\mathcal{T} \subseteq \mathcal{T}'$ (hence $\mathcal{T}'$ is finer than $\mathcal{T}$), F is clopen in $\mathcal{T}'$, and $\mathcal{B}(X, \mathcal{T}) = \mathcal{B}(X, \mathcal{T}')$.

(Recall that a set is *clopen* in a topological space iff it is both closed and open.)

PROOF Both F and $X \setminus F$ are Polish by Theorem 1.22, so the topological sum of these Polish spaces is Polish again by Lemma 1.17. The sum topology is the desired topology. ⬚

LEMMA 1.26
Let $(\mathcal{T}_n)_{n \in \mathbb{N}}$ be a sequence of Polish topologies $\mathcal{T}_n$ with $\mathcal{T} \subseteq \mathcal{T}_n$.

a. *The topology $\mathcal{T}_\infty$ generated by $\bigcup_{n \in \mathbb{N}} \mathcal{T}_n$ is Polish.*

b. *If $\mathcal{T}_n \subseteq \mathcal{B}(X, \mathcal{T})$, then $\mathcal{B}(X, \mathcal{T}_\infty) = \mathcal{B}(X, \mathcal{T})$.*

PROOF 1. The product $\prod_{n \in \mathbb{N}}(X_n, \mathcal{T}_n)$ is by Lemma 1.17 a Polish space, where $X_n = X$ for all n. Define the map $f : X \to \prod_{n \in \mathbb{N}} X_n$ through $x \mapsto \langle x, x, \ldots \rangle$, then f is $\mathcal{T}_\infty$-$\prod_{n \in \mathbb{N}} \mathcal{T}_n$-continuous by construction. One infers that

$f[X]$ is a closed subset of $\prod_{n \in \mathbb{N}} X_n$: if $(x_n)_{n \in \mathbb{N}} \notin f[X]$, take $x_i \neq x_j$ with $i < j$, and let G_i and G_j be disjoint open neighborhoods of x_i resp. x_j. Then

$$\prod_{\ell < i} X_\ell \times G_i \times \prod_{i < \ell < j} X_\ell \times G_j \times \prod_{\ell > j} X_\ell$$

is an open neighborhood of $(x_n)_{n \in \mathbb{N}}$ that is disjoint from $f[X]$. By Lemma 1.16, the latter set is Polish. On the other hand, f is a homeomorphism between (X, T_∞) and $f[X]$, which establishes part a.

2. T_n has a countable basis $\{U_{i,n} \mid i \in \mathbb{N}\}$, with $U_{i,n} \in \mathcal{B}(X, T)$. This implies that T_∞ has $\{U_{i,n} \mid i, n \in \mathbb{N}\}$ as a countable basis, which entails $\mathcal{B}(X, T_\infty) \subseteq \mathcal{B}(X, T)$. The other inclusion is obvious, giving part b. $\quad\Box$

As a consequence, we may add to a Polish topology a Borel set without destroying the property of the space to be Polish or changing the Borel sets. This is true as well for sequences of Borel sets, as we will see now.

PROPOSITION 1.27

If $(B_n)_{n \in \mathbb{N}}$ is a sequence of Borel sets in X, then there exists a Polish topology T_0 on X such that T_0 is finer than T, T and T_0 have the same Borel sets, and each B_n is clopen in T_0.

PROOF   1. We show first that we may add just one Borel set to the topology without changing the Borel sets. In fact, call a Borel set $B \in \mathcal{B}(X, T)$ *neat* if there exists a Polish topology T_B that is finer than T such that $B \in T_B$, and $\mathcal{B}(X, T) = \mathcal{B}(X, T_B)$.

$$\mathcal{H} := \{B \in \mathcal{B}(X, T) \mid B \text{ is neat}\}.$$

Then $T \subseteq \mathcal{H}$, and each closed set is a member of $\mathcal{H}$ by Lemma 1.25. Furthermore, $\mathcal{H}$ is closed under complements and under countable unions by Lemma 1.26. Thus we may now infer that $\mathcal{H} = \mathcal{B}(X, T)$, so that each Borel set is neat.

2. Now construct inductively Polish topologies T_n that are finer than T with $\mathcal{B}(T) = \mathcal{B}(T_n)$. Start with $T_0 := T$. Adding B_{n+1} to the Polish topology T_n according to the first part yields a finer Polish topology T_{n+1} with the same Borel sets. Thus the assertion follows from Lemma 1.26. $\quad\Box$

This permits turning a Borel map into a continuous one, whenever the domain is Polish and the range is a second countable metric space.

PROPOSITION 1.28

Let (Y, S) be a separable metric space with topology S. If $f : X \to Y$ is a $\mathcal{B}(X, T)$-$\mathcal{B}(Y, S)$-Borel measurable map, then there exists a Polish topology T'

on X such that $\mathcal{T}'$ is finer than $\mathcal{T}$, $\mathcal{T}$ and $\mathcal{T}'$ have the same Borel sets, and f is $\mathcal{T}'$-$\mathcal{S}$ continuous.

PROOF   The metric topology $\mathcal{S}$ is generated from the countable basis $(H_n)_{n \in \mathbb{N}}$. Construct from the Borel sets $f^{-1}[H_n]$ and from $\mathcal{T}$ a Polish topology $\mathcal{T}'$ according to Proposition 1.27. Because $\forall n \in \mathbb{N} : f^{-1}[H_n] \in \mathcal{T}'$, the inverse image of each open set from $\mathcal{S}$ is $\mathcal{T}'$-open, hence f is $\mathcal{T}'$-$\mathcal{S}$ continuous. The construction entails $\mathcal{T}$ and $\mathcal{T}'$ having the same Borel sets.   ▯

This property is most useful, because it permits rendering measurable maps continuous, when they go into a second countable metric space (thus in particular into a Polish space).

Having a countable dense subset for a metric space, we can use the corresponding base for a fairly helpful characterization of the Borel sets. The next Lemma says that the Borel sets are in this case countably generated.

LEMMA 1.29

Let Y be a separable metric space. Then

$$\mathcal{B}(Y) = \sigma(\{B_r(d) \mid r > 0 \text{ rational}, d \in D\}),$$

where D is countable and dense.

PROOF   Because an open ball is an open set, we infer that

$$\sigma(\{B_r(d) \mid r > 0 \text{ rational}, d \in D\}) \subseteq \mathcal{B}(Y).$$

Conversely, let G be open. Then there exists a sequence $(B_n)_{n \in \mathbb{N}}$ of open balls with rational radii such that $\bigcup_{n \in \mathbb{N}} B_n = G$, accounting for the other inclusion.   ▯

This representation implies that the Borel sets $\mathcal{B}(X)$ of our Polish space X are countably generated.

Also the characterization of Borel sets in a metric space as the closure of the open (closed) sets under countable unions and countable intersections will be occasionally helpful.

LEMMA 1.30

The Borel sets in a metric space Y are the smallest collection of sets that contains the open (closed) sets and that are closed under countable unions and countable intersections.

PROOF   The smallest collection $\mathcal{G}$ of sets that contains the open sets and that is closed under countable unions and countable intersections is closed under complementation. This is so since each closed set is a G_δ by Theorem 1.22. Thus $\mathcal{B}(Y) \subseteq \mathcal{G}$; on the other hand $\mathcal{G} \subseteq \mathcal{B}(Y)$ by construction.

$\Box$

As a preparation for dealing with analytic sets, we will show now that each Borel subset of the Polish space X is the continuous image of $\mathbb{N}^\infty$. We begin with a reduction of the problem space: it is sufficient to establish this property for closed sets.

LEMMA 1.31

Assume that each closed set in X is a continuous image of $\mathbb{N}^\infty$. Then each Borel set of X is a continuous image of $\mathbb{N}^\infty$.

PROOF   (Sketch) 1. Let

$$\mathcal{G} := \{B \in \mathcal{B}(X) \mid B = f\,[\mathbb{N}^\infty] \text{ for } f : \mathbb{N}^\infty \to X \text{ continuous}\}$$

be the set of all good guys. Then $\mathcal{G}$ contains by assumption all closed sets. We show that $\mathcal{G}$ is closed under countable unions and countable intersections. Then the assertion will follow from Lemma 1.30.

2. Suppose $B_n = f_n\,[\mathbb{N}^\infty]$ for the continuous map f_n, then

$$\mathbb{M} := \{\langle \tau_1, \tau_2, \dots \rangle \mid f_1(\tau_1) = f_2(\tau_2) = \dots\}$$

is a closed subset of $(\mathbb{N}^\infty)^\infty$, and defining $f : \langle \tau_1, \tau_2, \dots \rangle \mapsto f_1(\tau_1)$ yields a continuous map $f : \mathbb{M} \to X$ with $f\,[\mathbb{M}] = \bigcap_{n \in \mathbb{N}} B_n$. $\mathbb{M}$ is homeomorphic to $\mathbb{N}^\infty$. Thus $\mathcal{G}$ is closed under countable intersections.

3. Suppose $B_n = f_n\,[\mathbb{N}^\infty]$ for the continuous map f_n, then define $M_j := \Sigma_j = \{\sigma \in \mathbb{N}^\infty \mid \sigma_1 = j\}$. This set is clopen in $\mathbb{N}^\infty$ for each $j \in \mathbb{N}$. Adjust f_j to a continuous map $\widetilde{f}_j : M_j \to B_j$ with $B_j = \widetilde{f}_j\,[M_j]$, and define $f : \mathbb{N}^\infty \to X$ so that $f(\tau) = \widetilde{f}_j(\tau)$ if $\tau \in M_j$. Then f is continuous, and $f\,[\mathbb{N}^\infty] = \bigcup_{n \in \mathbb{N}} B_n$. Thus $\mathcal{G}$ is closed under countable unions. $\Box$

Thus it is sufficient to show that each closed subset of a Polish space is the continuous image on $\mathbb{N}^\infty$. But since a closed subset of a Polish space is Polish in its own right, we will restrict our attention to Polish spaces proper.

PROPOSITION 1.32

There exists a continuous map $f : \mathbb{N}^\infty \to X$ with $f\,[\mathbb{N}^\infty] = X$.

PROOF   0. We will define recursively a sequence of closed sets indexed by elements of $\mathbb{N}^*$ that will enable us to define a continuous map on $\mathbb{N}^\infty$.

1. Let d be a metric that makes X complete. Represent X as $\bigcup_{n \in \mathbb{N}} A_n$ with closed sets $A_n \neq \emptyset$ such that the diameter $\text{diam}(A_n) < 1$ for each $n \in \mathbb{N}$

(the diameter of a set was introduced in the proof of Lemma 1.21). Assume that for a word $\alpha \in \mathbb{N}^*$ of length k the closed set $A_\alpha \neq \emptyset$ is defined, and write $A_\alpha = \bigcup_{n \in \mathbb{N}} A_{\alpha n}$ with closed sets $A_{\alpha n} \neq \emptyset$ such that $\mathsf{diam}(A_{\alpha n}) < 1/(k+1)$ for $n \in \mathbb{N}$. This yields for every $\tau = \langle n_1, n_2, \dots \rangle \in \mathbb{N}^\infty$ a sequence of nonempty closed sets $(A_{n_1 n_2 .. n_k})_{k \in \mathbb{N}}$ with diameter $\mathsf{diam}(A_{n_1 n_2 .. n_k}) < 1/k$. Because the metric is complete, $\bigcap_{k \in \mathbb{N}} A_{n_1 n_2 .. n_k}$ contains exactly one point, which is defined to be $f(\tau)$. This construction renders $f : \mathbb{N}^\infty \to X$ well defined.

2. Because we can find for each $x \in X$ an index $n_1' \in \mathbb{N}$ with $x \in A_{n_1'}$, an index n_2' with $x \in A_{n_1' n_2'}$, etc.; the map just defined is onto, so that $f(\langle n_1', n_2', n_3', \dots \rangle) = x$ for some $\tau' := \langle n_1', n_2', n_3', \dots \rangle \in \mathbb{N}^\infty$. Suppose $\epsilon > 0$ is given, and since the diameters tend to 0, we can find $k_0 \in \mathbb{N}$ with $\mathsf{diam}(A_{n_1' n_2' .. n_k'}) < \epsilon$ for all $k > k_0$. Put $\alpha' := n_1' n_2' .. n_{k_0}'$, then $\Sigma_{\alpha'}$ is an open neighborhood of τ' with $f[\Sigma_{\alpha'}] \subseteq B_{\epsilon, d}(f(\tau'))$. Thus we find for an arbitrary open neighborhood V of $f(\tau')$ an open neighborhood U of τ' with $f[U] \subseteq V$, equivalently, $U \subseteq f^{-1}[V]$. Thus f is continuous. □

Proposition 1.32 permits sometimes the transfer of arguments pertaining to Polish spaces to arguments using infinite sequences. Thus a specific space is studied instead of an abstractly given one, the former permitting some rather special constructions. This will be seen in the investigation of some astonishing properties of analytic sets which we will study now.

1.3.4 Analytic Spaces

An *analytic set* B is the projection of a Borel subset of $X \times X$, where X is a Polish space; the complement of an analytic set is called a *co-analytic* set. One may wonder whether these projections are Borel sets, but we will show in a moment that there are strictly more analytic sets than Borel sets, whenever the underlying Polish space is uncountable. Thus analytic sets are a proper extension to Borel sets. On the other hand, analytic sets arise fairly naturally from factoring Polish spaces through equivalence relations that are generated from a countable collection of Borel sets; see Proposition 1.53. Consequently it is sometimes more adequate to consider analytic sets rather than their Borel cousins.

This is a first characterization of analytic sets (using π_X for the projection to X).

PROPOSITION 1.33

Let X be a Polish space. Then the following statements are equivalent for $A \subseteq X$:

a. *A is analytic.*

b. *There exists a Polish space Y and a Borel set $B \subseteq X \times Y$ with $A = \pi_X[B]$.*

c. *There exists a continuous map $f : \mathbb{N}^\infty \to X$ with $f[\mathbb{N}^\infty] = A$.*

d. $A = \pi_X\,[C]$ *for a closed subset* $C \subseteq X \times \mathbb{N}^\infty$.

PROOF   The implication $a \Rightarrow b$ is trivial, $b \Rightarrow c$ follows from Proposition 1.32: $B = g\,[\mathbb{N}^\infty]$ for some continuous map $g : \mathbb{N}^\infty \to X \times Y$, so put $f := \pi_X \circ g$. We obtain $c \Rightarrow d$ from the observation that the graph $\{\langle \tau, f(\tau)\rangle \mid \tau \in \mathbb{N}^\infty\}$ of f is a closed subset of $\mathbb{N}^\infty \times X$ the first projection of which equals A. Finally, $d \Rightarrow a$ is obtained again from Proposition 1.32.   ▯

As an immediate consequence we obtain that a Borel set is analytic. Just for the record:

COROLLARY 1.34
Each Borel set in a Polish space is analytic.

PROOF   Proposition 1.33 together with Proposition 1.32.   ▯

The converse does not hold, as we will show now. This statement is not only of interest in its own right. Historically it initiated the study of analytic and co-analytic sets as a separate discipline in set theory (what is called now Descriptive Set Theory).

PROPOSITION 1.35
Let X be an uncountable Polish space. Then there exists an analytic set that is not Borel.

We show as a preparation for the proof of Proposition 1.35 that analytic sets are closed under countable unions, intersections, direct and inverse images of Borel maps. Before doing that, we establish a simple but useful property of the graphs of measurable maps.

LEMMA 1.36
Let $(M, \mathcal{M})$ be a measurable space, $f : M \to Z$ be a $\mathcal{M}$-$\mathcal{B}(Z)$-measurable map, where Z is a separable metric space. The graph of f,

$$\mathsf{graph}(f) := \{\langle m, f(m)\rangle \mid m \in M\},$$

is a member if $\mathcal{M} \otimes \mathcal{B}(Z)$.

PROOF   1. Let $(V_n)_{n \in \mathbb{N}}$ be the basis for the metric topology of Z, then

$$(M \times Z) \setminus \mathsf{graph}(f) = \bigcup_{n \in \mathbb{N}} \left(f^{-1}\,[V_n] \times (Z \setminus V_n) \cup M \setminus f^{-1}\,[V_n] \times V_n \right),$$

which is plainly a member of $\mathcal{M} \otimes \mathcal{B}(Z)$. □

Analytic sets have closure properties that are similar to those of Borel sets, but not quite the same: they are closed under countable unions and intersections, and under the inverse image of Borel maps. They are closed under the direct image of Borel maps as well, but suspiciously missing is the closure under complementation (which will give rise to Souslin's Theorem).

PROPOSITION 1.37

Analytic sets in a Polish space X are closed under countable unions and countable intersections. If Y is another Polish space, with analytic sets $A \subseteq X$ and $B \subseteq Y$, and $f : X \to Y$ is a Borel map, then $f[A] \subseteq Y$ is analytic in Y, and $f^{-1}[B]$ is analytic in X.

PROOF 1. Using the characterization of analytic sets in Proposition 1.33, one shows exactly as in the proof to Lemma 1.31 that analytic sets are closed under countable unions and under countable intersections.

2. Note first that the set $Y \times A$ is analytic in the Polish space $Y \times X$ by Proposition 1.33. Since $y \in f[A]$ iff $\langle x, y \rangle \in \mathsf{graph}(f)$ for some $x \in A$, we write

$$f[A] = \pi_Y[Y \times A \cap \{\langle y, x \rangle \mid \langle x, y \rangle \in \mathsf{graph}(f)\}].$$

The set $\{\langle y, x \rangle \mid \langle x, y \rangle \in \mathsf{graph}(f)\}$ is Borel in $Y \times X$ by Lemma 1.36, so the assertion follows for the direct image. The assertion is proved in exactly the same way for the inverse image. □

PROOF (of Proposition 1.35) 1. We will deal with the case $X = \mathbb{N}^\infty$ first. Let $F \subseteq \mathbb{N}^\infty \times (\mathbb{N}^\infty \times \mathbb{N}^\infty)$ be a universal closed set according to Proposition 1.19. Thus each closed set $C \subseteq \mathbb{N}^\infty \times \mathbb{N}^\infty$ can be represented as $C = F_\alpha$ for some $\alpha \in \mathbb{N}^\infty$. Taking first projections, we conclude that there exists a universal analytic set $U \subseteq \mathbb{N}^\infty \times \mathbb{N}^\infty$ such that each analytic set $A \subseteq \mathbb{N}^\infty$ can be represented as U_τ for some $\tau \in \mathbb{N}^\infty$.

Now set

$$A := \{\zeta \mid \langle \zeta, \zeta \rangle \in U\}.$$

Because analytic sets are closed under inverse images by Proposition 1.37, A is an analytic set. Suppose that A is a Borel set, then $\mathbb{N}^\infty \setminus A$ is also a Borel set, hence analytic. Thus we find $\xi \in \mathbb{N}^\infty$ such that $\mathbb{N}^\infty \setminus A = U_\xi$. But now

$$\xi \in A \Leftrightarrow \langle \xi, \xi \rangle \in U \Leftrightarrow \xi \in U_\xi \Leftrightarrow \xi \in \mathbb{N}^\infty \setminus A.$$

This is a contradiction.

2. The general case is reduced to the one treated above by observing that an uncountable Polish space contains a homeomorphic copy on $\mathbb{N}^\infty$. But since we are interested mainly in showing that analytic sets are strictly more general

than Borel sets, we refrain from a discussion of this case and refer the reader
to (Srivastava, 1998, Remark 2.6.5).   ▯

The representation of an analytic set through a continuous map on $\mathbb{N}^\infty$
has the remarkable consequence that we can separate two disjoint analytic
sets by disjoint Borel sets (Lusin's Theorem). This in turn implies a pretty
characterization of Borel sets due to Souslin which says that an analytic set
is Borel iff it is co-analytic as well. Since the latter characterization will be
most valuable to us, we will discuss it in greater detail now.
    We start with Lusin's Theorem.

PROPOSITION 1.38

*Given disjoint analytic sets A and B in a Polish space X, there exist disjoint
Borel sets E and F with $A \subseteq E$ and $B \subseteq F$.*

PROOF   0. Call two analytic sets A and B separated by Borel sets iff
$A \subseteq E$ and $B \subseteq F$ for disjoint Borel sets E and F. Observe that if two
sequences $(A_n)_{n\in\mathbb{N}}$ and $(B_n)_{n\in\mathbb{N}}$ have the property that A_m and B_n can be
separated by Borel sets for all $m, n \in \mathbb{N}$, then $\bigcup_{n\in\mathbb{N}} A_n$ and $\bigcup_{m\in\mathbb{N}} B_m$ can
be separated by Borel sets. In fact, if $E_{m,n}$ and $F_{m,n}$ separate A_n and B_m,
then $E := \bigcap_{m\in\mathbb{N}} \bigcup_{n\in\mathbb{N}} E_{m,n}$ and $F := \bigcup_{m\in\mathbb{N}} \bigcap_{n\in\mathbb{N}} F_{m,n}$ separate $\bigcup_{n\in\mathbb{N}} A_n$
and $\bigcup_{m\in\mathbb{N}} B_m$.
    1. Now suppose that $A = f[\mathbb{N}^\infty]$ and $B = g[\mathbb{N}^\infty]$ cannot be separated
by Borel sets, where $f, g : \mathbb{N}^\infty \to X$ are continuous and chosen according to
Proposition 1.33. Because $\mathbb{N}^\infty = \bigcup_{j\in\mathbb{N}} \Sigma_j$, ($\Sigma_\alpha$ is defined in Example 1.18),
we find indices k_1 and ℓ_1 such that $f[\Sigma_{j_1}]$ and $g[\Sigma_{\ell_1}]$ cannot be separated
by Borel sets. For the same reason, there exist indices k_2 and ℓ_2 such that
$f[\Sigma_{j_1 j_2}]$ and $g[\Sigma_{\ell_1 \ell_2}]$ cannot be separated by Borel sets. Continuing with
this, we define infinite sequences $\kappa := \langle k_1, k_2, \ldots \rangle$ and $\lambda := \langle \ell_1, \ell_2, \ldots \rangle$ such
that for each $n \in \mathbb{N}$ the sets $f[\Sigma_{j_1 j_2 \ldots j_n}]$ and $g[\Sigma_{\ell_1 \ell_2 \ldots \ell_n}]$ cannot be separated
by Borel sets. Because $f(\kappa) \in A$ and $g(\lambda) \in B$, we know $f(\kappa) \neq g(\lambda)$, so we
find $\epsilon > 0$ with $d(f(\kappa), g(\lambda)) < 2 \cdot \epsilon$. But we may choose n large enough so
that both $f[\Sigma_{j_1 j_2 \ldots j_n}]$ and $g[\Sigma_{\ell_1 \ell_2 \ldots \ell_n}]$ have a diameter smaller than ϵ each.
This is a contradiction since we now have separated these sets by open balls.
▯

We obtain as a consequence Souslin's Theorem.

THEOREM 1.39

(Souslin) *Let A be an analytic set in a Polish space. If $X \setminus A$ is analytic,
then A is a Borel set.*

PROOF   Let A and $X \setminus A$ be analytic, then they can be separated by disjoint Borel sets E with $A \subseteq E$ and F with $X \setminus A \subseteq F$ by Lusin's Theorem Proposition 1.38. Thus $A = E$ is a Borel set. ⬛

Souslin's Theorem is important when one wants to show that a set is a Borel set that is given for example through the image of another Borel set. A typical scenario for its use is establishing for a Borel set B and a Borel map $f : X \to Y$ that both $A = f[B]$ and $Y \setminus A = f[X \setminus B]$ hold. Then one infers from Proposition 1.37 that both A and $Y \setminus A$ are analytic, and from Souslin's Theorem that A is a Borel set.

We make the properties of analytic sets a bit more widely available by introducing analytic spaces. Roughly, an analytic space is Borel isomorphic to an analytic set in a Polish space; to be more specific:

DEFINITION 1.40   *A measurable space* $(M, \mathcal{M})$ *is called an* analytic space *iff there exists a Polish space* X *and an analytic set* A *in* X *such that the measurable spaces* $(M, \mathcal{M})$ *and* $(A, \mathcal{B}(X) \cap A)$ *are Borel isomorphic. The elements of* $\mathcal{M}$ *are then called the* Borel sets *of* M. $\mathcal{M}$ *is denoted by* $\mathcal{B}(M)$.

We will omit the σ-algebra from the notation of an analytic space.

Analytic spaces share many favorable properties with analytic sets, and with Polish spaces, but they are a wee bit more general: whereas an analytic set lives in a Polish space, an analytic space does only require a Polish space to sit in the background somewhere and to be Borel isomorphic to it. This makes life considerably easier, since we are not always obliged to present a Polish space directly when dealing with properties of analytic spaces.

An immediate consequence is that the image of an analytic space under a Borel map into a Polish space is analytic again.

PROPOSITION 1.41
Let $f : X \to Y$ *be a Borel map from the analytic space* X *to the Polish space* Y, *then* $f[X]$ *is an analytic set in* Y.

PROOF   This is a mere reformulation from Proposition 1.37. ⬛

Take a Borel measurable bijection between two Polish spaces. It is not a priori clear whether or not this map is an isomorphism. Souslin's Theorem gives a helpful hand here as well. We will need this property in a moment for a characterization of countably generated sub-σ-algebras of Borel sets, but it appears to be interesting in its own right.

PROPOSITION 1.42
Let X *and* Y *be analytic spaces and* $f : X \to Y$ *be a bijection that is Borel measurable. Then* f *is a Borel isomorphism.*

PROOF   1. It is no loss of generality to assume that we can find Polish spaces P and Q such that X and Y are subsets of P resp. Q. We want to show that $f[X \cap B]$ is a Borel set in Y, whenever $B \in \mathcal{B}(P)$ is a Borel set. For this we need to find a Borel set $G \in \mathcal{B}(Q)$ such that $f[X \cap B] = G \cap G$.

2. Clearly, both $f[X \cap B]$ and $f[X \setminus B]$ are analytic sets in Q by Proposition 1.41, and because f is injective, they are disjoint. Thus we can find a Borel set $G \in \mathcal{B}(Q)$ with $f[X \cap B] \subseteq G \cap Y$, and $f[X \setminus B] \subseteq Q \setminus G \cap Y$. Because f is surjective, we have $f[X \cap B] \cup f[X \setminus B]$, thus $f[X \cap B] = G \cap Y$
⬜

Call a measurable space $(M, \mathcal{M})$ *separable* iff the σ-algebra $\mathcal{M}$ has a countable set $(A_n)_{n \in \mathbb{N}}$ of generators which separates points, i.e., given $x, x' \in M$ with $x \neq x'$ there exists A_n which contains exactly one of them. A Polish space is separable as a measurable space, so is an analytic space, as we will show now.

Separable measurable spaces are characterized through subsets of Polish spaces.

LEMMA 1.43
The measurable space $(M, \mathcal{M})$ is separable iff there exists a Polish space X and a subset $P \subseteq X$ such that the measurable spaces $(M, \mathcal{M})$ and $(P, \mathcal{B}(X) \cap P)$ are Borel isomorphic.

PROOF   1. Because $\mathcal{B}(X)$ is countably generated for a Polish space X by Lemma 1.29, the σ-algebra $\mathcal{B}(X) \cap P$ is countably generated. Since this property is not destroyed by Borel isomorphisms, the condition above is sufficient.

2. It is also necessary. Let $(A_n)_{n \in \mathbb{N}}$ be a generator for $\mathcal{M}$, and define

$$f(t) := \langle \chi_{A_1}(t), \chi_{A_2}(t), \dots \rangle$$

(χ_A is the indicator function for set A), then $f : M \to \{0,1\}^\infty$ is injective, because $(A_n)_{n \in \mathbb{N}}$ separates points. Put $X := \{0,1\}^\infty$ and equip X with the product topology, then X is compact, hence Polish. Put $P := f[M]$, and let

$$B_n := \{\tau \in P \mid \tau_n = 1\} = P \cap \chi_{A_n}^{-1}[\{1\}].$$

Since $\mathcal{B}(X)$ is generated from the sequence $(\{\tau \in X \mid \tau_n = 1\})_{n \in \mathbb{N}}$, we infer that f is a Borel isomorphism between $(M, \mathcal{M})$ and $(P, \mathcal{B}(X) \cap P)$.   ⬜

Thus analytic spaces are separable.

COROLLARY 1.44
An analytic space is a separable measurable space.

A second consequence is that separable measurable spaces are derived from separable metric spaces in a rather straightforward way.

LEMMA 1.45

For a separable measurable space $(X, \mathcal{A})$ there exists a separable metric topology $\mathcal{T}$ on X such that $\mathcal{B}(X, \mathcal{T}) = \mathcal{A}$.

By the way, this innocently looking statement has some remarkable consequences for our context. Just as an appetizer:

COROLLARY 1.46

Let $(M, \mathcal{M})$ be a separable measurable space. Then

a. The diagonal is measurable in the product, i.e.,

$$\Delta_{M \times M} := \{\langle t, t \rangle \mid t \in M\} \in \mathcal{M} \otimes \mathcal{M}.$$

b. If $f_i : X_i \to M$ is $\mathcal{A}_i - \mathcal{M}$-measurable, where $(X_i, \mathcal{A}_i)$ is a measurable space $(i = 1, 2)$, then

$$f_1^{-1}[\mathcal{M}] \otimes f_2^{-1}[\mathcal{M}] = (f_1 \times f_2)^{-1}[\mathcal{M} \otimes \mathcal{M}].$$

PROOF   1. Let $(A_n)_{n \in \mathbb{N}}$ be a generator for M that separates point, then

$$(M \times M) \setminus \Delta_{M \times M} = \bigcup_{n \in \mathbb{N}} (A_n \times M \setminus A_n \cup M \setminus A_n \times A_n),$$

which is a member of $\mathcal{M} \otimes \mathcal{M}$.

2. The product σ-algebra $\mathcal{M} \otimes \mathcal{M}$ is generated by the rectangles $B_1 \times B_2$ with B_i taken from some generator $\mathcal{B}_0$ for $\mathcal{B}$ $(i = 1, 2)$. Since

$$(f_1 \times f_2)^{-1}[B_1 \times B_2] = f_1^{-1}[B_1] \times f_2^{-1}[B_2],$$

we see that

$$(f_1 \times f_2)^{-1}[\mathcal{B} \otimes \mathcal{B}] \subseteq f_1^{-1}[\mathcal{B}] \otimes f_2^{-1}[\mathcal{B}].$$

This is true without the assumption of separability. Now let $\mathcal{T}$ be a second countable metric topology on Y with $\mathcal{B} = \mathcal{B}(Y, \mathcal{T})$ and let $\mathcal{T}_0$ be a countable base for the topology. Then

$$\mathcal{T}_p := \{T_1 \times T_2 \mid T_1, T_2 \in \mathcal{T}_0\}$$

is a countable base for the product topology $\mathcal{T} \otimes \mathcal{T}$, and (this is the crucial property)

$$\mathcal{B} \otimes \mathcal{B} = \mathcal{B}(Y \times Y, \mathcal{T} \otimes \mathcal{T})$$

holds: because the projections from $X \times Y$ to X and to Y are measurable, we observe $\mathcal{B} \otimes \mathcal{B} \subseteq \mathcal{B}(Y \times Y, \mathcal{T} \otimes \mathcal{T})$; because $\mathcal{T}_p$ is a countable base for the product topology $\mathcal{T} \otimes \mathcal{T}$, we infer the other inclusion.

3. Since for $T_1, T_2 \in \mathcal{T}_0$ clearly

$$f_1^{-1}[T_1] \times f_2^{-1}[T_2] \in (f_1 \times f_2)^{-1}[\mathcal{T}_p] \subseteq (f_1 \times f_2)^{-1}[\mathcal{B} \otimes \mathcal{B}]$$

holds, the nontrivial inclusion is inferred from the fact that the smallest σ-algebra containing $\{f_1^{-1}[T_1] \times f_2^{-1}[T_2] \mid T_1, T_2 \in \mathcal{T}_0\}$ equals $f_1^{-1}[\mathcal{B}] \otimes f_2^{-1}[\mathcal{B}]$.
⬜

COROLLARY 1.47

Let $f : M \to Z$ be a $\mathcal{M}$-$\mathcal{N}$-measurable map, where $(M, \mathcal{M})$ and $(N, \mathcal{N})$ are measurable spaces, the latter being separable. Then the kernel of f

$$\ker(f) := \{\langle m_1, m_2 \rangle \mid f(m_1) = f(m_2)\}$$

is a member of $\mathcal{M} \otimes \mathcal{M}$.

PROOF   The map $f \times f : \langle m_1, m_2 \rangle \mapsto \langle f(m_1), f(m_2) \rangle$ is $\mathcal{M} \otimes \mathcal{M}$-$\mathcal{N} \otimes \mathcal{N}$-measurable, and the diagonal $\Delta_{N \times N}$ is a member of $\mathcal{B}(Z) \otimes \mathcal{B}(Z)$. Since $\ker(f) = (f \times f)^{-1}[\Delta_{Z \times Z}]$, the assertion follows. ⬜

Returning to analytic spaces, have a brief look at countably generated sub-σ-algebras of an analytic space. This will help us to establish that the factor space for a particularly interesting and important class of equivalence relations is an analytic space.

PROPOSITION 1.48

Let X be an analytic space, $\mathcal{B}_0$ a countably generated sub-σ-algebra of $\mathcal{B}(X)$ that separates points. Then $\mathcal{B}_0 = \mathcal{B}(X)$.

PROOF   1. $(X, \mathcal{B}_0)$ is a separable measurable space, so there exists a Polish space P and a subset $Y \subseteq P$ of P such that $(X, \mathcal{B}_0)$ is Borel isomorphic to $P, \mathcal{B}(P) \cap Y$ by Lemma 1.43. Let f be this isomorphism, then $\mathcal{B}_0 = f^{-1}[\mathcal{B}(P) \cap Y]$.

2. f is a Borel map from $(X, \mathcal{B}(X))$ to $(Y, \mathcal{B}(P) \cap Y)$, thus Y is an analytic set with $\mathcal{B}(Y) = \mathcal{B}(X) \cap P$ by Proposition 1.41. By Proposition 1.42, f is an isomorphism, hence $\mathcal{B}(X) = f^{-1}[\mathcal{B}(P) \cap Y]$. But this establishes the assertion. ⬜

This gives an interesting characterization of measurable spaces to be analytic, provided they have a separating sequence of sets, to be specific:

LEMMA 1.49
Let X be analytic, $f : X \to Y$ $\mathcal{B}(X)$-$\mathcal{N}$-measurable and onto for a measurable space $(Y, \mathcal{N})$ which has a sequence of sets in $\mathcal{N}$ that separate points. Then $(Y, \mathcal{N})$ is analytic.

PROOF   1. Let $(B_n)_{n \in \mathbb{N}}$ be the sequence of sets that separates points, take an arbitrary set $N \in \mathcal{N}$ and define the σ-algebra $\mathcal{B}_0 := \sigma(\{B_n \mid n \in \mathbb{N}\} \cup \{N\})$. Then $(Y, \mathcal{B}_0)$ is a separable measurable space, so by Lemma 1.43 we can find a Polish space P with $Y \subseteq P$ and $\mathcal{B}_0$ is the trace of $\mathcal{B}(P)$ on Y. Proposition 1.41 tells us that $Y = f[X]$ is analytic with $\mathcal{B}_0 = \mathcal{B}(Y) = \sigma(\{B_n \mid n \in \mathbb{N}\})$, the latter equality being implied by Proposition 1.48. Because N was arbitrary, and because $B_n \in \mathcal{N}$, this yields $\mathcal{N} = \sigma(\{B_n \mid n \subset \mathbb{N}\}) = \mathcal{B}(Y)$.   □

We will use Lemma 1.49 for demonstrating that the factor space of an analytic space for a smooth equivalence relation is analytic again. This class of relations will be defined now and briefly characterized here; later chapters will use them extensively.

We give a definition in terms of a determining sequence of Borel sets and relate other characterizations of smoothness in Lemma 1.52.

DEFINITION 1.50   Let X be an analytic space and ρ an equivalence relation on X. Then ρ is called smooth iff there exists a sequence $(A_n)_{n \in \mathbb{N}}$ of Borel sets such that

$$x \, \rho \, x' \Leftrightarrow \forall n \in \mathbb{N} : [x \in A_n \Leftrightarrow x' \in A_n].$$

$(A_n)_{n \in \mathbb{N}}$ is said to determine the relation ρ.

We obtain immediately from the definition that a smooth equivalence relation — seen as a subset of the Cartesian product — is a Borel set:

COROLLARY 1.51
Let ρ be a smooth equivalence relation on the analytic space X, then ρ is a Borel subset of $X \times X$.

PROOF   Suppose that $(A_n)_{n \in \mathbb{N}}$ determines ρ. Since $x \, \rho \, x'$ is false iff there exists $n \in \mathbb{N}$ with $\langle x, x' \rangle \in (A_n \times (X \setminus A_n)) \cup ((X \setminus A_n) \times A_n)$, we obtain

$$(X \times X) \setminus \rho = \bigcup_{n \in \mathbb{N}} (A_n \times (X \setminus A_n)) \cup ((X \setminus A_n) \times A_n).$$

This is clearly a Borel set in $X \times X$.   □

The following characterization of smooth equivalence relations is sometimes helpful and shows that it is not necessary to look only at sequences of sets.

It indicates that the kernels of Borel measurable maps and smooth relations are intimately related.

LEMMA 1.52

Let ρ be an equivalence relation on an analytic set X. Then these conditions are equivalent:

a. ρ is smooth.

b. There exists a sequence $(f_n)_{n \in \mathbb{N}}$ of Borel maps $f_n : X \to Z$ into an analytic space Z such that $\rho = \bigcap_{n \in \mathbb{N}} \ker(f_n)$.

c. There exists a Borel map $f : X \to Y$ into an analytic space Y with $\rho = \ker(f)$.

PROOF   1. $a \Rightarrow b$: Let $(A_n)_{n \in \mathbb{N}}$ determine ρ, then

$$x \, \rho \, x' \Leftrightarrow \forall n \in \mathbb{N} : [x \in A_n \Leftrightarrow x' \in A_n]$$
$$\Leftrightarrow \forall n \in \mathbb{N} : \chi_{A_n}(x) = \chi_{A_n}(x').$$

Thus take $Z = \{0, 1\}$ and $f_n := \chi_{A_n}$.

2. $b \Rightarrow c$: Put $Y := Z^\infty$. This is an analytic space in the product σ-algebra, and

$$f : \begin{cases} X & \to Y \\ x & \mapsto (f_n(x))_{n \in \mathbb{N}} \end{cases}$$

is Borel measurable with $f(x) = f(x')$ iff $\forall n \in \mathbb{N} : f_n(x) = f_n(x')$.

3. $c \Rightarrow a$: Since Y is analytic, it is separable; hence the Borel sets are generated through a sequence $(B_n)_{n \in \mathbb{N}}$ which separates points. Put $A_n := f^{-1}[B_n]$, then $(A_n)_{n \in \mathbb{N}}$ is a sequence of Borel sets, because the base sets B_n are Borel in Y, and because f is Borel measurable. We claim that $(A_n)_{n \in \mathbb{N}}$ determines ρ:

$$f(x) = f(x') \Leftrightarrow \forall n \in \mathbb{N} : [f(x) \in B_n \Leftrightarrow f(x') \in B_n]$$
$$\text{(since } (B_n)_{n \in \mathbb{N}} \text{ separates points in } Z)$$
$$\Leftrightarrow \forall n \in \mathbb{N} : [x \in A_n \Leftrightarrow x' \in A_n].$$

⊔

Thus each smooth equivalence relation may be represented as the kernel of a Borel map, and vice versa.

The interest in analytic spaces comes from the fact that factoring an analytic space through a smooth equivalence relation will result in an analytic space again. This requires first and foremost the definition of a measurable structure induced by the relation. The natural choice is the structure imposed by the factor map. The final σ-algebra on X/ρ with respect to the Borel sets on X and the natural projection η_ρ will be chosen; it is denoted by $\mathcal{B}(X)/\rho$. Recall that $\mathcal{B}(X)/\rho$ is the largest σ-algebra $\mathcal{C}$ on X/ρ rendering η_ρ a $\mathcal{B}(X)$-$\mathcal{C}$-measurable map. Then it turns out that $\mathcal{B}(X/\rho)$ coincides with $\mathcal{B}(X)/\rho$:

PROPOSITION 1.53

Let X be an analytic space, and assume that α is a smooth equivalence relation on X. Then X/α is an analytic space.

PROOF   In accordance with the characterization of smooth relations in Lemma 1.52 we assume that α is given through a sequence $(f_n)_{n\in\mathbb{N}}$ of measurable maps $f_n : X \to \mathbb{R}$. The factor map is measurable and onto. Put $E_{n,r} := \{[x]_\alpha \mid x \in X, f_n(x) < r\}$, then $\mathcal{E} := \{E_{n,r} \mid n \in \mathbb{N}, r \in \mathbb{Q}\}$ is a countable set of element of the factor σ-algebra that separates points. The assertion now follows without difficulties from Lemma 1.49.   ⬜

The Blackwell-Mackey-Theorem analyzes those Borel sets that are unions of $\mathcal{A}$-atoms for a sub-σ-algebra $\mathcal{A} \subseteq \mathcal{B}(X)$. Recall that a set $W \in \mathcal{A}$ is an $\mathcal{A}$-*atom* (or simply an atom) iff for each $V \in \mathcal{A}$ with $V \subseteq W$ either $V = \emptyset$ or $V = W$ holds. If $\mathcal{A}$ is countably generated by, say, $(A_n)_{n\in\mathbb{N}}$, then it is not difficult to see that an atom in $\mathcal{A}$ can be represented as

$$\bigcap_{i\in T} A_i \cap \bigcap_{i\in\mathbb{N}\setminus T} (X \setminus A_i)$$

for a suitable subset $T \subseteq \mathbb{N}$.

THEOREM 1.54

(Blackwell-Mackey) *Let X be an analytic space and $\mathcal{A} \subseteq \mathcal{B}(X)$ be a countably generated sub-σ-algebra of the Borel sets of X. If $B \subseteq X$ is a Borel set that is a union of atoms of $\mathcal{A}$, then $B \in \mathcal{A}$.*

PROOF   Let $\mathcal{A}$ be generated by $(A_n)_{n\in\mathbb{N}}$, and define

$$f : X \to \{0,1\}^\infty$$

through

$$x \mapsto \langle \chi_{A_1}(x), \chi_{A_2}(x), \chi_{A_3}(x), \ldots \rangle.$$

Then f is $\mathcal{A}$-$\mathcal{B}(\{0,1\}^\infty)$-measurable. We claim that $f[B]$ and $f[X \setminus B]$ are disjoint. Suppose not, then we find $t \in \{0,1\}^\infty$ with $t = f(x) = f(x')$ for some $x \in B, x' \in X \setminus B$. Because B is the union of atoms, we find a subset $T \subseteq \mathbb{N}$ with $x \in A_n$, provided $n \in T$, and $x \notin A_n$, provided $n \notin T$. But since $f(x) = f(x')$, the same holds for x' as well, which means that $x' \in B$, contradicting the choice of x'.

Because $f[B]$ and $f[X \setminus B]$ are disjoint analytic sets, we find through Souslin's Theorem 1.39 a Borel set C with

$$f[B] \subseteq C, f[X \setminus B] \cap C = \emptyset.$$

Thus $f[B] = C$, and we are done.   □

Sometimes one starts not with a topological space and its Borel sets but rather with a measurable space: A *Standard Borel* space $(X, \mathcal{A})$ is a measurable space such that the σ-algebra $\mathcal{A}$ equals $\mathcal{B}(X, \mathcal{T})$ for some Polish topology $\mathcal{T}$ on X. But we will not use this construction extensively.

1.4   Measurable Selectors

Assume that X and Z are sets. Consider a set valued map $R : X \to \mathfrak{Pow}(Z)$, equivalently, a relation $R \subseteq X \times Z$. We will not distinguish too narrowly between relations and set valued maps, so that for a relation R the set $R(x)$ will be defined as well. We define for R and a set $G \subseteq Z$ the *weak inverse*

$$\exists R(G) := \{x \in X \mid R(x) \cap G \neq \emptyset\}$$

and the *strong inverse* as

$$\forall R(G) := \{x \in X \mid R(x) \subseteq G\}.$$

EXAMPLE 1.55
Both weak and strong inverse of a relation invoke an analogy to modal logic. The formulas of this logic are given in its simplest form through

$$\varphi ::= \top \mid p \mid \neg\varphi \mid \varphi_1 \wedge \varphi_2 \mid \Diamond\varphi$$

with p an atomic sentence. Assume that $\mathcal{M} = (S, R, V)$ is a Kripke model for the basic modal logic, S being the state space, $R \subseteq S \times S$ the transition relation, and V the valuation for the atomic sentences. We define the semantics of formulas inductively, starting with $[\![p]\!]_\mathcal{M} := V(p)$. The meaning of negation and conjunction are given as usual, and the semantics for $\Diamond\varphi$ is defined through

$$\mathcal{M}, s \models \Diamond\varphi \Leftrightarrow \exists s' \in S : \langle s, s' \rangle \in R \wedge \mathcal{M}, s' \models \varphi.$$

Thus

$$s \in [\![\Diamond\varphi]\!]_\mathcal{M} \Leftrightarrow \exists s' : s' \in R(s) \wedge s' \in [\![\varphi]\!]_\mathcal{M}$$
$$\Leftrightarrow R(s) \cap [\![\varphi]\!]_\mathcal{M} \neq \emptyset$$
$$\Leftrightarrow s \in \exists R([\![\varphi]\!]_\mathcal{M}).$$

Similarly, the semantics for $\Box\varphi$ which is as usual defined to equal $\neg\Diamond\neg\varphi$ is shown to be $\forall R(\llbracket\varphi\rrbracket) = \llbracket\Box\varphi\rrbracket$.

This relationship between both inverses and modal logics will be exploited in Chapter 6 in the context of stochastic relations. Nondeterministic refinements will be investigated as well; see Section 6.2.2.

We will discuss here the relationship of measurable relations and those measurable maps that always select an element from the range of such a relation. For this, we endow X and Z with a measurable resp. a Polish structure. Assume that X is a measurable space, and that Z is Polish.

Assume that $R(x)$ always takes closed and nonempty values. If the *weak inverse* $\exists R(G)$ is a measurable set, whenever $G \subseteq Z$ is open, then R is called a *weakly measurable relation* on $X \times Z$. Relation R is called a *measurable relation* iff the *strong inverse* $\forall R(F)$ is measurable, whenever $F \subseteq Z$ is closed. R is called $\mathcal{C}$-*measurable* iff for any compact set $C \subseteq Z$ the weak inverse $\exists R(C)$ of C is a Borel set in X.

Weakly measurable relations can be represented through measurable selectors (sometimes called a *Castaing representation*). This representation implies in particular that a weakly measurable set valued map has a measurable selector. Formally:

DEFINITION 1.56   *Given the measurable space X and the Polish space Z, assume that $\emptyset \neq R(x) \subseteq Z$ takes always closed values.*

a. *A measurable map $f : X \to Z$ is called a* measurable selector *for R iff $f(x) \in R(x)$ holds for all $x \in X$.*

b. *The sequence $(f_n)_{n\in\mathbb{N}}$ of measurable selectors $f_n : X \to Z$ is called a* Castaing representation *for the relation $R \subseteq X \times Z$ iff*

$$R(x) = \mathsf{cl}\left(\{f_n(x) \mid n \in \mathbb{N}\}\right)$$

holds for all $x \in X$.

Thus for relation R to have a Castaing representation it is a necessary condition that $R(x)$ is nonempty and closed for each $x \in X$. We will establish the following characterization.

PROPOSITION 1.57

Given the measurable space X and the Polish space Z. Let $R \subseteq X \times Z$ be a relation with $\emptyset \neq R(x) \subseteq Z$ is closed for every $x \in$. Then the following conditions are equivalent:

a. *R is weakly measurable.*

b. *There exists Castaing representation for R.*

c. *If R is C-measurable.*

This statement will be proved through a sequence of auxiliary statements.

LEMMA 1.58

Assume that X is a measurable space, Z is a Polish space, and assume that $R \subseteq X \times Z$ is weakly measurable. Then there exists a measurable selector f for R.

PROOF   1. Fix a complete metric d for the topology on Z, and assume that $(z_n)_{n \in \mathbb{N}}$ is a dense sequence in Z. We define inductively a sequence of measurable maps $(f_n)_{n \in \mathbb{N}}$ with the following properties:

i. $d(f_n(x), R(x)) < 2^{-n}$,

ii. $d(f_{n+1}(x), f_n(x)) < 2 \cdot 2^{-n}$.

Put $f_0(x) := z_n$, where $n = n(x)$ is the smallest integer such that $R(x) \cap B_1(z_n) \neq \emptyset$. Because

$$\{x \in X \mid f_1(x) = x_n\} = \exists R(B_1(x_n)) \cap \bigcup_{m < n} \exists R(B_1(x_m)),$$

this defines a measurable map.

2. Suppose that $f_0, \ldots, f_n$ is chosen with the desired properties, then let

$$X_i := \{x \in X \mid f_n(x) = z_i\}.$$

This is a measurable set, and if $x \in X_i$, we know that $R(x) \cap B_{2^{-i}}(x_i) \neq \emptyset$. This is so since $f_n(x) = z_i$ implies $d(f_n(x), R(x)) < 2^{-n}$. If $x \in X_i$ is given, define $f_{n+1}(x) := z_k$ iff k is the smallest index with

$$R(x) \cap B_{2^{-n}}(z_i) \cap B_{2^{-(n+1)}}(x_k) \neq \emptyset.$$

Thus f_{n+1} is measurable, we have $d(f_{n+1}(x), R(x)) < 2^{-(n+1)}$, and

$$d(f_{n+1}(x), f_n(x)) \leq d(f_{n+1}(x), z_i) + d(z_i, f_n(x)) < 2^{-(n+1)} + 2^{-n} = 2 \cdot 2^{-n}.$$

This implies that $(f_n(x))_{n \in \mathbb{N}}$ is a Cauchy sequence, which converges, because (Z, d) is complete. The limit f is a measurable map with $f(x) \in R(x)$, as desired.   ∎

A closer analysis of the arguments in the previous Lemma will show now that we even get a Castaing representation.

LEMMA 1.59

Under the conditions of Lemma 1.58 there exists a Castaing representation for R.

PROOF 1. Let again $(z_n)_{n \in \mathbb{N}}$ be dense in Z, and put for $n, k \in \mathbb{N}$

$$R_{n,k}(x) := \begin{cases} R(x) \cap B_{2^{-k}}(z_n), x \in \exists R(B_{2^{-k}}(z_n)) \\ R(x), \text{otherwise.} \end{cases}$$

If $G \subseteq Z$ is an open set, then

$$\{x \in X \mid \mathsf{cl}\,(R_{n,k}(x)) \cap G \neq \emptyset\} = \{x \in X \mid R_{n,k}(x) \cap G \neq \emptyset\}$$
$$= \exists R(B_{2^{-k}}(z_n) \cap U) \cup$$
$$(\exists R(U) \cap (X \setminus \exists R(B_{2^{-k}}(z_n)))).$$

This is a measurable set, thus $x \mapsto \mathsf{cl}\,(R_{n,k}(x))$ constitutes a weakly measurable relation, for which there exists a measurable selector $f_{n,k}$ according to Lemma 1.58.

2. We claim that

$$R(x) = \mathsf{cl}\,(\{f_{n,k}(x) \mid n, k \in \mathbb{N}\})$$

holds. In fact, take $\varepsilon > 0$ and select $k \in \mathbb{N}$ so that $2 \cdot 2^{-k} < \varepsilon$. Given $x \in X, z \in R(x)$, we can find n with $d(z_n, z) < 2^{-k}$, thus $x \in \exists R(B_{2^{-k}}(z_n))$, and $f_{n,k}(x) \in \mathsf{cl}\,(B_{2^{-k}}(z_n))$. Hence $d(f_{n,k}(x), z) \le d(f_{n,k}(x), z_n) + d(z_n, z) < \varepsilon$. $\Box$

PROOF (of Proposition 1.57) 1. The implication $b \Rightarrow a$ follows from Lemma 1.59. For $a \Rightarrow b$ we argue as follows. Let $(f_n)_{n \in \mathbb{N}}$ be a Castaing representation for R, $G \subseteq Z$ open, then

$$\exists R(G) = \{x \in X \mid \mathsf{cl}\,(\{f_n(x) \mid n \in \mathbb{N}\}) \cap G \neq \emptyset\}$$
$$= \{x \in X \mid \{f_n(x) \mid n \in \mathbb{N}\} \cap G \neq \emptyset\}$$
$$= \bigcup_{n \in \mathbb{N}} f_n^{-1}[G].$$

This is a measurable set.

2. For $b \Rightarrow c$, we assume that $R(x) = \mathsf{cl}\,(f_n(x) \mid n \in \mathbb{N}\})$ holds, where the f_n are measurable selectors for R. Let $C \subseteq Z$ be compact, and define for a compatible and complete metric d on Z the open set $K_m := \{z \in Z \mid d(z, C) < 1/m\}$. Clearly, $C = \bigcap_{m \in \mathbb{N}} K_m$, and we show that

$$\exists R(C) = \bigcap_{m \in \mathbb{N}} \exists R(K_m)$$

holds. If $x \in \exists R(C)$, then clearly $x \in \exists R(K_m)$ for any $m \in \mathbb{N}$. Take conversely $z_m \in R(x) \cap K_m$, and select $y_m \in C$ with $d(y_m, z_m) < 1/m$. Since C is compact, we can find a convergent subsequence y_{m_k} and $y \in C$ with $y_{m_k} \to y$, thus $z_{m_k} \to y$, hence $y \in R(x) \cap C$, which means $x \in \exists R(C)$. Consequently, we can represent $\exists R(C)$ as

$$\exists R(C) = \bigcap_{m \in \mathbb{N}} \bigcup_{n \in \mathbb{N}} f_n^{-1}[K_m],$$

so that R is $\mathcal{C}$-measurable.

3. It remains to establish $c \Rightarrow a$. We know that Z is a dense subset of a compact metric space W by Alexandrov's Theorem 1.23. Thus $R_0(x) := \mathsf{cl}\,(R(x))^W$ maps X to the nonempty closed subsets of W, $\mathsf{cl}\,(\cdot)^W$ denoting the closure in W. Since each open set in a metric space is the countable union of closed sets, and since each closed set in W is compact, $\mathcal{C}$-measurability implies that R_0 is weakly measurable, and since

$$\mathsf{cl}\,(R(x))^W \cap U \neq \emptyset \Leftrightarrow R(x) \cap U \neq \emptyset,$$

whenever $U \subseteq Z$ is open, it follows that R is weakly measurable. $\square$

1.5   Probability Measures

Stochastic relations and Markov transition systems are based eventually on subprobability measures. We will discuss these measures now and point out some salient features from which we will develop some algebraic properties.

Dealing first with properties of individual measures, the attention shifts soon to the set of all subprobabilities on a Polish space. This space is endowed with a topology, rendering it a Polish space again, and some properties of this topology will be investigated. We need in the sequel some particular constructions (like projective limits for the interpretation of logics operating on infinite sequences); these constructions will be provided here as well.

A *probability measure* on the measurable space $(N, \mathcal{N})$ is a monotone and σ-additive map $\mu : \mathcal{N} \to [0, 1]$ with $\mu(\emptyset) = 0$ and $\mu(N) = 1$. That μ is σ-additive means that

$$\mu\left(\bigcup_{i \in \mathbb{N}} D_i\right) = \sum_{i \in \mathbb{N}} \mu(D_i)$$

holds whenever $(D_n)_{n \in \mathbb{N}}$ is a countable family of mutually disjoint sets in $\mathcal{N}$. Denote by $\mathfrak{P}\,(N, \mathcal{N})$ the set of all probability measures on $(N, \mathcal{N})$. We will use *subprobability measures* as well: they are defined like probability measures with the exception that the entire spaces is assigned a mass which does not exceed unity; $\mathfrak{S}\,(N, \mathcal{N})$ is the set of all subprobability measures on $(N, \mathcal{N})$.

A rather important tool is the well-known Monotone Convergence Theorem (Halmos, 1950, Theorem 27.B), which yields the analogue to σ-additivity for the integral. Recall that the elements of $\mathcal{F}\,(M, \mathcal{M})$ are bounded.

PROPOSITION 1.60

Let $f \in \mathcal{F}\,(M, \mathcal{M})$ for the measurable space $(M, \mathcal{M})$ be a nonnegative and bounded measurable function with $f \geq 0$, assume that $0 \leq f_1 \leq f_2 \leq \ldots$ is a *monotonically increasing sequence* $(f_n)_{n\in\mathbb{N}} \subseteq \mathcal{F}(M, \mathcal{M})$ *with* $f = \sup_{n\in\mathbb{N}} f_n$, *and let* $\mu \in \mathfrak{S}(N, \mathcal{M})$ *be a subprobability measure. Then*

$$\int_M f \, d\mu = \lim_{n\to\infty} \int_M f_n \, d\mu.$$

An easy and occasionally very practical first consequence of Proposition 1.60 is the representation of an integral with respect to an arbitrary measure through an integral on the real line, what is sometimes called the Choquet representation. This representation builds a bridge between the classical Riemann integral and Lebesgue integral, since it permits computing a Lebesgue integral for a nonnegative function f through the area

$$\{\langle x, t \rangle \in M \times \mathbb{R} \mid 0 \leq t < f(x)\}$$

under its graph.

PROPOSITION 1.61
Let $f \in \mathcal{F}(M, \mathcal{M})$ *for the measurable space* $(M, \mathcal{M})$ *be a nonnegative and bounded measurable function with* $f \geq 0$, *then*

$$\int_M f \, d\mu = \int_0^\infty \mu(\{x \in M \mid f(x) > t\}) \, dt.$$

PROOF Define for $f \geq 0$ the set

$$C(f) := \{\langle x, t \rangle \in M \times \mathbb{R} \mid 0 \leq t < f(x)\},$$

we claim that $C(f) \in \mathcal{M} \otimes \mathcal{B}(\mathbb{R})$ and

$$\int_M f \, d\mu = (\mu \otimes \lambda)(C(f))$$

holds, where λ is Lebesgue measure, and $\mu \otimes \lambda$ is the product measure. Consider these cases.

1. If $f = \chi_A$ with $A \in \mathcal{M}$, then $C(f) = M \setminus A \times \{0\} \cup A \times [0, 1[\in \mathcal{M} \otimes \mathcal{B}(\mathbb{R})$ thus

$$\int_M \chi_A \, d\mu = \mu(A) = (\mu \otimes \lambda)(C(f)).$$

2. If f is represented as a step function with a finite number of mutually disjoint steps, say, $f = \sum_{i=1}^k r_i \cdot \chi_{A_i}$ with $r_i \geq 0$ and all $A_i \in \mathcal{M}$, then

$$C(f) = \left(M \setminus \bigcup_{i=1}^k A_i \right) \times \{0\} \cup \bigcup_{i=1}^k A_i \times [0, r_i[\in \mathcal{M} \otimes \mathcal{B}(\mathbb{R}),$$

and

$$\int_M f \, d\mu = \sum_{i=1}^k \int_{A_i} r_i \, d\mu = (\mu \otimes \lambda)(C(f)).$$

3. If f is represented as a monotone limit of step function $(f_n)_{n \in \mathbb{N}}$ with $f_n \geq 0$ according to Proposition 1.5, then $C(f) = \bigcup_{n \in \mathbb{N}} C(f_n)$, thus $C(f) \in \mathcal{M} \otimes \mathcal{B}(\mathbb{R})$, and, by Proposition 1.60,

$$\begin{aligned} \int_M f \, d\mu &= \lim_{n \to \infty} \int_M f_n \, d\mu \\ &= \lim_{n \to \infty} (\mu \otimes \lambda)(C(f_n)) \\ &= (\mu \otimes \lambda)(C(f)). \end{aligned}$$

Thus we have for $f \geq 0$ the representation

$$\int_M f \, d\mu = (\mu \otimes \lambda)(C(f)) = \int_0^\infty \mu(C(f)_t) \, dt = \int_0^\infty \mu(\{x \in M \mid f(x) > t\}) \, dt,$$

the latter equality being derived from Fubini's Theorem for product integration. $\square$

Here $\mu \otimes \lambda$ is the *product measure* with factors μ and λ. Recall that given $\mu_i \in \mathfrak{S}(N_i, \mathcal{N}_i)$ there exists a unique measure $\mu_1 \otimes \mu_2 \in \mathfrak{S}(N_1 \times N_2, \mathcal{N}_1 \otimes \mathcal{N}_2)$ such that $\mu_1 \otimes \mu_2(b_1 \times B_2) = \mu_1(B_1) \cdot \mu_2(B_2)$ whenever $B_1 \in \mathcal{N}_1, B_2 \in \mathcal{N}_2$.

1.5.1   Regularity and Tightness

Now let X be a Polish space. We write $\mathfrak{S}(X)$ for $\mathfrak{S}(X, \mathcal{B}(X))$, similarly for $\mathfrak{P}$. Each $\mu \in \mathfrak{S}(X)$ is *regular* in the following sense: given a Borel set B, $\mu(B)$ can be approximated from within through closed sets; symmetrically for open sets. This property is called *regularity* and will help us to show that in a Polish space each Borel set can be approximated from the inside even through compact sets.

LEMMA 1.62

For each $\mu \in \mathfrak{S}(X)$ and each Borel set $B \in \mathcal{B}(X)$,

$$\mu(B) = \sup\{\mu(G) \mid G \supseteq B \text{ open}\} = \sup\{\mu(F) \mid F \subseteq B \text{ closed}\}.$$

PROOF   0. The assertion says that we can find for each $B \in \mathcal{B}(X)$ and for each $\varepsilon > 0$ a closed set $F_\varepsilon \subseteq B$ and an open set G_ε with $B \subseteq G_\varepsilon$ such that $\mu(G_\varepsilon \setminus F_\varepsilon) < \varepsilon$.

1. Let

$$\mathcal{R} := \{B \in \mathcal{B}(X) \mid \mu(B) = \inf_{G \supseteq B \text{ open}} \mu(G) = \sup_{F \subseteq B \text{ closed}} \mu(F)\}.$$

It is clear that $\mathcal{R}$ is closed under complementation.

2. Let F be a closed set, then $F = \bigcap_{n \in \mathbb{N}} G_n$, where $(G_n)_{n \in \mathbb{N}}$ is a decreasing sequence of open sets. This is so since each closed set is a G_δ-set by Section 1.3.1. Thus $\mu(F) = \inf_{n \in \mathbb{N}} \mu(G_n)$ by σ-additivity, so that $F \in \mathcal{R}$.

3. $\mathcal{R}$ is closed under countable disjoint unions as well. Let $(B_n)_{n \in \mathbb{N}}$ be a countable disjoint sequence of sets in R, select for $\varepsilon > 0$ a closed set $F_n \subseteq B_n$ with $\mu(B_n \setminus F_n) < \varepsilon/2^{-(n+1)}$, and choose an open set $G_n \supseteq B_n$ with $\mu(G_n \setminus B_n) < \varepsilon/2^{-n}$. Then there exists $k \in \mathbb{N}$ so that

$$\mu(\bigcup_{n=1}^{k} F_n) > \mu(\bigcup_{n=1}^{\infty} F_j) - \varepsilon/2;$$

thus we have for $B := \bigcup_{n \in \mathbb{N}} B_n$ and the closed set $F := \bigcup_{n=1}^{k} F_n$ that $\mu(B \setminus F) < \varepsilon$ holds. Similarly, $G := \bigcup_{n \in \mathbb{N}} G_n$ is an open set that contains B, and we have

$$\mu(G \setminus B) \leq \sum_{n \in \mathbb{N}} \mu(G_n \setminus B_n) < \varepsilon.$$

Thus $B = \bigcup_{n \in \mathbb{N}} B_n \in \mathcal{R}$.

4. Thus R is a λ-system containing the closed sets. The π-λ-Theorem 1.1 now shows that $\mathcal{R} = \mathcal{B}(X)$. $\Box$

Consequently, we can approximate the probability for an arbitrary event by a topologically closed event implying it and by an topologically open one implied by it up to arbitrary precision. The approximation from the inside can be rendered considerably more convenient by taking compact rather than closed sets. Thus we will show that given $\varepsilon > 0$ there exists for each Borel set B a compact set $K \subseteq B$ with $\mu(B \setminus K) < \varepsilon$. This means that the measure lives essentially on a compact set, which in turn will permit us to capitalize on the properties of compact sets, at least under favorable circumstances.

We first need an auxiliary characterization of compact sets for showing this.

LEMMA 1.63

Let C be a closed subset of a complete metric space X, and assume that for each $n \in \mathbb{N}$ there exists a finite number $K_{n,1}, \ldots, K_{n,k_n}$ of closed balls with radius not exceeding $1/n$ with $C \subseteq \bigcup_{j=1}^{k_n} K_{n,j}$. Then C is compact.

PROOF It is by Theorem 1.24 enough to show that each sequence $(x_n)_{n \in \mathbb{N}}$ in C has a convergent subsequence.

We find an index n_1 such that K_{1,n_1} contains infinitely many elements of the sequence; since K_{1,n_1} is covered by a subset of the closed balls $K_{2,1}, \ldots, K_{2,k_2}$, we find an index n_2 such that K_{2,n_2} contains an infinite number of members of the sequence, and $K_{2,n_2} \subseteq K_{1,n_1}$. Continuing this process, we find for each m an index k_m so that $K_{1,n_1} \supseteq K_{2,n_2} \supseteq \cdots \supseteq K_{m,n_m}$, and K_{m,n_m} contains

infinitely many elements of the sequence. Since $\mathsf{diam}(K_{m,n_m}) \leq 1/m$, and since the space is complete, the intersection $\bigcap_{m \in \mathbb{N}} K_{m,n_m}$ contains exactly one point x. Each neighborhood of x contains by construction infinitely many elements of the sequence, so that we can find a subsequence converging to x.
☐

We derive from this observation that each subprobability lives essentially on a compact set, provided the base space is Polish. We will use the criterion for compactness from Lemma 1.63 together with the fact that a Polish space has a countable dense subset for constructing a compact set the probability of which will be sufficiently close to that of the space proper.

PROPOSITION 1.64
Let X be a Polish space, $\mu \in \mathfrak{S}(X)$ be a subprobability measure. Given $\varepsilon > 0$, there exists a compact subset $C_\varepsilon \subseteq X$ with $\mu(X \setminus C_\varepsilon) < \varepsilon$.

PROOF   Because X is Polish, it contains a countable dense subset. Thus we can cover X for each $n \in \mathbb{N}$ by a countable number of closed balls $(K_{n,\ell})_{\ell \in \mathbb{N}}$ with a radius not greater than $1/n$. Fix $\varepsilon > 0$, then we can find for each n an index k_n such that

$$\mu\left(X \setminus \bigcup_{j=1}^{k_n} K_{n,j}\right) \geq 1 - \frac{\varepsilon}{2^n}.$$

Then $X_n := \bigcup_{j=1}^{k_n} K_{n,j}$ is closed, being the finite union of closed sets, and $C_\varepsilon := \bigcap_{n \in \mathbb{N}} X_n$ is compact by Lemma 1.63. We obtain

$$\mu(X \setminus C_\varepsilon) \leq \sum_{n \in \mathbb{N}} \mu(X \setminus X_n) \leq \sum_{n \in \mathbb{N}} \frac{\varepsilon}{2^n} = \varepsilon.$$

☐

As a consequence we obtain that each Borel set can be approximated through a compact set from the inside; this property is usually called *tight*.

PROPOSITION 1.65
Given $\mu \in \mathfrak{S}(X)$ for Polish X and $\varepsilon > 0$ there exists a compact subset $C \subseteq X$ with $\mu(X \setminus C) < \varepsilon$.

PROOF   This follows immediately from Proposition 1.64 together with Lemma 1.62, since a closed subset of a compact set is compact again.   ☐

1.5.2 Weak Topology

We will define the topology of weak convergence on $\mathfrak{S}(X)$, and we show that we can find a metric for it. If X is Polish, the weak topology will be shown to be Polish as well. The Borel sets for this topology will also be investigated. They will be a valuable tool for the investigations to follow even without the weak topology being present explicitly.

Let X be a metric space, then $\mathfrak{S}(X) = \mathfrak{S}(X, \mathcal{B}(X))$ is equipped with the *topology of weak convergence*. This is the smallest topology on $\mathfrak{S}(X)$ which makes the map $\mu \mapsto \int_X f \, d\mu$ continuous for each continuous and bounded $f : X \to \mathbb{R}$. Denote by $\mathcal{C}(X)$ the linear space of all these functions, and by $\longrightarrow_w$ convergence in this topology.

A base for the weak topology is furnished through sets of the form

$$U(\mu_0, \varepsilon, f_1, \ldots, f_n) :=$$

$$\{\mu \in \mathfrak{S}(X) \mid \left| \int_X f_i \, d\mu - \int_X f_i \, d\mu_0 \right| < \varepsilon \text{ for } 1 \leq i \leq n\}$$

with $\varepsilon > 0, \mu_0 \in \mathfrak{S}(X), f_1, \ldots, f_n \in \mathcal{C}(X)$. This topology is characterized through the famous Portmanteau Theorem :

PROPOSITION 1.66
The following conditions are equivalent for a sequence $(\mu_n)_{n \in \mathbb{N}}$ and a measure $\mu \in \mathfrak{S}(X)$, whenever X is a Polish space:

a. $\mu_n \longrightarrow_w \mu$.

b. $\int_X f \, d\mu_n \to \int_X f \, d\mu$ *for each bounded and continuous $f : X \to \mathbb{R}$.*

c. $\int_X f \, d\mu_n \to \int_X f \, d\mu$ *for each bounded and uniformly continuous $f : X \to \mathbb{R}$.*

d. $\liminf_{n \to \infty} \mu_n(G) \geq \mu(G)$ *for each open subset $G \subseteq X$, and $\mu_n(X) \to \mu(X)$.*

e. $\limsup_{n \to \infty} \mu_n(F) \leq \mu(F)$ *for each closed subset $F \subseteq X$, and $\mu_n(X) \to \mu(X)$.*

PROOF 1. The equivalence $a \Leftrightarrow b$ is a mere restatement of the definition of the weak topology; the equivalence $d \Leftrightarrow e$ is obvious as well. The implication $b \Rightarrow c$ is trivial, since each uniformly continuous function is continuous.

2. In order to establish $d \Rightarrow c$, we may and do assume that $f \geq 0$, because the integral is linear. Then we can represent the integral through its Choquet representation(Proposition 1.61)

$$\int_X f \, d\nu = \int_0^\infty \nu(\{x \in X \mid f(x) > t\}) \, dt.$$

Since f is continuous, the set $\{x \in X \mid f(x) > t\}$ is open. By Fatou's Lemma (Halmos, 1950, Theorem 27.F) we obtain from the assumption

$$\liminf_{n \to \infty} \int_X f \, d\mu_n = \liminf_{n \to \infty} \int_0^\infty \mu_n(\{x \in X \mid f(x) > t\}) \, dt$$

$$\geq \int_0^\infty \liminf_{n \to \infty} \mu_n(\{x \in X \mid f(x) > t\}) \, dt$$

$$\geq \int_0^\infty \mu(\{x \in X \mid f(x) > t\}) \, dt$$

$$= \int_X f \, d\mu.$$

Because $f \geq 0$ is bounded, we find $T \in \mathbb{R}$ such that $f(x) \leq T$ for all $x \in X$, hence $g(x) := T - f(x)$ defines a nonnegative and bounded function. Then by the preceding argument $\liminf_{n \to \infty} \int_X g \, d\mu_n \geq \int_X g \, d\mu$. Since $\mu_n(X) \to \mu(X)$, we infer

$$\limsup_{n \to \infty} \int_X f \, d\mu_n \leq \int_X f \, d\mu,$$

which implies the desired equality.

3. Now take for $c \Rightarrow d$ an open set G, let d be a suitable metric on X, and define for $k \in \mathbb{N}$ the uniformly continuous map $f_k(x) := \min\{1, k \cdot d(x, X \setminus G)\}$ (f_k is uniformly continuous because $|d(x_1, X \setminus G) - d(x_2, X \setminus G)| \leq d(x_1, x_2)$, as noted above). Then

$$0 \leq f_1(x) \leq f_2(x) \leq \ldots \chi_G(x),$$

so that

$$\int_X f_k \, d\mu_n \leq \int_X \chi_G \, d\mu_n = \mu_n(G).$$

Moreover $f_k(x) \to \chi_G(x)$. Since the convergence is monotone, we have

$$\lim_{k \to \infty} \int_X f_k \, d\mu = \int_X \chi_G \, d\mu$$

by Proposition 1.60, and we know from the assumption that

$$\lim_{n \to \infty} \int_X f_k \, d\mu_n = \int_X f_k \, d\mu.$$

But this implies

$$\lim_{n \to \infty} \int_X f_k \, d\mu_n \leq \liminf_{n \to \infty} \mu_n(G),$$

hence

$$\mu(G) \leq \liminf_{n \to \infty} \mu_n(G).$$

□

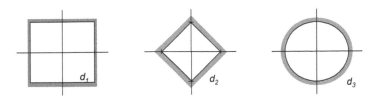

Figure 1.2: Neighborhood of Unit Balls

It is clear that $\mu_n \to_w \mu$ does not imply that $\mu_n(A) \to \mu_0(A)$ for all Borel sets holds (this would be a case for the topology of strong convergence, which is by far not that interesting — for us, that is). The following simple example illustrates the point. Recall that the *Dirac measure* δ_x is defined through $\delta_x(A) := \chi_A(x)$, so $\delta_x(A) = 1$ iff $x \in A$, and $\delta_x(A) = 0$ otherwise. Topologically, $x \mapsto \delta_x$ yields a homeomorphic embedding of X into $\mathfrak{S}(X)$.

EXAMPLE 1.67
Take $X :- [0,1]$, let $\mu_n := \delta_{1/n}$, and consider the Borel set $A :=]0,1]$. Since $\int_X f \, d\mu_n = f(1/n)$, we see that $\mu_n \to_w \delta_0$, but $\mu_n(A) = 1$, $\delta_0(A) = 0$.

Denote by $\mathsf{bd}\,(A)$ the *boundary* of a set $A \subseteq X$, thus

$$\mathsf{bd}\,(A) = \mathsf{cl}\,(A) \setminus \left(\bigcup \{ G \mid G \subseteq A \text{ open} \} \right)$$

$(\bigcup \{ G \mid G \subseteq A$ open$\}$ is just the largest open set that is contained in A). Then the Portmanteau Theorem entails

COROLLARY 1.68
$\mu_n \to_w \mu$ *iff* $\lim_{n\to\infty} \mu_n(A) = \mu(A)$ *for each Borel set A with* $\mu(\mathsf{bd}\,(A)) = 0$.

Hence in order to learn something about the weak limit of a sequence $(\mu_n)_{n\in\mathbb{N}}$ of measures, we could compute the setwise limit of $(\mu_n(A))_{n\in\mathbb{N}}$, but only for those sets A to which the boundary is assigned the value zero by the limiting measure. Consequently, determining the limit probability through the limit of sequences of probabilities for sets is close to hopeless.

Let (X, d) be a metric space, and define for $A \subseteq X, r > 0$ the r-neighborhood of A by

$$A^r := \{ x \in X \mid d(x, A) < r \}.$$

Thus $\{x\}^r = B_{r,d}(x)$, so A^r builds a kind of a measured cloud around A. Figure 1.2 gives small neighborhoods for the unit balls from Figure 1.1.

We have the following elementary properties.

LEMMA 1.69

Let $r, s > 0, \mu_1, \mu_2 \in \mathfrak{S}(X)$, then

a. $A^r = (\mathrm{cl}\,(A))^r$ for $A \subseteq X$,

b. $(A^r)^s \subseteq A^{r+s}$ for $A \subseteq X$,

c. The inequality $\mu_1(A) \leq \mu_2(A^r) + r$ holds for all $A \in \mathcal{B}(X)$ iff it holds for all closed sets $A \subseteq X$ iff it holds for all open sets $A \subseteq X$.

PROOF   1. Part a. follows from $d(x, A) = d(x, \mathrm{cl}\,(A))$.

2. $x \in (A^r)^s$ iff there exists $y \in A^r$ with $d(x, y) < s$. Thus there exists $z \in A$ with $d(y, z) < r$. Consequently, $d(x, z) < r + s$, hence $x \in A^{r+s}$. This establishes part b.

3. The statement in part c referring to closed sets follows from $A \subseteq \mathrm{cl}\,(A)$ and from part a. The assertion for open sets follows from the observation that $F = \bigcap_{n \in \mathbb{N}} F^{1/n}$ holds, and since $F^{1/n}$ is open.   ☐

Now define the *Prohorov metric* $\mathbf{d}_P$ on $\mathfrak{S}(X)$ is through

$$\mathbf{d}_P(\mu_1, \mu_2) :=$$
$$\inf\{\varepsilon > 0 \mid \forall A \in \mathcal{B}(X) : \mu_1(A) \leq \mu_2(A^\varepsilon) + \varepsilon \text{ and } \mu_2(A) \leq \mu_1(A^\varepsilon) + \varepsilon\}.$$

We see from Lemma 1.69 that we can restrict our attention to closed (or to open) sets.

LEMMA 1.70

$(\mathfrak{S}(X), \mathbf{d}_P)$ is a metric space.

PROOF   1. Let $\mathbf{d}_P(\mu_1, \mu_2) = 0$, then $\mu_1(F) \leq \mu_2(F^\varepsilon) + \varepsilon$ and $\mu_2(F) \leq \mu_1(F^\varepsilon) + \varepsilon$ holds for all $\varepsilon > 0$ and for all closed sets F. Letting $\varepsilon \to 0$, we obtain $\mu_1(F) = \mu_2(F)$ for all closed sets (making use of the observation that $F = \{x \in X \mid d(x, F) = 0\}$ holds for closed sets; this equality is not available for all Borel sets, of course). The π-λ-Theorem 1.1 now shows that $\mu_1(B) = \mu_2(B)$ is true for any Borel set B.

2. Symmetry is trivial. The triangle inequality follows from Lemma 1.69, part b.   ☐

Relating the Prohorov metric and the topology of weak convergence, we note first that metric convergence implies weak convergence. This is a simple application of the Portmanteau Theorem 1.66.

LEMMA 1.71

If $\mathbf{d}_P(\mu_n, \mu) \to 0$, then $\mu_n \to_w \mu$. Thus each set which is open in the metric topology is open in the weak topology.

PROOF Let $F \subseteq X$ be closed, $\varepsilon < 0$. There exists $\delta > 0$ so that $\mu(F^\delta) < \mu(F) + \varepsilon$ (this is so since $\mu(F) = \inf_{n \in \mathbb{N}} \mu(F^{1/n})$, because μ is σ-additive). Let $\eta > 0$ be smaller than both ε and δ, then we find $m \in \mathbb{N}$ with $\mu_n(F) \leq \mu(F^\eta) + \eta$, whenever $n \geq m$. Thus

$$\liminf_{n \to \infty} \mu_n(F) = \inf_{m \in \mathbb{N}} \sup_{n \geq m} \mu_m(F) \leq \mu(F^\eta) + \eta \leq \mu(F) + 2 \cdot \varepsilon.$$

Because $|\mu_n(X) - \mu(X)| \leq \mathbf{d}_P(\mu_n, \mu)$, we see that also $\mu_n(X) \to \mu(X)$. This establishes weak convergence through the Portmanteau Theorem. ⬚

The converse is considerably more involved. Recall that the Boolean algebra generated from some finite set $\{A_1, \ldots, A_n\}$ of subsets has at most 2^n elements. This is so since this algebra is just the set

$$\{\bigcap_{i \in T} A_i \cap \bigcap_{i \notin T} X \setminus A_i \mid T \subseteq \{1, \ldots, n\}\}$$

The second point which will be helpful is the observation that the sets A with $\mu(\mathsf{bd}\,(A)) = 0$ form a Boolean algebra. A point x is in the boundary of a set A iff each neighborhood of x intersects both A and $X \setminus A$. Consequently, $\mathsf{bd}\,(X \setminus A) = \mathsf{bd}\,(A)$, and $\mathsf{bd}\,(A_1 \cup A_2) \subseteq \mathsf{bd}\,(A_1) \cup \mathsf{bd}\,(A_2)$. Hence these sets form a Boolean algebra indeed.

Armed with these tools, we enter the discussion of

PROPOSITION 1.72

Let X be a metric space with a countable dense subset. Then each open set in the weak topology is open in the metric topology.

PROOF 1. Let $(x_n)_{n \in \mathbb{N}}$ be a countable dense subset of X. We can find for an arbitrary $\varepsilon > 0$ a sequence $(A_n)_{n \in \mathbb{N}}$ of subsets with these properties

i. $(A_n)_{n \in \mathbb{N}}$ covers X, hence $X = \bigcup_{n \in \mathbb{N}} A_n$,

ii. $\mathsf{diam}(A_n) < \varepsilon$ for all $n \in \mathbb{N}$,

iii. $\mu_0(\mathsf{bd}\,(A_n)) = 0$ for all $n \in \mathbb{N}$.

This is so since $(x_n)_{n \in \mathbb{N}}$ is dense in X, so that we may cover X with balls of radius ε. The boundary $\mathsf{bd}\,(B_r(x))$ of a ball of radius r is contained in the set $\{y \mid d(x, y) = r\}$; all these sets are disjoint, so there is always a radius the corresponding set of which has measure zero.

Because the sets the boundary of which has mass zero form a Boolean algebra, we may and do assume that the sequence $(A_n)_{n \in \mathbb{N}}$ consists of mutually disjoint sets (otherwise define a new sequence $(A'_n)_{n \in \mathbb{N}}$ with $A'_1 := A_1, \ldots, A'_{n+1} := A_{n+1} \setminus (A_1 \cup \cdots \cup A_n)$ without losing the properties above).

We can find an index $n \in \mathbb{N}$ such that $\mu(X \setminus (\bigcup_{i=1}^n A_i)) < \varepsilon/3$. Denote for simplicity the Boolean algebra generated by $\{A_1, \ldots, A_n\}$ by $\mathcal{A}$.

2. We claim that whenever $\mu \in \mathfrak{S}(X)$ with $|\mu(A) - \mu_0(A)| < \varepsilon/3$ for all $A \in \mathcal{A}$, then we have $\mu(B) \leq \mu_0(B^\varepsilon) + \varepsilon$ and $\mu_0(B) \leq \mu(B^\varepsilon) + \varepsilon$ for any Borel set $B \in \mathcal{B}(X)$. Well, put

$$A := \bigcup \{A_k \in \mathcal{A} \mid A_k \cap B \neq \emptyset, 1 \leq k \leq n\} \in \mathcal{A},$$

then

   i. $A \subseteq B^\varepsilon$, because if $x \in A$, there exists exactly one index k with $x \in A_k$, and there exists $b \in B$ with $b \in A_k$, hence $d(x, B) \leq d(x, b) < \varepsilon$,

   ii. $B \subseteq A \cup X \setminus (A_1 \cup \cdots \cup A_n)$ follows directly from the construction.

But now

$$\begin{aligned}
\mu(B) &\leq \mu(A) + \mu(X \setminus (A_1 \cup \cdots \cup A_n)) \\
&\leq \mu_0(A) + \varepsilon/3 + \mu_0(X \setminus (A_1 \cup \cdots \cup A_n)) + \varepsilon/3 \\
&\leq \mu_0(A) + \varepsilon \\
&\leq \mu_0(B^\varepsilon) + \varepsilon.
\end{aligned}$$

Interchanging the rôles of μ_0 and μ yields $\mu_0(B) \leq \mu(B^\varepsilon) + \varepsilon$.

3. Recall the construction in the third part of the Portmanteau Theorem Proposition 1.66, where we approximated the indicator function of an open set from below by continuous functions. We find for each $A \in \mathcal{A}$ two continuous maps $f_\ell \leq \chi_A \leq f_u$ with $\int_X (f_u - f_\ell) \, d\mu_0 < \varepsilon/6$. This is so since we can approximate the indicator function for the largest open set contained in A from below, and similarly $\chi_{\mathrm{cl}(A)}$ from above by continuous functions to arbitrary precision; taking into account that the boundary has μ_0-mass zero, the assertion follows.

Thus, if we have $\left| \int_X f \, d\mu - \int_X f \, d\mu_0 \right| < \varepsilon/6$, for $f = f_\ell$ and for $f = f_u$, then

$$|\mu(A) - \mu_0(A)| < \varepsilon/3$$

follows: since

$$\left| \int_X f \, d\mu - \mu_0(A) \right| \leq \left| \int_X f \, d\mu - \int_X f \, d\mu_0 \right| + \left| \int_X f \, d\mu_0 - \mu_0(A) \right|,$$

the assertion follows from $\int_X f_\ell \, d\mu \leq \mu(A) \leq \int_X f_u \, d\mu$.

4. Now we are done: select bounding functions $f_{A,\ell} \leq \chi_A \leq f_{A,u}$ for each $A \in \mathcal{A}$ according to part 3, then $\left| \int_X f_k \, d\mu - \int_X f_k \, d\mu_0 \right| < \varepsilon/6$ for all $f_k \in \{f_{A,\ell}, f_{A,u} \mid A \in \mathcal{A}\}$ implies $\mathbf{d}_P(\mu_n, \mu_0) < \varepsilon$. ⬚

Thus we have established that the metric topology and the topology of weak convergence are the same.

THEOREM 1.73
Let X be a separable metric space, then the Prohorov metric is a metric for the topology of weak convergence.

It is easy to find a dense subset in $\mathfrak{S}(X)$. As one might expect, the measures living on discrete subsets are dense.

PROPOSITION 1.74
Let X be a separable metric space. The set of

$$\{\sum_{k \in \mathbb{N}} r_k \cdot \delta_{x_k} \mid x_k \in X, r_k \geq 0, \sum_{k \in \mathbb{N}} r_k \leq 1\}$$

discrete measures is dense in the topology of weak convergence.

PROOF 1. Fix $\mu \in \mathfrak{S}(X)$. Cover X for each $k \in \mathbb{N}$ with mutually disjoint Borel sets $(A_{n,k})_{n \in \mathbb{N}}$, each of which has a diameter not less that $1/k$. Select an arbitrary $x_{n,k} \in A_{n,k}$. We claim that $\mu_n := \sum_{k \in \mathbb{N}} \mu(A_{n,k}) \cdot \delta_{x_{n,k}}$ converges weakly to μ.
2. Let $f : X \to \mathbb{R}$ be a uniformly continuous and bounded map. Since f is uniformly continuous,

$$\eta_n := \sup_{k \in \mathbb{N}} \left(\sup_{x \in A_{n,k}} f(x) - \inf_{x \in A_{n,k}} f(x) \right)$$

tends to 0, as $n \to \infty$. Thus

$$\left| \int_X f \, d\mu_n - \int_X f \, d\mu \right| = \left| \sum_{k \in \mathbb{N}} \left(\int_{A_{n,k}} f \, d\mu_n - \int_{A_{n,k}} f \, d\mu \right) \right|$$
$$\leq \eta_n \cdot \sum_{k \in \mathbb{N}} \mu(A_{n,k})$$
$$\leq \eta_n$$
$$\to 0.$$

☐

COROLLARY 1.75
If X is a separable metric space, then $\mathfrak{S}(X)$ is a separable metric space in the topology of weak convergence.

PROOF Because

$$\sum_{1 \leq k \leq n} r_k \cdot \delta_{x_k} \to_w \sum_{k \in \mathbb{N}} r_k \cdot \delta_{x_k},$$

as $n \to \infty$, and because the rationals $\mathbb{Q}$ are dense in the reals, we obtain from Proposition 1.74 that

$$\{ \sum_{1 \le k \le n} r_k \cdot \delta_{x_k} \mid x_k \in D, 0 \le r_k \in \mathbb{Q}, n \in \mathbb{N}, \sum_{1 \le k \le n} r_k \le 1 \}$$

is a countable and dense subset of $\mathfrak{S}(X)$, whenever $D \subseteq X$ is a countable and dense subset of X. ⬛

We will show now that $\mathfrak{S}(X)$ is a Polish space, provided X is one; thus applying the $\mathfrak{S}$-functor to a Polish space does not leave the realm of Polish spaces. We will show later that this functor is actually an endofunctor on the category of Polish spaces; for the time being, however, we will lay the ground work for this.

In order to establish this property, we will need to recall the celebrated Riesz Representation Theorem (Parthasarathy, 1967, Theorem II.5.8). It states that a linear and positive map from the continuous functions on a compact metric space X to the reals that maps the constant function 1 to a value not exceeding 1 can be represented uniquely through a subprobability measure on the Borel sets on X. Since we need it later on anyway, we quote it here for easier reference.

THEOREM 1.76

(Riesz Representation Theorem) *Let X be a compact metric space, and $\Lambda : \mathcal{C}(X) \to \mathbb{R}$ be a map with the following properties*

a. *$\Lambda(r_1 \cdot f_1 + r_2 \cdot f_2) = r_1 \cdot \Lambda(f_1) + r_2 \cdot \Lambda(f_2)$ for all $f_1, f_2 \in \mathcal{C}(X)$ and all $r_1, r_2 \in \mathbb{R}$,*

b. *If $f \ge 0$, then $\Lambda(f) \ge 0$,*

c. *$\Lambda(1) \le 1$.*

Then there exists a unique subprobability measure $\mu \in \mathfrak{S}(X)$ such that

$$\Lambda(f) = \int_X f \, d\mu.$$

We know by Alexandrov's Theorem 1.23 that a separable metrizable space is Polish iff it can be embedded as a G_δ-set into the Hilbert cube. We show first that for compact metric X the space $\mathfrak{S}(X)$ with the topology of weak convergence is itself a compact metric space. This is established by embedding it as a closed subspace into $[-1, +1]^\infty$.

LEMMA 1.77

Let X be a compact metric space. Then $\mathfrak{S}(X)$ is a compact metric space.

PROOF   (Sketch) 1. The space $\mathcal{C}(X)$ of continuous maps into the reals is for compact metric X a Banach space. The closed unit ball

$$\mathbf{C}_1 := \{f \in \mathcal{C}(X) \mid \|f\|_\infty \leq 1\}$$

is a separable metric space in its own right. Here

$$\|f\|_\infty := \sup_{x \in X} |f(x)|$$

is as usual the sup-norm. Let $(g_n)_{n \in \mathbb{N}}$ be a countable sense subset in $\mathbf{C}_1$, and define

$$\Theta : \mathfrak{S}(X) \ni \mu \mapsto \langle \int_X g_1 \, d\mu, \int_X g_2 \, d\mu, \dots \rangle \in [-1, 1]^\infty.$$

Then Θ is injective, and we show that $\Theta[\mathfrak{S}(X)]$ is closed, hence compact.

2. In fact, let $(\mu_n)_{n \in \mathbb{N}}$ be a sequence in $\mathfrak{S}(X)$ such that $(\Theta(\mu_n))_{n \in \mathbb{N}}$ converges in $[-1, 1]^\infty$, put $\alpha_i := \lim_{n \to \infty} \int_X g_i \, d\mu_n$. For each $f \in \mathbf{C}_1$ there exists a subsequence $(g_{n_k})_{k \in \mathbb{N}}$ such that $\|f - g_{n_k}\|_\infty \to 0$ as $k \to \infty$. Thus

$$\Lambda(f) := \lim_{n \to \infty} \int_X f \, d\mu_n$$

exists. Define for $r \in \mathbb{R}$ $\Lambda(r \cdot f) := r \cdot \Lambda(f)$, then it is not difficult to show that $\Lambda : \mathcal{C}(X) \to \mathbb{R}$ is linear and that $\Lambda(f) \geq 0$, provided $f \geq 0$.

3. The Riesz Representation Theorem 1.76 now gives a unique $\mu \in \mathfrak{S}(X)$ with

$$\Lambda(f) = \int_X f \, d\mu,$$

and the construction shows that

$$\lim_{n \to \infty} \Theta(\mu_n) = \langle \int_X g_1 \, d\mu, \int_X g_2 \, d\mu, \dots \rangle.$$

Thus $\Theta[\mathfrak{S}(X)]$ is closed, hence compact. ⬜

This is the decisive step, the next step being nearly canonic. If X is a Polish space, it may be embedded as a G_δ-set into a compact space $\widetilde{X}$, the subprobabilities of which are topologically a closed subset of $[-1, +1]^\infty$, as we have just seen. Wouldn't it be wonderful if $\mathfrak{S}(X)$ would be a G_δ in $\mathfrak{S}(\widetilde{X})$ as well? Well, it is, as the proof below demonstrates.

PROPOSITION 1.78

Let X be a Polish space. Then $\mathfrak{S}(X)$ is a Polish space in the topology of weak convergence.

PROOF   1. Embed X as a G_δ-subset into a compact metric space $\widetilde{X}$, hence $X \in \mathcal{B}(\widetilde{X})$. Put

$$\mathfrak{S}_0 := \{\mu \in \mathfrak{S}\left(\widetilde{X}\right) \mid \mu(\widetilde{X} \setminus X) = 0\},$$

so $\mathfrak{S}_0$ contains exactly those subprobabilities on $\widetilde{X}$ that are concentrated on X. Then $\mathfrak{S}_0$ is homeomorphic to $\mathfrak{S}(X)$.

2. Represent $X = \bigcap_{n \in \mathbb{N}} G_n$, where $(G_n)_{n \in \mathbb{N}}$ is a sequence of open sets in $\widetilde{X}$. Given $r > 0$, the set

$$\Gamma_{k,r} := \{\mu \in \mathfrak{S}\left(\widetilde{X}\right) \mid \mu(\widetilde{X} \setminus G_k) < r\}$$

is open in $\mathfrak{S}\left(\widetilde{X}\right)$. In fact, if $\mu_n \notin \Gamma_{k,r}$ with $\mu_n \rightharpoonup_w \mu_0$, then

$$\mu_0(\widetilde{X} \setminus G_k) \geq \limsup_{n \to \infty} \mu_n(\widetilde{X} \setminus G_k) \geq r$$

by the Portmanteau Theorem 1.66, since $\widetilde{X} \setminus G_k$ is closed. Consequently, $\mu_0 \notin \Gamma_{k,r}$. This shows that $\Gamma_{k,r}$ is open, because its complement is closed. Thus

$$\mathfrak{S}_0 = \bigcap_{n \in \mathbb{N}} \bigcap_{k \in \mathbb{N}} \Gamma_{n,1/k}$$

is a G_δ-set, and the assertion follows.   □

This proof rested on the embedding of a Polish space into a compact metric space, solving the problem there, and transporting the solution back into the original environment. The solution for the compact case then depends on a classical tool like the Riesz Representation Theorem. We will use this technique again in Chapter 4 for the construction of a semi-pullback.

It can actually be shown that the Prohorov metric is complete. The proof of this is considerably more complicated. It rests also on the Riesz Theorem. But since we need merely the fact that $\mathfrak{S}(X)$ is Polish for Polish X, we do not want to delve into the complexities of that proof.

The σ-algebra of Borel sets for the topology of weak convergence is just the weak*-σ-algebra. The latter σ-algebra can be described for each measurable space.

DEFINITION 1.79   *Let $(M, \mathcal{M})$ be a measurable space. The initial σ-algebra $\mathcal{M}^\bullet$ which makes all evaluation maps $\mu \mapsto \mu(E)$ for $E \in \mathcal{M}$ measurable is called the* weak-*-σ-algebra.

Two remarks are in order. First, to show for a measurable space $(N, \mathcal{N})$ that a map $\Phi : N \to \mathfrak{S}(M, \mathcal{M})$ is $\mathcal{N}$-$\mathcal{M}^\bullet$-measurable it is sufficient to show that for $B \in \mathcal{M}$ the set $\{n \in N \mid \Phi(n)(B) \bowtie q\}$ is a member of $\mathcal{N}$ by Lemma 1.2, when $\bowtie$ is taken from the relational operators $\leq, <, >, \geq$; the set B may even be taken from a generator for $\mathcal{M}$. This observation makes handling measurable maps into $\mathfrak{S}(M, \mathcal{M})$ more versatile than meets the eye at first. Second, if $\mathcal{M}_0$ is a generator from $\mathcal{M}$, then we may deduce from Lemma 1.2

$$\mathcal{M}^\bullet = \sigma(\{\{\mu \in \mathfrak{S}(M, \mathcal{M}) \mid \mu(B) \bowtie q\} \mid q \in \mathbb{Q}, B \in \mathcal{M}_0\}),$$

so that $\mathcal{M}^\bullet$ is countably generated whenever $\mathcal{M}_0$ is.

The weak-*-σ-algebra constitute the Borel sets for the topology of weak convergence.

PROPOSITION 1.80
If X is a separable metric space, then $\mathcal{B}(X)^\bullet = \mathcal{B}(\mathfrak{S}(X))$.

PROOF   1. By the Portmanteau Theorem 1.66 the set $\{\mu \in \mathfrak{S}(M, \mathcal{M}) \mid \mu(B) > q\}$ is closed, whenever F is closed: if $(\mu_n)_{n\in\mathbb{N}}$ is a set with $\mu_n(B) \geq q$ for all n, then $q \leq \liminf_{n\to\infty} \mu_n(B) \leq \mu(B)$. This implies that $\mathcal{B}(X)^\bullet \subseteq \mathcal{B}(\mathfrak{S}(X))$.

2. On the other hand, $\mu \mapsto \int_X f \, d\mu$ is $\mathcal{B}(X)^\bullet$-measurable, whenever $f : X \to \mathbb{R}$ is Borel measurable. This is clear if $f = \chi_A$ by the definition of the weak-*-σ-algebra, because $\mu(A) = \int_X \chi_A \, d\mu$. Thus it is true if f is a measurable step function, hence we may deduce it for all f by approximating it through step functions Proposition 1.60 (decompose $f = f^+ - f^-$ with $f^+, f^- \geq 0$ and approximate each map separately).

Thus $\mu \mapsto \int_X f \, d\mu$ is also measurable for continuous f. Consequently each element of a base $U(\mu_0, \varepsilon, f_1, \ldots, f_n)$ is an element of $\mathcal{B}(X)^\bullet$; hence each open set, being a countable union of base elements, is in $\mathcal{B}(X)^\bullet$. This implies $\mathcal{B}(\mathfrak{S}(X)) \subseteq \mathcal{B}(X)^\bullet$. $\qquad \square$

1.5.3   Disintegration

We will occasionally encounter the situation that we need to decompose a measure on a product of two spaces (Section 4.4.1, Section 5.8). This problem is of course easiest dealt with when one can deduce that the measure is the product of measures on the coordinate spaces; probabilistically, this would correspond to the distribution of two independent random variables. But sometimes one is not so lucky, and there is some hidden dependence, or one simply cannot assess the degree of independence. Then one has to live with a somewhat weaker result: in this case one can decompose the measure into a measure on one component and a transition probability. This will be made specific in the discussion to follow.

Because it will not cost substantially more attention, we will treat the question a bit more generally. Let $(X, \mathcal{A})$, $(Y, \mathcal{B})$, and $(Z, \mathcal{C})$ be measurable spaces, assume that $\mu \in \mathfrak{S}(X, \mathcal{A})$, and let $f : X \to Y$ and $g : X \to Z$ be measurable maps. Then $\mu_f(B) := \mu(f^{-1}[B])$ and $\mu_g(C) := \mu(g^{-1}[C])$ define subprobabilities on $(Y, \mathcal{B})$ resp. $(Z, \mathcal{C})$. This is so since the inverse image is compatible with the set operations, and because the maps are measurable. Consider the probability for the event that $f(x) \in B$ and $g(x) \in C$. μ_f and μ_g can be interpreted as the probability distribution of f resp. g under μ.

We will show that we can represent the joint distribution as

$$\mu(\{x \in X \mid f(x) \in B, g(x) \in C\}) = \int_B K(y)(C)\, \mu_f(dy),$$

where $K : Y \times \mathcal{C} \to [0, 1]$ is a measurable map in one component, and a subprobability on $\mathcal{C}$ in the other. This requires Z to be a Polish space with $\mathcal{C} = \mathcal{B}(Z)$.

Let us see how this corresponds to the initially stated problem. Suppose $X := Y \times Z$ with $\mathcal{A} = \mathcal{B} \otimes \mathcal{C}$, and let $f := \pi_Y$, $g := \pi_Z$, then $\mu_f(B) = \mu(B \times Z)$, $\mu_g(C) = \mu(Y \times Z)$, and $\mu(\{x \in X \mid f(x) \in B, g(x) \in C\}) = \mu(B \times C)$. Granted that we have established the decomposition, we can then write

$$\mu(B \times C) = \int_B K(y)(C)\, \mu_f(dy);$$

thus we have decomposed the probability on the product into a probability on the first component, and, conditioned on the value the first component may take, a probability on the second factor.

DEFINITION 1.81   *Using the notation from above, K is called a* regular conditional distribution of g given f *iff*

$$\mu(\{x \in X \mid f(x) \in B, g(x) \in C\}) = \int_B K(y)(C)\, \mu_f(dy)$$

holds for each $B \in \mathcal{B}, C \in \mathcal{C}$, where $K : Y \times \mathcal{C} \to [0, 1]$ is a sub-Markov kernel, on $(X, \mathcal{A})$ and $(Z, \mathcal{C})$. This means that K has these properties

1. *$y \mapsto K(y)(C)$ is $\mathcal{B}$-measurable for all $C \in \mathcal{C}$,*

2. *$K(y) \in \mathfrak{S}(Z, \mathcal{C})$ for all $y \in Y$.*

If K satisfies only property 1, then it will be called a conditional distribution of g given f.

Sub-Markov kernels are also called *transition subprobability functions*. We will discuss them later as stochastic relations.

The existence of regular conditional distribution will be established, provided Z is Polish with $\mathcal{C} = \mathcal{B}(Z)$. This will happen in several steps: first the existence of a conditional distribution will be shown using the well known Radon-Nikodym Theorem. The latter construction will then be scrutinized. It will turn out that there exists a set of measure zero outside of which the conditional distribution behaves like a regular one, but at first sight only on an algebra of sets, not on the entire σ-algebra. But don't worry, using a classical extension argument will then do the job and yield a regular conditional distribution on the Borel sets, just as we want it. The proofs are actually a kind of a round trip through the first principles of measure theory, where the Radon-Nikodym Theorem together with the classical Hahn Extension Theorem are the main vehicles. It displays also some nice and helpful proof techniques.

We fix $(X, \mathcal{A})$, $(Y, \mathcal{B})$, and $(Z, \mathcal{C})$ as measurable spaces, assume that $\mu \in \mathfrak{S}(X, \mathcal{A})$, and take $f : X \to Y$ and $g : X \to Z$ to be measurable maps. The measures μ_f and μ_g are defined as above as the distribution of f resp. g under μ.

The existence of a conditional distribution of g given f is established first, and it is shown that it is essentially unique.

LEMMA 1.82
Using the notation from above, then

a. *there exists a conditional distribution K_0 of g given f,*

b. *if there is another conditional distribution K_0' of g given f, then there exists for any $C \in \mathcal{C}$ a set $N_C \in \mathcal{B}$ with $\mu_f(N_C) = 0$ such that $K_0(y)(C) = K_0'(C)$ for all $y \notin C$.*

PROOF   1. Fix $C \in \mathcal{C}$, then

$$\varpi_C(B) := \mu(f^{-1}[B] \cap g^{-1}[C])$$

defines a subprobability measure ϖ_C on $\mathcal{B}$ which is absolutely continuous with respect to μ_g, because $\mu_g(B) = 0$ implies $\varpi_C(B) = 0$. The classic Radon-Nikodym Theorem(Halmos, 1950, Theorem 31.A) now gives a density $h_C \in \mathcal{F}(Y, \mathcal{B})$ with

$$\varpi_C(B) = \int_B h_C \, d\mu_f$$

for all $B \in \mathcal{B}$. Setting $K_0(y)(C) := h_C(y)$ yields the desired conditional distribution.

2. Suppose K_0' is another conditional distribution of g given f, then we have for all $C \in \mathcal{C}$

$$\forall B \in \mathcal{B} : \int_B K_0(y)(C) \, \mu_f(dy) = \int_B K_0(y)(C) \, \mu_f(dy),$$

which implies that the set on which $K_0(\cdot)(C)$ disagrees with $K_0'(\cdot)(C)$ is μ_f-null. $\quad\square$

Essential uniqueness may strengthened if the σ-algebra $\mathcal{C}$ is countably generated, and if the conditional distribution is regular.

LEMMA 1.83

Assume that K and K' are regular conditional distributions of g given f, and that $\mathcal{C}$ has a countable generator. Then there exists a set $N \in \mathcal{B}$ with $\mu_f(N) = 0$ such that $K(y)(C) = K'(y)(C)$ for all $C \in \mathcal{C}$ and all $y \notin N$.

PROOF   If $\mathcal{C}_0$ is a countable generator of $\mathcal{C}$, then

$$\mathcal{C}_f := \{\bigcap \mathcal{E} \mid \mathcal{E} \subseteq \mathcal{C}_0 \text{ is finite}\}$$

is a countable generator of $\mathcal{C}$ well, and $\mathcal{C}_f$ is closed under finite intersections; note that $Z \in \mathcal{C}_f$. Construct for $D \in \mathcal{C}_f$ the set $N_D \in \mathcal{B}$ outside of which $K(\cdot)(D)$ and $K'(\cdot)(D)$ coincide, and define

$$N := \bigcup_{D \in \mathcal{C}_f} N_D \in \mathcal{B}.$$

Evidently, $\mu_f(N) = 0$. We claim that $K(y)(C) = K'(y)(C)$ holds for all $C \in \mathcal{C}$, whenever $y \notin N$. In fact, fix $y \notin N$, and let

$$\mathcal{C}_1 := \{C \in \mathcal{C} \mid K(y)(C) = K'(y)(C)\},$$

then $\mathcal{C}_1$ contains $\mathcal{C}_f$ by construction, and is a π-λ-system. This is so since it is closed under complements and countable disjoint unions. Thus $\mathcal{C} = \sigma(\mathcal{C}_f) \subseteq \mathcal{C}_1$, by the π-λ-Theorem 1.1, and we are done.   □

We will show now that a regular conditional distribution of g given f exists. This will be done through several steps, given the construction of a conditional distribution K_0:

A. A set $N_a \in \mathcal{B}$ is constructed with $\mu_f(N_a) = 0$ such that $K_0(y)$ is additive on a countable generator $\mathcal{C}_z$ for $\mathcal{C}$.

B. We construct a set $N_z \in B$ with $\mu_f(N_z) = 0$ such that $K_0(y)(Z) \leq 1$ for $y \notin N_z$.

C. For each element G of $\mathcal{C}_z$ we will find a set $N_G \in \mathcal{B}$ with $\mu_f(N_G) = 0$ such that $K_0(y)(G)$ can be approximated from inside through compact sets, whenever $y \notin N_G$.

D. Then we will combine all these sets of μ_f-measure zero to produce a set $N \in \mathcal{B}$ with $\mu_f(N) = 0$ outside of which $K_0(y)$ is a premeasure on the generator $\mathcal{C}_z$, hence can be extended to a measure on all of $\mathcal{C}$.

Well, this looks like a full program, so let us get on with it.

THEOREM 1.84

Given measurable spaces $(X, \mathcal{A})$ and $(Y, \mathcal{B})$, a Polish space Z, a subprobability $\mu \in \mathfrak{S}(X, \mathcal{A})$, and measurable maps $f : X \to Y$, $g : X \to Z$, there exists a regular conditional distribution K of g given f. K is uniquely determined up to a set of μ_f-measure zero.

PROOF   0. Since Z is a Polish space, its topology has a countable base. We infer from Lemma 1.29 that $\mathcal{B}(Z)$ has a countable generator $\mathcal{C}$. Then the Boolean algebra $\mathcal{C}_1$ generated by $\mathcal{C}$ is also a countable generator of $\mathcal{B}(Z)$.

1. Given $C_n \in \mathcal{C}_1$, we find by Proposition 1.65 a sequence $(L_{n,k})_{k \in \mathbb{N}}$ of compact sets in Z with

$$L_{n,1} \subseteq L_{n,2} \subseteq L_{n,3} \ldots \subseteq C_n$$

such that

$$\mu_g(C_n) = \sup_{k \in \mathbb{N}} \mu_g(L_{n,k}).$$

Then the Boolean algebra $\mathcal{C}_z$ generated by $\mathcal{C} \cup \{L_{n,k} \mid n, k \in \mathbb{N}\}$ is also a countable generator of $\mathcal{B}(Z)$.

2. From the construction of the conditional distribution of g given f we infer that for disjoint $C_1, C_2 \in \mathcal{C}_z$

$$
\begin{aligned}
\int_Y K_0(y)(C_1 \cup C_2)\, \mu_f(dy) &= \mu(\{x \in X \mid f(x) \in B, g(x) \in C_1 \cup C_2\}) \\
&= \mu(\{x \in X \mid f(x) \in B, g(x) \in C_1\}) + \\
&\qquad \mu(\{x \in X \mid f(x) \in B, g(x) \in C_2\}) \\
&= \int_Y K_0(y)(C_1)\, \mu_f(dy) + \int_Y K_0(y)(C_2)\, \mu_f(dy).
\end{aligned}
$$

Thus there exists $N_{C_1, C_2} \in \mathcal{B}$ with $\mu_f(N_{C_1, C_2}) = 0$ such that

$$K_0(y)(C_1 \cup C_2) = K_0(y)(C_1) + K_0(y)(C_2)$$

for $y \notin N_{C_1, C_2}$. Because $\mathcal{C}_z$ is countable, we may deduce (by taking the union of N_{C_1, C_2} over all pairs C_1, C_2) that there exists a set $N_a \in \mathcal{B}$ such that K_0 is additive outside N_a, and $\mu_f(N_a) = 0$. This accounts for part A in the plan above.

3. It is by the previous arguments easy to construct a set $N_z \in \mathcal{B}$ with $\mu_f(N_z) = 0$ such that $K_0(y)(Z) \leq 1$ for $y \notin N_z$ (part B).

4. Because

$$\int_Y K_0(y)(C_n)\,\mu_f(dy) = \mu(f^{-1}[Y] \cap g^{-1}[C_n])$$

$$= \mu_g(C_n)$$

$$= \sup_{k \in \mathbb{N}} \mu_g(L_{n,k})$$

$$= \sup_{k \in \mathbb{N}} \int_Y K_0(y)(L_{n,k})\,\mu_f(dy)$$

$$= \int_Y \sup_{k \in \mathbb{N}} K_0(y)(L_{n,k})\,\mu_f(dy)$$

we find for each $n \in \mathbb{N}$ a set $N_n \in \mathcal{B}$ with

$$\forall y \notin N_n : K_0(y)(C_n) = \sup_{k \in \mathbb{N}} K_0(y)(L_{n,k})$$

and $\mu_f(N_n) = 0$. This accounts for part C.

5. Now we may begin to work on part D. Put

$$N := N_a \cup N_z \cup \bigcup_{n \in \mathbb{N}} N_n,$$

then $N \in \mathcal{B}$ with $\mu_f(N) = 0$. We claim that $K_0(y)$ is a premeasure on $\mathcal{C}_z$ for each $y \notin N$. It is clear that $K_0(y)$ is additive on $\mathcal{C}_z$, hence monotone, so merely σ-additivity has to be demonstrated: let $(D_\ell)_{\ell \in \mathbb{N}}$ be a sequence in $\mathcal{C}_z$ that is monotonically decreasing with

$$\eta := \inf_{\ell \in \mathbb{N}} K_0(y)(D_\ell) > 0,$$

then we have to show that

$$\bigcap_{\ell \in \mathbb{N}} D_\ell \neq \emptyset.$$

We approximate the sets D_ℓ now by compact sets, so we assume that $D_\ell = C_{n_\ell}$ for some n_ℓ (otherwise the sets are compact themselves). By construction we find for each $\ell \in \mathbb{N}$ a compact set $L_{n_\ell,k_\ell} \subseteq C_\ell$ with

$$K_0(y)(C_{n_\ell} \setminus L_{n_\ell,k_\ell}) < \eta \cdot 2^{\ell+1},$$

then

$$L_r := \bigcap_{i=\ell}^{r} L_{n_\ell,k_\ell} \subseteq C_{n_r} = D_r$$

defines a decreasing sequence of compact sets with

$$K_0(y)(L_r) \geq K_0(y)(C_{n_r}) - \sum_{i=\ell}^{r} K_0(y)(L_{n_\ell,k_\ell}) > \eta/2,$$

thus $L_r \neq \emptyset$. Since L_r is compact and decreasing, we know that the sequence has a nonempty intersection (otherwise one of the L_r would already be empty). We may infer

$$\bigcap_{\ell \in \mathbb{N}} D_\ell \supseteq \bigcap_{r \in \mathbb{N}} L_r \neq \emptyset.$$

6. The classic Hahn Extension Theorem (Halmos, 1950, Theorem 13.A) now tells us that there exists a unique extension of $K_0(y)$ from $\mathcal{C}_z$ to a measure $K(y)$ on $\sigma(\mathcal{C}_z) = \mathcal{B}(Z)$, whenever $y \notin N$. If, however, $y \in N$, then we define $K(y) := \nu$, where $\nu \in \mathfrak{S}(Z)$ is arbitrary. Because

$$\int_B K(y)(C)\, \mu_f(dy) =$$

$$\int_B K_0(y)(C)\, \mu_f(dy) = \mu(\{x \in X \mid f(x) \in B, g(x) \in C\})$$

holds for $C \in \mathcal{C}_z$, the π-λ-Theorem 1.1 asserts that this equality is valid for all $C \in \mathcal{B}(Z)$ as well.

Measurability of $y \mapsto K(y)(C)$ needs to be shown, and then we are done. Put

$$\mathcal{E} := \{C \in \mathcal{B}(Z) \mid y \mapsto K(y)(C) \text{ is } \mathcal{B} - \text{measurable}\}.$$

Then $\mathcal{E}$ is a σ-algebra, and $\mathcal{E}$ contains the generator $\mathcal{C}_z$ by construction, thus $\mathcal{E} = \mathcal{B}(Z)$. □

The scenario in which the space $X - Y \times Z$ with a measurable space $(Y, \mathcal{B})$ and a Polish space Z with $\mathcal{A} = \mathcal{B} \otimes \mathcal{B}(Z)$ with f and g as projections deserves particular attention. In this case we decompose a measure on A into its projection onto Z and a conditional distribution for the projection onto Z given the projection onto Y. This is sometimes called the *disintegration* of a measure $\mu \in \mathfrak{S}(Y \times Z)$.

We state the corresponding Proposition explicitly, since we will use it in this specialized form.

PROPOSITION 1.85

Given a measurable space $(Y, \mathcal{B})$ and a Polish space Z, there exists for every subprobability $\mu \in \mathfrak{S}(Y \times Z, \mathcal{B} \otimes \mathcal{B}(Z))$ a regular conditional distribution of π_Z given π_Y.

1.5.4 Applications of the π-λ-Theorem

The π-λ-Theorem is used typically in the following scenario: we have a property P for which we know the following

i. $P(A)$ holds for all elements A of a generator $\mathcal{A}$ of a σ-algebra $\mathcal{B}$.

ii. $P(A)$ implies $P(X \setminus A)$ with X as the basic set.

iii. if $P(A_n)$ holds for all $n \in \mathbb{N}$, and $(A_n)_{n \in \mathbb{N}}$ is mutually disjoint, then $P(\bigcup_{n \in \mathbb{N}} A_n)$ holds.

We then have a look at

$$\mathcal{G} := \{A \in \mathcal{B} \mid P(A)\}$$

($\mathcal{G}$ stands of course for the *good guys*), then $P(A)$ holding for all $A \in \mathcal{A}$ translates into $\mathcal{A} \subseteq \mathcal{G}$, and the other two properties make sure that G is a π-λ-system. We conclude then from the π-λ-Theorem 1.1 that $\mathcal{G} = \sigma(\mathcal{A}) = \mathcal{B}$, provided $\mathcal{A}$ is closed under finite intersections. This was the argumentation for example in the proofs of Lemma 1.62 and Lemma 1.70. It shows that this Theorem is a rather versatile tool.

We want to demonstrate its application when exploring measure extensions, and for settling questions of measurability, which is also somewhat typical for getting it to work later on. The existence of projective limits which is established in Proposition 1.88 may also be listed among its useful applications.

Suppose that μ_n is a probability measure on the measurable space $(X_n, \mathcal{A}_n)$ for each $n \in \mathbb{N}$, and define for a cylinder set

$$\hat{\mu} \left(\prod_{n \in \mathbb{N}} A_n \right) := \prod_{n \in \mathbb{N}} \mu_n(A_n).$$

Observe that in this infinite product all but a finite number of factors equal unity. Then $\hat{\mu}$ extends uniquely to a probability measure $\mu^{\#}$ on the product $(\prod_{n \in \mathbb{N}} A_n, \bigotimes_{n \in \mathbb{N}} \mathcal{A}_n)$; in particular,

$$\mu^{\#}(A_1 \times \ldots A_n \times \prod_{j > n} X_j) = \mu_1(A_1) \cdot \ldots \cdot \mu_n(A_n)$$

holds. Accordingly, $\mu^{\#}$ is called the product measure of $(\mu_n)_{n \in \mathbb{N}}$ and denoted by $\bigotimes_{n \in \mathbb{N}} \mu_n$. Of course, a finite product is also available. The π-λ-Theorem assures us that the extension is unique.

Another application is given when applying horizontal or vertical cuts from a measurable set in a product and then asking about measurability of associated maps.

LEMMA 1.86

Let $(X, \mathcal{A})$ and $(Y, \mathcal{B})$ be measurable spaces, and fix $D \in \mathcal{A} \otimes \mathcal{B}$. The map $\langle \nu, x \rangle \mapsto \nu(D_x)$ is a $\mathcal{B}^{\bullet} \otimes \mathcal{A}$-measurable map on $\mathfrak{S}(Y, \mathcal{B}) \times X$.

PROOF   Consider

$$\mathcal{D} := \{D \in \mathcal{A} \otimes \mathcal{B} \mid \langle \nu, x \rangle \mapsto \nu(D_x) \text{ is } \mathcal{B}^{\bullet} \otimes \mathcal{A} - \text{measurable}\}.$$

Since $((X \times Y) \setminus D)_x = Y \setminus (D_x)$ and

$$\left(\bigcup_{n \in \mathbb{N}} D_n \right)_x = \bigcup_{n \in \mathbb{N}} (D_n)_x,$$

it is clear that $\mathcal{D}$ is closed under taking complements and countable disjoint unions. Now let $D = A \times B$ with $A \in \mathcal{A}, B \in \mathcal{B}$. Then $\nu(D_x) = \chi_A(x) \cdot \nu(B)$, thus $\langle \nu, x \rangle \mapsto \nu(D_x)$ is evidently $\mathcal{B}^\bullet \otimes \mathcal{A}$-measurable. But this implies that all measurable rectangles are members of $\mathcal{D}$, and since the set of all these rectangles is closed under finite intersections, $\mathcal{D}$ equals the σ-algebra generated from them, which coincides with $\mathcal{A} \otimes \mathcal{B}$. The assertion is hence true for all measurable subsets of the product. ⬚

Lemma 1.86 entails that both $x \mapsto \nu(D_x)$ and $\nu \mapsto \nu(D_x)$ are measurable (but the Lemma says considerably more: it establishes joint measurability).

1.5.5 Projective Systems

We need for the interpretation of the continuous time stochastic logics **CSL** and μ**CSL** in Chapter 6 the projective limit of a projective family of stochastic relations. Denote by $X^\infty := \prod_{k \in \mathbb{N}} X$ the infinite product of X with itself.

DEFINITION 1.87 *Let X be a Polish space, and $(\mu_n)_{n \in \mathbb{N}}$ a sequence of probability measures $\mu_n \in \mathfrak{P}(X^n)$. This sequence is called a* projective system *iff $\mu_n(A) = \mu_{n+1}(A \times X)$ for all $n \in \mathbb{N}$ and all Borel sets $A \in \mathcal{B}(X^n)$. A probability measure $\mu_\infty \in \mathfrak{P}(X^\infty)$ is called the* projective limit *of the projective system $(\mu_n)_{n \in \mathbb{N}}$ iff*

$$\mu_n(A) = \mu_\infty(A \times \prod_{j > n} X)$$

for all $n \in \mathbb{N}$ and $A \in \mathcal{B}(X^n)$.

Thus a sequence of measures is a projective system iff each measure is the projection of the next one; its projective limit is characterized through the property that its values on cylinder sets coincides with the value of a member of the sequence, after taking projections.

It is not immediately obvious that a projective limit exists. The basic idea is to define the limit on the cylinder sets and then to extend this premeasure — but it has to be established that it is indeed a premeasure. The crucial property is that $\mu_{n_k}(A_k) \to 0$ whenever $(A_n)_{n \in \mathbb{N}}$ is a sequence of cylinder sets A_k (with at most n_k components that do not equal X) that decreases to $\emptyset$. This property is difficult to establish without topological assumptions. This is why we did postulate the base space X to be Polish.

The central statement is

PROPOSITION 1.88

Let X be a compact metric space. Then a unique projective limit μ_∞ exists for the projective system $(\mu_n)_{n\in\mathbb{N}}$.

PROOF   1. Let $A = A'_k \times \prod_{j>k} X$ be a cylinder set with $A'_k \in \mathcal{B}(X^k)$; then define $\mu^*(A) := \mu_k(A'_k)$. Then μ^* is well defined, since the sequence forms a projective system. In order to show that μ^* is a premeasure on the cylinder sets, we have to take a decreasing sequence $(A_n)_{n\in\mathbb{N}}$ of cylinder sets with $\bigcap_{n\in\mathbb{N}} A_n = \emptyset$ and show that $\inf_{n\in\mathbb{N}} \mu^*(A_n) = 0$. In fact, suppose that $(A_n)_{n\in\mathbb{N}}$ is decreasing with $\mu^*(A_n) \geq \delta$ for all $n \in \mathbb{N}$, then we show that $\bigcap_{n\in\mathbb{N}} A_n \neq \emptyset$.

We can write $A_n = A'_n \times \prod_{j>k_n} X$ for some $A'_n \in \mathcal{B}(X^{k_n})$. From Proposition 1.65 we get for each n a compact set $K'_n \subseteq A'_n$ such that $\mu_{k_n}(A'_n \setminus K'_n) < \delta/2^n$. Because X^∞ is compact by Tichonov's Theorem,

$$K''_n := K'_n \times \prod_{j>k_n} X$$

is a compact set, and $K_n := \bigcap_{j=1}^n K''_j \subseteq A_n$ is compact as well, with

$$\mu^*(A_n \setminus K_n) \leq \mu^*(\bigcup_{j=1}^n A''_n \setminus K''_j) \leq$$

$$\sum_{j=i}^n \mu^*(A''_j \setminus K''_j) = \sum_{j=1}^n \mu_{k_j}(A'_j \setminus K'_j) \leq \sum_{j=1}^\infty \delta/2^j = \delta.$$

Thus $(K_n)_{n\in\mathbb{N}}$ is a decreasing sequence of nonempty compact sets; consequently,

$$\emptyset \neq \bigcap_{n\in\mathbb{N}} K_n \subseteq \bigcap_{n\in\mathbb{N}} A_n.$$

2. Since the cylinder sets generate the Borel sets of X^∞, and since μ^* is a premeasure, we know that there exists a unique extension $\mu_\infty \in \mathfrak{P}(X^\infty)$ to it. Clearly, if $A \subseteq X^n$ is a Borel set, then

$$\mu_\infty(A \times \prod_{j>n} X) = \mu^*(A \times \prod_{j>n} X) = \mu_n(A),$$

so we have constructed a projective limit.

3. Suppose that μ' is another probability measure in $\mathfrak{P}(X^\infty)$ that has the desired property. Consider

$$\mathcal{D} := \{D \in \mathcal{B}(X^\infty) \mid \mu_\infty(D) = \mu'(D)\}.$$

It is clear the $\mathcal{D}$ contains all cylinder sets, that it is closed under complements, and under countable disjoint unions. By the π-λ-Theorem 1.1 $\mathcal{D}$ contains the

σ-algebra generated by the cylinder sets, hence all Borel subset of X^∞. This establishes uniqueness of the extension. □

The proof makes critical use of the tightness property for finite measures on Polish spaces that says that we can approximate the measure of a Borel set arbitrarily well by compact sets; see Proposition 1.65. It is also important that compact sets have the finite intersection property: if each finite intersection of a family of compact sets is nonempty, the intersection of the entire family cannot be empty (see the proof of Theorem 1.84, where exactly this property was crucial as well). Consequently the proof given above works in general Hausdorff spaces, provided the measures under consideration are tight.

The construction from above can be made use of when we work in Proposition 1.88 in a compact scenario. We liberate us from that restrictive assumption using the Alexandrov embedding of Polish spaces into compact metric spaces that we will also put to good use in Section 4.3.2, when we transport a measure extension from a compact to a general Polish space.

PROPOSITION 1.89

Let X be a Polish space, $(\mu_n)_{n\in\mathbb{N}}$ be a projective system on X. Then there exists a unique projective limit $\mu_\infty \in \mathfrak{P}(X^\infty)$ for $(\mu_n)_{n\in\mathbb{N}}$.

PROOF X is a dense measurable subset of a compact metric space $\widetilde{X}$ by (Kechris, 1994, Theorem 4.14). Defining $\widetilde{\mu}_n(B) := \mu_n(B \cap X^n)$ for the Borel set $B \subseteq \widetilde{X}^n$ yields a projective system $(\widetilde{\mu}_n)_{n\in\mathbb{N}}$ on $\widetilde{X}$ with a projective limit $\widetilde{\mu}_\infty$ by Proposition 1.88. Since by construction $\widetilde{\mu}_\infty(X^\infty) = 1$, restrict $\widetilde{\mu}_\infty$ to the Borel sets of X^∞, then the assertion follows. □

Our interest in this construction comes from sub-Markov kernels that may form a projective system. We will show now that there exists such a kernel which may be thought as the (pointwise) projective limit.

COROLLARY 1.90

Let X and Y be Polish spaces, and assume that $J^{(n)}$ is a sub-Markov kernel on X and Y^n for each $n \in \mathbb{N}$ such that the sequence $(J^{(n)}(x))_{n\in\mathbb{N}}$ forms a projective system on Y for each $x \in X$, in particular $J^{(n)}(x)(Y^n) = 1$ for all $x \in X$. Then there exists a unique sub-Markov kernel J_∞ on X and Y^∞ such that $J_\infty(x)$ is the projective limit of $(J^{(n)}(x))_{n\in\mathbb{N}}$ for each $x \in X$.

PROOF 0. Let for x fixed $J_\infty(x)$ be the projective limit of the projective system $(J^{(n)}(x))_{n\in\mathbb{N}}$. By the definition of a sub-Markov kernel we need to show that the map $x \mapsto J_\infty(x)(B)$ is measurable for every $B \in \mathcal{B}(Y^\infty)$.

1. In fact, consider

$$\mathcal{D} := \{B \in \mathcal{B}(Y^\infty) \mid x \mapsto J_\infty(x)(B) \text{ is measurable}\}$$

then the general properties of measurable functions imply that $\mathcal{D}$ is a σ-algebra on Y^∞. Take a cylinder set $B = B_0 \times \prod_{j>k} Y$ with $B_0 \in \mathcal{B}(Y^k)$ for some $k \in \mathbb{N}$, then, by the properties of the projective limit, we have $J_\infty(x)(B) = J^{(k)}(x)(B_0)$. But $x \mapsto J^{(k)}(x)(B_0)$ constitutes a measurable function on X. Consequently, $B \in \mathcal{D}$, and so $\mathcal{D}$ contains the cylinder sets which generate $\mathcal{B}(Y^\infty)$. Thus measurability is established for each Borel set $B \subseteq Y^\infty$, arguing with the π-λ-Theorem as in the last part of the proof for Proposition 1.88.   ⬚

This construction will be needed when interpreting a path logic over infinite paths in Chapter 6.

1.6   Categories

A *category* $\mathfrak{C}$ consists of a collection of objects, and for any objects a and b a set of arrows $f : a \to b$ from a to b; a is called the *domain*, b the *codomain* or the *range* of f; an arrow is also called a *morphism*. Given arrows $f : a \to b$ and $g : b \to c$, there is an operation $\circ$ called *composition* that composes arrows, so that $g \circ f : a \to c$ is an arrow in $\mathfrak{C}$ from a to c. Composition is associative, for each object a there exists an identity arrow $id_a : a \to a$ so that $f \circ id_a = f = id_b \circ f$ holds, whenever $f : a \to b$ is an arrow of $\mathfrak{C}$. The collection $\mathfrak{C}(a, b)$ denotes all morphisms $a \to b$ of $\mathfrak{C}$, the *hom-set of a and b*.

EXAMPLE 1.91
Let us consider some simple example categories.

1. The objects in category $\mathfrak{Set}$ are sets; morphisms between sets are maps with composition as the usual composition of maps. The identity id_a is just the identity map $a \to a$.

The example illustrates that the collection of objects is not necessarily a set, but that the morphisms between two objects may form one.

2. The objects of a category $\mathfrak{NoName}_P$ are the elements of a fixed ordered set $(P, \le)$; there exists a morphism between objects a and b iff $a \le b$ holds. Reflexivity corresponds to the existence of an identity arrow, transitivity of the order relation translates into associativity of the composition.

The example illustrates that the morphisms in a category do not need to be maps.

3. The category $\mathfrak{Top}$ has as objects topological spaces. Morphisms are the continuous maps.

4. $\mathfrak{Meas}$ has measurable spaces as objects. Morphisms are the measurable maps.

5. $\mathfrak{Borel}$ has as objects the Borel sets of Polish spaces; a morphism $\mathcal{B}(S) \to \mathcal{B}(T)$ is a map $m_f : \mathcal{B}(S) \to \mathcal{B}(T)$ corresponding to the inverse image f^{-1} of a Borel map $f : T \to S$ with $m_f(B) := f^{-1}[B]$, whenever $B \in \mathcal{B}(S)$. m_f is a map because measurability of f implies that the inverse image of a Borel set is a Borel set again. Note that the transition from f to m_f reverses the direction of the arrows.

The category $\mathfrak{C}^{op}$ *opposite* or *dual* to $\mathfrak{C}$ has just the objects from $\mathfrak{C}$, and a morphism $a \to b$ in $\mathfrak{C}^{op}$ is a morphism $b \to a$ in $\mathfrak{C}$, so that the arrows are reversed.

Let $f : a \to b, g : b \to d$ and $h : a \to c, i : c \to d$ be arrows in $\mathfrak{C}$ such that $g \circ f = i \circ h$, then the pictorial representation

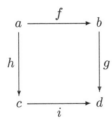

yields a commutative diagram.

DEFINITION 1.92   *Given two categories $\mathfrak{C}$ and $\mathfrak{D}$.*

a. *A covariant functor $\mathfrak{F} : \mathfrak{C} \to \mathfrak{D}$ maps each object a in $\mathfrak{C}$ into an object $\mathfrak{F}(a)$ in $\mathfrak{D}$, and maps each morphism $f : a \to b$ in $\mathfrak{C}$ into a morphism $\mathfrak{F}(f) : \mathfrak{F}(a) \to \mathfrak{F}(b)$ with these properties:*

   i. *$\mathfrak{F}(id_a) = id_{\mathfrak{F}(a)}$ holds for all objects a in $\mathfrak{C}$,*

   ii. *$\mathfrak{F}(f \circ_{\mathfrak{C}} g) = \mathfrak{F}(f) \circ_{\mathfrak{D}} \mathfrak{F}(g)$ holds, whenever $f \circ_{\mathfrak{C}} g$ is defined in $\mathfrak{C}$.*

b. *A contravariant functor $\mathfrak{F} : \mathfrak{C} \to \mathfrak{D}$ is a (covariant) functor $\mathfrak{F} : \mathfrak{C} \to \mathfrak{D}^{op}$*

Thus a covariant functor preserves identities as well as the composition of arrows (we will of course continue writing $\circ$ rather than $\circ_{\mathfrak{C}}$; the reader should be aware of the composition being specific for the category), a contravariant functor reverses the order of the composition. When talking about functors without characterizing them as co- or contravariant, we always have a covariant functor in mind.

If the categories coincide, then $\mathfrak{F}$ is called an *endofunctor*; the identity $\mathbb{1}_{\mathfrak{C}}$ is a trivial example for an endofunctor. For defining a functor $\mathfrak{F}$ one has to define how $\mathfrak{F}$ acts on objects. Additionally one has to say what $\mathfrak{F}$ does to morphisms.

EXAMPLE 1.93

1. Assign to each topological space the set it is based on, and assign each continuous map itself. Then this yields a functor $\mathfrak{U} : \mathfrak{Top} \to \mathfrak{Set}$ which forgets the structure (hence is called *forgetful*).

2. Assign to each topological space $(X, \mathcal{S})$ the associated measurable space $(X, \mathcal{B}(X, \mathcal{S}))$, and map each $\mathcal{S}\text{-}\mathcal{T}$-continuous map $f : X \to Y$ to itself. Because f is $\mathcal{B}(X, \mathcal{S})\text{-}\mathcal{B}(Y, \mathcal{T})$-measurable by Lemma 1.6, this yields a functor $\mathfrak{Top} \to \mathfrak{Meas}$.

3. Let $f : P \to Q$ an order morphism between the ordered sets $(P, \leq)$ and $(Q, \sqsubseteq)$, thus $f(a) \sqsubseteq b(b)$ provided $a \leq b$. Then $a \mapsto f(a)$ induces a functor $\mathfrak{NoName}_P \to \mathfrak{NoName}_Q$ (which for reasons of consistency remains anonymous as well).

4. Let $\mathfrak{B}(X) := \mathcal{B}(X)$, whenever X is a Polish space, and define for the Borel map $f : X \to Y$ the map $\mathcal{B}(Y) \to \mathcal{B}(X)$ through $\mathfrak{B}(f)(D) := f^{-1}[D]$ (this is m_f from Example 1.91). Then $\mathfrak{B} : \mathfrak{Pol} \to \mathfrak{Borel}^{op}$ is a covariant functor, equivalently, $\mathfrak{B} : \mathfrak{Pol} \to \mathfrak{Borel}$ is a contravariant functor.

5. Let a and b objects of $\mathfrak{C}$, then $\mathfrak{C}(a, -) : x \mapsto \mathfrak{C}(a, x)$ defines a covariant functor $\mathfrak{C}(a, -) : \mathfrak{C} \to \mathfrak{Set}$ upon defining for the morphism $f : c \to d$ the map

$$\mathfrak{C}(a, f) := f_* : \begin{cases} \mathfrak{C}(a, c) & \to \mathfrak{C}(a, d) \\ g & \mapsto f \circ g. \end{cases}$$

The properties of a functor are elementary. Similarly, $\mathfrak{C}(-, b) : x \mapsto \mathfrak{C}(a, x)$ defines a contravariant functor $\mathfrak{C}(-, b) : \mathfrak{C} \to \mathfrak{Set}$ upon defining for the morphism $f : c \to d$ the map

$$\mathfrak{C}(f, b) := f^* : \begin{cases} \mathfrak{C}(d, b) & \to \mathfrak{C}(c, b) \\ g & \mapsto g \circ f. \end{cases}$$

This functor is commonly called the covariant *hom-set functor*, its contravariant cousin $\mathfrak{C}(b, -)$ is defined similarly.

If $\mathfrak{F}$ is an endofunctor on category $\mathfrak{C}$, then the pair $\langle c, f \rangle$ consisting of an object c and a morphism $f : \mathfrak{F}(c) \to c$ is called an $\mathfrak{F}$-*algebra*. Suppose that $\langle d, g \rangle$ is another $\mathfrak{F}$-algebra, and $\phi : c \to d$ is a morphism in $\mathfrak{C}$ such that the diagram

$$
\begin{array}{ccc}
c & \xrightarrow{\ \phi\ } & d \\
{\scriptstyle f}\downarrow & & \downarrow{\scriptstyle g} \\
\mathfrak{F}(c) & \xrightarrow[\mathfrak{F}(\phi)]{} & \mathfrak{F}(d)
\end{array}
$$

commutes, then $\phi : \langle c, f \rangle \to \langle d, g \rangle$ is an $\mathfrak{F}$-algebra morphism. Dually, a pair $\langle c, f \rangle$ consisting of an object c and a morphism $f : c \to \mathfrak{F}(c)$ is called an $\mathfrak{F}$-*coalgebra*. A $\mathfrak{F}$-coalgebra morphism $\phi : \langle c, f \rangle \to \langle d, g \rangle$ is a morphism $\psi : c \to d$

in $\mathfrak{C}$ such that $\mathfrak{F}(\psi) \circ f = g \circ \psi$ holds. Both algebras and coalgebras for a given functor form a category with these morphisms.

1.6.1 The Subprobability Functor

Recall that $(\mathfrak{S}(X, \mathcal{A}), \mathcal{A}^\bullet)$ is a measurable space, whenever $(X, \mathcal{A})$ is one (the weak-*-σ-algebra $\mathcal{A}^\bullet$ is defined in Definition 1.79). Let $f : X \to Y$ be $\mathcal{A}$-$\mathcal{B}$-measurable, where $(Y, \mathcal{B})$ is another measurable space, and define

$$\mathfrak{S}(f)(\mu)(B) := \mu(f^{-1}[B]).$$

PROPOSITION 1.94
$\mathfrak{S}$ *is an endofunctor on the category* $\mathfrak{Meas}$ *of measurable spaces with measurable maps as morphisms.*

PROOF Given a measurable map f, the induced map $\mathfrak{S}(f) : \mathfrak{S}(X, \mathcal{A}) \to \mathfrak{S}(Y, \mathcal{B})$ is $\mathcal{A}^\bullet$-$\mathcal{B}^\bullet$-measurable. For this, we have to establish

$$\mathfrak{S}(f)^{-1}[\mathcal{B}^\bullet] \subseteq \mathcal{A}^\bullet,$$

hence we have to show that for each $W \in \mathcal{B}^\bullet$ its inverse image under $\mathfrak{S}(f)$ is in $\mathcal{A}^\bullet$. The construction of the weak-*-σ-algebra entails that by Lemma 1.2 we may assume $W = \{\nu \in \mathfrak{S}(Y) \mid \nu(B) < t\}$ for some measurable $B \subseteq Y, t \in \mathbb{R}$, since sets of this form generate the weak-*-σ-algebra $\mathcal{B}^\bullet$. But then

$$\mu \in \mathfrak{S}(f)^{-1}[W] \Leftrightarrow \mu \in \{\mu' \in \mathfrak{S}(X) \mid \mu'(f^{-1}[B]) < t\},$$

because of the assumption on f's measurability, $f^{-1}[B] \subseteq X$ is measurable. This establishes the measurability of $\mathfrak{S}(f)$ □

$\mathfrak{S}(f)(\mu)$ is the image of measure μ under measurable map f; it is probabilistically interpreted as the distribution of random variable f under μ. Lacking this notation, we introduced the distribution of f under μ as μ_f in Section 1.5.3 for discussing disintegration and regular conditional distributions. Integration with respect to the image measure may be captured through the *Change of Variable formula* which will be somewhat helpful in the sequel.

PROPOSITION 1.95
(Change of Variables) Let $g \in \mathcal{F}(Y, \mathcal{B})$ *be a bounded and measurable function, then*

$$(\ddagger) \quad \int_Y g(y) \, \mathfrak{S}(f)(\mu)(dy) = \int_X (g \circ f)(x) \, \mu(dx).$$

PROOF (Sketch) We have a look at all g for which the assertion is true:

$$\mathcal{F}_0 := \{g \in \mathcal{F}(Y, \mathcal{B}) \mid (\ddagger) \text{ holds for } g\}.$$

Then $\chi_B \in \mathcal{F}_0$, provided $B \in \mathcal{B}$ is measurable. This is so since

$$\int_Y \chi_B \, d\mathfrak{S}\,(f)\,(\mu) = \mathfrak{S}\,(f)\,(\mu)(B)$$

$$= \mu(f^{-1}\,[B])$$

$$= \int_X \chi_B \circ f \, d\mu.$$

It is clear from the integral's additivity that $\mathcal{F}_0$ is a linear space, so that measurable step functions are contained in $\mathcal{F}_0$. Since for each $g \in \mathcal{F}\,(Y, \mathcal{B})$ with $g \geq 0$ there exists an increasing sequence $(g_n)_{n \in \mathbb{N}}$ of measurable step functions such that $g = \sup_{n \in \mathbb{N}} g_n$ (Proposition 1.5), we obtain

$$\int_Y g \, d\mathfrak{S}\,(f)\,(\mu) = \lim_{n \to \infty} \int_Y g_n \, d\mathfrak{S}\,(f)\,(\mu)$$

$$= \lim_{n \to \infty} \int_X g_n \circ f \, d\mu$$

$$= \int_Y g \circ f \, d\mu.$$

Consequently, $\mathcal{F}_0$ contains each nonnegative measurable and bounded function. Since each function g can be written as $g = \max(g, 0) + \min(g, 0)$, it follows that $\mathcal{F}_0 = \mathcal{F}\,(Y, \mathcal{B})$, hence the assertion is true for all measurable and bounded functions on Y. $\quad\Box$

The reader is probably more familiar with a version that permits changing real variables. It says that for a monotone and continuous differentiable map g with domain $[a, b]$ and range $[\alpha, \beta]$ the equality

$$\int_\alpha^\beta f(y) \, dy = \int_a^b (f \circ g)(x) \cdot |\, g'(x)\,| \, dx$$

holds, whenever f is integrable over $[\alpha, \beta]$. This is the classical version of Calculus, and it is in fact a special case of the Proposition above. It is discussed at length in (Hewitt and Stromberg, 1965, Chapter 20.2).

Returning to properties of functor $\mathfrak{S}$, we infer from Proposition 1.95 that $\mathfrak{S}$ is an endofunctor on the category of Polish spaces. Denote by $\mathfrak{cPol}$ the category of Polish spaces with continuous maps as morphisms, and by $\mathfrak{BPol}$ the category of Polish spaces with Borel maps as morphisms.

COROLLARY 1.96
Equip $\mathfrak{S}\,(X)$ with the topology of weak convergence, whenever X is a Polish space. Then $\mathfrak{S} : \mathfrak{cPol} \to \mathfrak{cPol}$ is an endofunctor.

PROOF   We know from Proposition 1.78 that $\mathfrak{S}\,(X)$ is Polish, provided X is Polish, so we are left to show that for continuous $f : X \to Y$ the map

$\mathfrak{S}(f) : \mathfrak{S}(X) \to \mathfrak{S}(X)$ is continuous. Let $(\mu_n)_{n \in \mathbb{N}}$ be a sequence in $\mathfrak{S}(X)$ with $\mu_n \rightharpoonup_w \mu$, and let $g \in \mathcal{C}(Y)$ be continuous and bounded. Then an application of the Change of Variables Formula yields

$$\int_Y g \, d\mathfrak{S}(f)(\mu_n) = \int_X g \circ f \, d\mu_n$$

$$\to \int_X g \circ f \, d\mu$$

$$= \int_Y g \, d\mathfrak{S}(f)(\mu).$$

Thus $\mathfrak{S}(f)(\mu_n) \rightharpoonup_w \mathfrak{S}(f)(\mu)$. ⬚

Strictly speaking, one should distinguish the subprobability functor operating on $\mathfrak{Meas}$ from the one working on $\mathfrak{cPol}$ or on other categories. But this would lead to a flurry of definitions and notations for functors that do essentially the same, so we use our mathematical licence and are a bit negligent as far as notation is concerned.

COROLLARY 1.97

Equip $\mathfrak{S}(X)$ with the topology of weak convergence, whenever X is a Polish space. Then $\mathfrak{S} : \mathfrak{BPol} \to \mathfrak{BPol}$ is an endofunctor.

We will work often in a category of Polish or analytic spaces with morphisms that are surjective Borel maps, and we require the image of a surjective Borel map under the subprobability functor to be surjective again. The proof of this fact is far from being trivial and requires the concept of universal measurability.

But let us have a closer look at the problem. Given $f : X \to Y$ as a surjective Borel map, we want to find for each $\nu \in \mathfrak{S}(Y)$ a subprobability $\mu \in \mathfrak{S}(X)$ with $\mathfrak{S}(f)(\mu) = \nu$. Hence we want $\nu(B) = \mu(f^{-1}[B])$ for all $B \in \mathcal{B}(Y)$. Now suppose that we can find $g : Y \to X$ so that $f \circ g = id_Y$, then $B = g^{-1}[f^{-1}[B]]$, so that $\mu := \mathfrak{S}(g)(\nu)$ would do the job. This requires in turn that g is well behaved. We will not be able to guarantee that g is a Borel map, but we will come quite close to it, in fact so close that the difference will not be discernible to us. Thus we are looking for a right inverse to f that is suitably close to being a Borel map, and we will see that universal measurability is the concept to work with here.

First we will recall the σ-algebra of universal measurable sets (Halmos, 1950, § 13). Let $\mu \in \mathfrak{S}(X, \mathcal{A})$ be a subprobability on the measurable space $(X, \mathcal{A})$, then $A \subseteq X$ is called μ-*measurable* iff there exist $M_1, M_2 \in \mathcal{A}$ with $M_1 \subseteq A \subseteq M_2$ and $\mu(M_1) = \mu(M_2)$. The μ-measurable subsets of X form a σ-algebra $\mathcal{M}_\mu(\mathcal{A})$; this is easy to see. The measure μ is extended silently from $\mathcal{A}$ to a measure on the σ-algebra $\mathcal{M}_\mu(\mathcal{A})$ upon setting $\mu(A) := \mu(M_1)$, if $A \in \mathcal{M}_\mu(\mathcal{A})$ is sandwiched between $M_1, M_2 \in \mathcal{A}$ with $\mu(M_1) = \mu(M_2)$.

The σ-algebra $\mathcal{U}(\mathcal{A})$ of universally measurable sets is defined by

$$\mathcal{U}(\mathcal{A}) := \bigcap \{\mathcal{M}_\mu(\mathcal{A}) \mid \mu \in \mathfrak{S}(X, \mathcal{A})\}$$

(in fact, one considers usually all finite or σ-finite measures; these definitions lead to the same universally measurable sets). If $f : X_1 \to X_2$ is an $\mathcal{A}_1$-$\mathcal{A}_2$-measurable map between the measurable spaces $(X_1, \mathcal{A}_1)$ and $(X_2, \mathcal{A}_2)$, then it is well known that f is also $\mathcal{U}(\mathcal{A}_1)$-$\mathcal{U}(\mathcal{A}_2)$-measurable; the converse does not hold, and one usually cannot conclude that a map $g : X_1 \to X_2$ which is $\mathcal{U}(\mathcal{A}_1)$-$\mathcal{A}_2$-measurable is also $\mathcal{A}_1$-$\mathcal{A}_2$-measurable.

LEMMA 1.98
Let X be a Polish space.

a. $\mathcal{B}(X) \subseteq \mathcal{U}(\mathcal{B}(X))$

b. If $A \subseteq X$ is analytic, then $A \in \mathcal{U}(\mathcal{B}(X))$.

PROOF   Part *a* is trivial. Part *b* requires a quite elaborate construction. The reader is referred to (Srivastava, 1998, Theorem 4.3.1) or to (Kechris, 1994, Theorem 21.10).   ⬚

We will first formulate a sequence of auxiliary statements that deal with finding for a surjective map $f : X \to Y$ a map $g : Y \to X$ such that $f \circ g = id_Y$. This map g should have some sufficiently pleasant properties, otherwise we could just pick arbitrarily for each $y \in Y$ an element $x \in X$ with $f(x) = y$ and put $g(y) := x$ (this is brought to you by the Axiom of Choice). Hence this simple approach does not work.

Thus in order to make the first step in the strategy outlined above it turns out to be helpful focussing the attention to analytic sets being the continuous images of $\mathbb{N}^\infty$. The latter space is ordered, as we have seen in the discussion of structural issues in Section 1.3.2. We will capitalize on this order, to be more precise, on the interplay between the order and the topology.

LEMMA 1.99
Let X be Polish, $Y \subseteq X$ analytic with $Y = f[\mathbb{N}^\infty]$ for some continuous $f : \mathbb{N}^\infty \to X$. Then there exists $g : Y \to \mathbb{N}^\infty$ such that

a. $f \circ g = id_Y$,

b. g is $\mathcal{U}(\mathcal{B}(Y))$-$\mathcal{U}(\mathcal{B}(\mathbb{N}^\infty))$-measurable.

PROOF   1. Since f is continuous, the inverse image $f^{-1}[\{y\}]$ for each $y \in Y$ is a closed and nonempty set in $\mathbb{N}^\infty$. Thus this set contains a minimal element $g(y)$ in the lexicographic order $\preceq$ by Lemma 1.20. It is clear that $f(g(y)) = y$ holds for all $y \in Y$.

2. Denote by
$$A(\tau') := \{\tau \in \mathbb{N}^\infty \mid \tau \prec \tau'\},$$

then $A(\tau')$ is open: let $\tau \prec \tau'$ and k be the first component in which τ differs from τ', then $\Sigma_{\tau_1 \dots \tau_{k-1}}$ is an open neighborhood of τ that is entirely contained in $A(\tau')$. It is easy to see that $\{A(\tau') \mid \tau' \in \mathbb{N}^\infty\}$ is a generator for the Borel sets of $\mathbb{N}^\infty$.

3. We claim that
$$g^{-1}[A(\tau')] = f[A(\tau')]$$

holds. In fact, let $y \in g^{-1}[A(\tau')]$, so that $g(y) \in A(\tau')$, then $y = f(g(y)) \in f[A(\tau')]$. If, on the other hand, $y = f(\tau)$ with $\tau \prec \tau'$, then by construction $\tau \in f^{-1}[\{y\}]$, thus $g(y) \preceq \tau \prec \tau'$, settling the other inclusion.

This equality implies that $g^{-1}[A(\tau')]$ is an analytic set, because it is the image of an open set under a continuous map. Consequently, $g^{-1}[A(\tau')]$ is universally measurable for each $A(\tau')$ by Lemma 1.98. Thus g is a universally measurable map. ⬜

This will help us to establish that a right inverse exists for surjective Borel maps between an analytic space and a separable measurable space.

PROPOSITION 1.100
Let X be an analytic space, $(Y, \mathcal{B})$ a separable measurable space and $f : X \to Y$ a surjective measurable map. Then there exists $g : Y \to X$ with these properties:

a. $f \circ g = id_Y$,

b. g is $\mathcal{U}(\mathcal{B})$-$\mathcal{U}(\mathcal{B}(X))$-measurable.

PROOF   1. We may and do assume by Lemma 1.49 that Y is an analytic subset of a Polish space Q, and that X is an analytic subset of a Polish space P. $x \mapsto \langle x, f(x) \rangle$ is a bijective Borel map from X to the graph of f, so $\mathsf{graph}(f)$ is an analytic set by Proposition 1.37. Thus we can find a continuous map $F : \mathbb{N}^\infty \to P \times Q$ with $F[\mathbb{N}^\infty] = \mathsf{graph}(f)$. Consequently, $\pi_Q \circ F$ is a continuous map from $\mathbb{N}^\infty$ to Q with

$$(\pi_Q \circ F)[\mathbb{N}^\infty] = \pi_Q[\mathsf{graph}(f)] = Y.$$

Now let $G : Y \to \mathbb{N}^\infty$ be chosen according to Lemma 1.98 for $\pi_Q \circ F$. Then $g := \pi_P \circ F \circ G : Y \to X$ is the map we are looking for:

i. g is universally measurable, because G is, and because $\pi_P \circ F$ are continuous, hence universally measurable as well,

ii. $f \circ g = f \circ (\pi_P \circ F \circ G) = (f \circ \pi_P) \circ F \circ G = \pi_Q \circ F \circ G = id_Y$, so g is right inverse to f.

☐

Now we are in a position to show that the image of a surjective map under the subprobability functor is onto again.

PROPOSITION 1.101
Let X be an analytic space, Y a second countable metric space. If $f : X \to Y$ is a surjective Borel map, so is $\mathfrak{S}(f) : \mathfrak{S}(X) \to \mathfrak{S}(Y)$.

PROOF   1. From Proposition 1.100 we find a map $g : Y \to X$ such that $f \circ g = id_Y$ and g is $\mathcal{U}(\mathcal{B}(Y)) - \mathcal{U}(\mathcal{B}(X))$-measurable.

2. Let $\nu \in \mathfrak{S}(Y)$, and define $\mu := \mathfrak{S}(g)(\nu)$, then $\mu \in \mathfrak{S}(X, \mathcal{U}(\mathcal{B}(X)))$ by construction. Restrict μ to the Borel sets on X, obtaining $\mu_0 \in \mathfrak{S}(X, \mathcal{B}(X))$. Since we have for each set $B \subseteq Y$ the equality $g^{-1}[f^{-1}[B]] = B$, we see that for each $B \in \mathcal{B}(Y)$

$$\mathfrak{S}(f)(\mu_0)(B) = \mu_0(f^{-1}[B]) = \mu(f^{-1}[B]) = \nu(g^{-1}[f^{-1}[B]]) = \nu(B)$$

holds.

☐

This has as a consequence that $\mathfrak{S}$ is an endofunctor on the category of Polish or analytic spaces with surjective Borel maps as morphisms; it displays a pretty interaction of reasoning in measurable spaces and arguing in categories. We will use this fact throughout without further reference. The category driven investigation of probabilistic structures will of course be continued as well.

We will continue now with the discussion of general constructions in categories.

1.6.2   Natural Transformations

Let $\mathfrak{F}, \mathfrak{G} : \mathfrak{C} \to \mathfrak{D}$ be functors. Then $\eta : \mathfrak{F} \overset{\bullet}{\to} \mathfrak{G}$ is called a *natural transformation* of $\mathfrak{F}$ and $\mathfrak{G}$ iff

i. $\eta_C : \mathfrak{F}(C) \to \mathfrak{G}(C)$ is a morphism in $\mathfrak{D}$ for each object C in $\mathfrak{C}$,

ii. this diagram commutes whenever $f : C_1 \to C_2$ is a morphism in $\mathfrak{C}$:

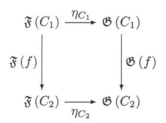

The morphism η_C is sometimes called the *component* of η at C.

Let $\mathfrak{H} : \mathfrak{C} \to \mathfrak{D}$ be another functor with a natural transformation $\theta : \mathfrak{G} \overset{\bullet}{\to} \mathfrak{H}$. Define

$$(\theta \circ \eta)_C := \theta_C \circ \eta_C,$$

then it is immediate that $\theta \circ \eta : \mathfrak{F} \overset{\bullet}{\to} \mathfrak{H}$ by pasting diagrams for the morphism $f : C_1 \to C_2$:

$$
\begin{array}{ccccc}
\mathfrak{F}(C_1) & \xrightarrow{\eta_{C_1}} & \mathfrak{G}(C_1) & \xrightarrow{\theta_{C_1}} & \mathfrak{H}(C_1) \\
\downarrow{\scriptstyle\mathfrak{F}(f)} & & \downarrow{\scriptstyle\mathfrak{G}(f)} & & \downarrow{\scriptstyle\mathfrak{H}(f)} \\
\mathfrak{F}(C_2) & \xrightarrow{\eta_{C_2}} & \mathfrak{G}(C_2) & \xrightarrow{\theta_{C_2}} & \mathfrak{H}(C_2)
\end{array}
$$

This entails that natural transformations are the morphisms in the category of all functors, which has as objects all functors, and for the functors $\mathfrak{F}, \mathfrak{G} : \mathfrak{C} \to \mathfrak{D}$ the natural transformations $\mathfrak{F} \overset{\bullet}{\to} \mathfrak{G}$ as morphisms.

Given $\eta : \mathfrak{F} \overset{\bullet}{\to} \mathfrak{G}$ for the functors $\mathfrak{F}, \mathfrak{G} : \mathfrak{C} \to \mathfrak{D}$ and $\mathfrak{K} : \mathfrak{D} \to \mathfrak{E}$, it is clear that $\mathfrak{K} \circ \mathfrak{F} : \mathfrak{C} \to \mathfrak{E}$ and $\mathfrak{K} \circ \mathfrak{G} : \mathfrak{C} \to \mathfrak{E}$ are functors. Define for the object C in $\mathfrak{C}$ the morphism

$$(\mathfrak{K}\eta)_C := \mathfrak{K}(\eta_C),$$

then $\mathfrak{K}\eta : \mathfrak{K} \circ \mathfrak{F} \overset{\bullet}{\to} \mathfrak{K} \circ \mathfrak{G}$. For the functor $\mathfrak{L} : \mathfrak{B} \to \mathfrak{C}$ a natural transformation $\eta\mathfrak{L} : \mathfrak{F} \circ \mathfrak{L} \overset{\bullet}{\to} \mathfrak{G} \circ \mathfrak{L}$. is defined through

$$(\eta\mathfrak{L})_B := \eta_{\mathfrak{L}(B)}.$$

EXAMPLE 1.102

Define for the Polish space X, the Borel set $C \in \mathcal{B}(X)$ and the real number $r \in \mathbb{R}$ the set

$$\lambda_X^r(C) := \{\mu \in \mathfrak{S}(X) \mid \mu(C) \leq r\}.$$

We claim that $\lambda^r : \mathfrak{B} \overset{\bullet}{\to} \mathfrak{B} \circ \mathfrak{S}$ is a natural transformation ($\mathfrak{B} : \mathfrak{Pol} \to \mathfrak{Borel}^{op}$ is defined in Example 1.93).

From Proposition 1.80 it is inferred that $\lambda_X^r(C) \in \mathcal{B}(\mathfrak{S}(X))$, whenever $C \in \mathcal{B}(X)$, thus $\lambda_X^r : \mathcal{B}(X) \to \mathcal{B}(\mathfrak{S}(X))$. Let $f : X_1 \to X_2$ be a morphism in $\mathfrak{Pol}$, then so is $\mathfrak{S}(f) : \mathfrak{S}(X_1) \to \mathfrak{S}(X_2)$, and the $\mathfrak{Borel}^{op}$-morphism $\mathfrak{B}(f) : \mathcal{B}(X_1) \to \mathcal{B}(X_2)$ corresponds to the map $\overline{\mathfrak{B}(f)} : \mathcal{B}(X_2) \to \mathcal{B}(X_1)$, similar for the $\mathfrak{Borel}^{op}$-morphism $\mathfrak{B}(\mathfrak{S}(f)) : \mathcal{B}(\mathfrak{S}(X_1)) \to \mathcal{B}(\mathfrak{S}(X_2))$. Hence we have to show that this diagram of maps

$$
\begin{array}{ccc}
\mathcal{B}(X_1) & \xrightarrow{\lambda_{X_1}} & \mathcal{B}(\mathfrak{S}(X_1)) \\
\Big\uparrow{\scriptstyle\overline{\mathfrak{B}(f)}} & & \Big\uparrow{\scriptstyle\overline{\mathfrak{B}(\mathfrak{S}(f))}} \\
\mathcal{B}(X_2) & \xrightarrow[\lambda_{X_2}]{} & \mathcal{B}(\mathfrak{S}(X_2))
\end{array}
$$

commutes (note the direction of the arrows). Now let $C \in \mathcal{B}(X_2)$, then we have

$$
\begin{aligned}
\overline{\mathfrak{B}(\mathfrak{S}(f))}\left(\lambda^r_{X_2}(C)\right) &= \mathfrak{S}(f)^{-1}\left[\lambda^r_{X_2}(C_2)\right] \\
&= \{\mu \in \mathfrak{S}(X_1) \mid \mathfrak{S}(f)\,\mu(C) \le r\} \\
&= \{\mu \in \mathfrak{S}(X_1) \mid \mu(f^{-1}[C]) \le r\} \\
&= \lambda^r_{X_1}(f^{-1}[C]) \\
&= (\lambda^r_{X_1} \circ \overline{\mathfrak{B}(f)})(C).
\end{aligned}
$$

1.6.3   Adjunctions, Monads, Algebras and the Kleisli Construction

We will define in this section monads and the corresponding Kleisli construction. Each monad is based on an endofunctor for a category, so each monad has algebras associated with it. The Eilenberg-Moore algebras are particularly interesting, and they will be investigated later on for the monad which is defined somewhat naturally through the subprobability functor.

On first sight, all these constructions appear somewhat unrelated, but it is well known that they are tied together through adjunctions. Thus we will define adjunctions as well, state the fundamental characterization through a pair of natural transformations and relate monads and Eilenberg-Moore algebras to adjunctions.

Adjunctions.   We define the basic notion of an adjunction and show that an adjunction defines a pair of natural transformations through universal arrows (which is sometimes taken as the basis for adjunctions).

DEFINITION 1.103   *Let $\mathfrak{X}$ and $\mathfrak{A}$ be categories. Then $(\mathfrak{F}, \mathfrak{G}, \varphi)$ is called an* adjunction *iff*

a. *$\mathfrak{F} : \mathfrak{X} \to \mathfrak{A}$ and $\mathfrak{G} : \mathfrak{A} \to \mathfrak{X}$ are functors,*

b. *for each object a in* $\mathfrak{A}$ *and x in* $\mathfrak{X}$ *there is a bijection*

$$\varphi_{x,a} : \mathfrak{A}(\mathfrak{F}(x), a) \to \mathfrak{X}(x, \mathfrak{G}(a))$$

which is natural in x and a.

$\mathfrak{F}$ *is called the* left adjoint *to* $\mathfrak{G}$, $\mathfrak{G}$ *is called the* right adjoint *to* $\mathfrak{F}$.

That $\varphi_{x,a}$ is natural for each x, a means that for all morphisms $f : a \to b$ in $\mathfrak{A}$ and $g : x \to y$ in $\mathfrak{X}$ both diagrams commute:

$$
\begin{array}{ccc}
\mathfrak{A}(\mathfrak{F}(x), a) & \xrightarrow{\varphi_{x,a}} & \mathfrak{X}(x, \mathfrak{G}(a)) \\
{\scriptstyle f_*}\downarrow & & \downarrow{\scriptstyle (\mathfrak{G}(f))_*} \\
\mathfrak{A}(\mathfrak{F}(x), b) & \xrightarrow{\varphi_{x,b}} & \mathfrak{X}(x, \mathfrak{G}(b))
\end{array}
\qquad
\begin{array}{ccc}
\mathfrak{A}(\mathfrak{F}(x), a) & \xrightarrow{\varphi_{x,a}} & \mathfrak{X}(x, \mathfrak{G}(a)) \\
{\scriptstyle (\mathfrak{F}(g))^*}\downarrow & & \downarrow{\scriptstyle g^*} \\
\mathfrak{A}(\mathfrak{F}(y), a) & \xrightarrow{\varphi_{y,a}} & \mathfrak{X}(y, \mathfrak{G}(a))
\end{array}
$$

Here $f_* := \mathfrak{A}(\mathfrak{F}(x), f)$ and $g^* := \mathfrak{X}(g, \mathfrak{G}(a))$ are the hom-set functors associated with f resp. g, similar for $(\mathfrak{G}(f))_*$ and for $(\mathfrak{F}(g))^*$; for the hom-set functors see Example 1.93.

An adjunction induces natural transformations which make this important construction easier to handle, and which helps indicating connections of adjunctions to monads and Eilenberg-Moore algebras in the sequel. Before entering the discussion, universal arrows are introduced.

DEFINITION 1.104   *Let* $\mathfrak{G} : \mathfrak{C} \to \mathfrak{D}$ *be a functor, and c an object in* $\mathfrak{C}$.

a. $u : c \to \mathfrak{G}(r)$ *is called a* universal arrow *from c to* $\mathfrak{G}$ *iff for any arrow* $f : c \to \mathfrak{G}(d)$ *there exists a unique arrow* $f' : r \to d$ *in* $\mathfrak{C}$ *such that* $f = \mathfrak{G}(f') \circ u$.

b. $v : \mathfrak{G}(r) \to c$ *is called a* universal arrow *from* $\mathfrak{G}$ *to c iff for any arrow* $f : \mathfrak{G}(d) \to c$ *there exists a unique arrow* $f' : d \to r$ *in* $\mathfrak{C}$ *such that* $f = v \circ \mathfrak{G}(f')$.

Thus, if $c \to \mathfrak{G}(r)$ is universal from c to $\mathfrak{G}$, then each arrow $c \to \mathfrak{G}(d)$ in $\mathfrak{D}$ factors uniquely through the $\mathfrak{G}$-image of an arrow $r \to d$ in $\mathfrak{C}$. Similarly, if $\mathfrak{G}(r) \to c$ is universal from $\mathfrak{G}$ to c, then each $\mathfrak{D}$-arrow $\mathfrak{G}(d) \to c$ factors uniquely through the $\mathfrak{G}$-image of an $\mathfrak{C}$-arrow $d \to r$.

Universal arrows will be used now for a characterization of adjunctions in terms of natural transformations (we will omit the indices for the natural transformation φ that comes with an adjunction).

THEOREM 1.105

Let $(\mathfrak{F}, \mathfrak{G}, \varphi)$ be an adjunction for the functors $\mathfrak{F} : \mathfrak{X} \to \mathfrak{A}$ and $\mathfrak{G} : \mathfrak{A} \to \mathfrak{X}$. Then there exist natural transformations $\eta : \mathbb{1}_{\mathfrak{X}} \overset{\bullet}{\to} \mathfrak{G} \circ \mathfrak{F}$ and $\varepsilon : \mathfrak{F} \circ \mathfrak{G} \to \mathbb{1}_{\mathfrak{A}}$ with these properties:

a. the arrow η_x is universal from $\mathfrak{G}$ to x for each x in $\mathfrak{X}$, and $\varphi(f) = \mathfrak{G}(f) \circ \eta_x$ holds for each $f : \mathfrak{F}(x) \to a$,

b. the arrow ε_a is universal from a to $\mathfrak{F}$ for each a in $\mathfrak{A}$, and $\varphi^{-1}(g) = \varepsilon_a \circ \mathfrak{F}(g)$ holds for each $g : x \to \mathfrak{G}(a)$,

c. the composites

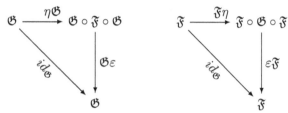

are the identities for $\mathfrak{G}$ resp. $\mathfrak{F}$.

PROOF   (Sketch) 1. Define $\eta_x := \varphi(id_{\mathfrak{F}(x)})$, then the well known Yoneda Lemma (MacLane, 1997, Proposition III.2.1) entails that η_x is a universal arrow from x to $\mathfrak{G}$. Let $h : x \to y$ be a morphism in $\mathfrak{X}$, then

$$\mathfrak{G}(\mathfrak{F}(h)) \circ \varphi(id_{\mathfrak{F}(x)}) = \varphi(\mathfrak{F}(h) \circ id_{\mathfrak{F}(x)}) = \varphi(id_{\mathfrak{F}(y)} \circ \mathfrak{F}(h)) = \varphi(id_{\mathfrak{F}(y)}) \circ h,$$

because φ is natural. This implies the commutativity of the diagram

$$
\begin{array}{ccc}
x & \xrightarrow{\eta_x} & \mathfrak{G}(\mathfrak{F}(x)) \\
{\scriptstyle h}\downarrow & & \downarrow{\scriptstyle \mathfrak{G}(\mathfrak{F}(h))} \\
y & \xrightarrow[\eta_y]{} & \mathfrak{G}(\mathfrak{F}(y))
\end{array}
$$

Thus $\eta : \mathbb{1}_{\mathfrak{X}} \overset{\bullet}{\to} \mathfrak{G} \circ \mathfrak{F}$ is natural.

2. Let $f : \mathfrak{F}(x) \to a$ be a morphism in $\mathfrak{A}$, then the naturality of φ implies

$$\varphi(f) = \varphi(f \circ id_{\mathfrak{F}(x)}) = \mathfrak{G}(f) \circ \varphi(id_{\mathfrak{F}(x)}) = \mathfrak{G}(f) \circ \eta_x.$$

3. Define for the object a in $\mathfrak{A}$ the morphism $\varepsilon_a := \varphi^{-1}(id_{\mathfrak{G}(a)})$, then an argumentation quite similar to the one above shows that $\varepsilon : \mathfrak{F} \circ \mathfrak{G} \overset{\bullet}{\to} \mathbb{1}_{\mathfrak{A}}$ is a natural transformation, and that for each morphism $g : x \to \mathfrak{G}(a)$ the equality $\varphi^{-1}(g) = \varepsilon_a \circ \mathfrak{F}(g)$ holds.

4. From $\varphi(f) = \mathfrak{G}(f) \circ \eta_x$ we obtain

$$id_{\mathfrak{G}(a)} = \varphi(\varepsilon_a) = \mathfrak{G}\varepsilon_a \circ \eta_{\mathfrak{G}(a)},$$

so that $\mathfrak{G}\varepsilon \circ \eta\mathfrak{G}$ is the identity transformation on $\mathfrak{G}$. Similarly, $\eta\mathfrak{F} \circ \mathfrak{F}\varepsilon$ is the identity for $\mathfrak{F}$. ▯

The transformation η is sometimes called the *unit* of the adjunction, whereas ε is called its *counit*. The converse to Theorem 1.105 holds as well: from two transformations η and ε with the signatures as above one can construct an adjunction. The proof is a straightforward verification.

PROPOSITION 1.106
Let $\mathfrak{F} : \mathfrak{X} \to \mathfrak{A}$ and $\mathfrak{G} : \mathfrak{A} \to \mathfrak{X}$. be functors, and assume that natural transformations $\eta : \mathbb{1}_{\mathfrak{X}} \overset{\bullet}{\to} \mathfrak{G} \circ \mathfrak{F}$ and $\varepsilon : \mathfrak{F} \circ \mathfrak{G} \to \mathbb{1}_{\mathfrak{A}}$ are given so that $(\mathfrak{G}\varepsilon) \circ (\eta\mathfrak{G})$ is the identity of $\mathfrak{G}$, and $(\varepsilon\mathfrak{F}) \circ (\mathfrak{F}\eta)$ is the identity of $\mathfrak{F}$. Define $\varphi(f) := \mathfrak{G}(f) \circ \eta_x$, whenever $f : \mathfrak{F}(x) \to a$ is a morphism in $\mathfrak{A}$. Then $(\mathfrak{F}, \mathfrak{G}, \varphi)$ defines an adjunction.

Thus for identifying an adjunction it is usually sufficient to identify its unit and its counit; this includes verifying the identity laws of the functors for the corresponding compositions. In Section 3.6 we will, however, take the direct road and verify the laws of an adjunction according to Definition 1.103.

Monads. An endofunctor $\mathfrak{T} : \mathcal{C} \to \mathcal{C}$ together with the natural transformations $\mathfrak{e} : \mathbb{1}_{\mathcal{C}} \overset{\bullet}{\to} \mathfrak{T}$ (the unit) and $\mathfrak{m} : \mathfrak{T}^2 \overset{\bullet}{\to} \mathfrak{T}$ (the multiplication) is a *monad* iff these diagrams commute

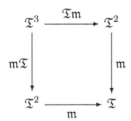
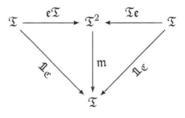

The commutativity of the leftmost diagram is expressed for an object x of $\mathcal{C}$ through

$$\mathfrak{m}_x \circ \mathfrak{T}(\mathfrak{m}_x) = \mathfrak{m}_x \circ \mathfrak{m}_{\mathfrak{T}(x)},$$

while the commutativity of the rightmost diagram is written down as

$$\mathfrak{m}_x \circ \mathfrak{e}_{\mathfrak{T}(x)} = id_x = \mathfrak{m}_x \circ \mathfrak{T}(\mathfrak{e}_x).$$

These expressions are sometimes easier to handle than the purely functorial notation in the diagrams above.

Two monads will be discussed extensively: the one based on the power set functor in the category of sets, and the one based on the subprobability functor on the category of Polish spaces. Since the former monad is well known, it will serve mainly as a paragon for proceeding in the latter. Chapter 3 discusses these monads together with an application to software architecture.

Now let $(\mathfrak{F}, \mathfrak{G}, \varphi)$ be an adjunction with functors $\mathfrak{F} : \mathfrak{X} \to \mathfrak{A}$ and $\mathfrak{G} : \mathfrak{A} \to \mathfrak{X}$, the unit η and the counit ε. Define the functor $\mathfrak{T}$ through $\mathfrak{T} := \mathfrak{G} \circ \mathfrak{F}$. Then $\mathfrak{T} : \mathfrak{X} \to \mathfrak{X}$ defines an endofunctor on category $\mathfrak{X}$ with $\mathfrak{m}_a := (\mathfrak{G}\varepsilon\mathfrak{F})(a) = \mathfrak{G}(\varepsilon_{\mathfrak{F}(a)})$ as a morphism $\mathfrak{m}_a : \mathfrak{T}^2(a) \to \mathfrak{T}(a)$. Because $\varepsilon_a : \mathfrak{F}(\mathfrak{G}(a)) \to a$ is a m morphism in $\mathfrak{A}$, and because $\varepsilon : \mathfrak{F} \circ \mathfrak{G} \overset{\bullet}{\to} \mathbb{1}_{\mathfrak{A}}$ is natural, the diagram

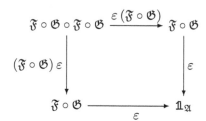

is commutative. This means that in the functor category the diagram

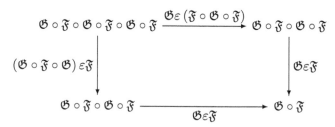

commutes. Multiplying from the left with $\mathfrak{G}$ and from the right with $\mathfrak{F}$ yields this diagram.

$$
\begin{array}{ccc}
\mathfrak{G} \circ \mathfrak{F} \circ \mathfrak{G} \circ \mathfrak{F} \circ \mathfrak{G} \circ \mathfrak{F} & \xrightarrow{\;\mathfrak{G}\varepsilon\,(\mathfrak{F} \circ \mathfrak{G} \circ \mathfrak{F})\;} & \mathfrak{G} \circ \mathfrak{F} \circ \mathfrak{G} \circ \mathfrak{F} \\
{\scriptstyle (\mathfrak{G} \circ \mathfrak{F} \circ \mathfrak{G})\,\varepsilon\mathfrak{F}} \Big\downarrow & & \Big\downarrow {\scriptstyle \mathfrak{G}\varepsilon\mathfrak{F}} \\
\mathfrak{G} \circ \mathfrak{F} \circ \mathfrak{G} \circ \mathfrak{F} & \xrightarrow[\;\mathfrak{G}\varepsilon\mathfrak{F}\;]{} & \mathfrak{G} \circ \mathfrak{F}
\end{array}
$$

When rotated along the left-to-right diagonal, this diagram shows that $\mathfrak{m} : \mathfrak{T}^2 \overset{\bullet}{\to} \mathfrak{T}$ satisfies the laws of a multiplication in a monad. Define $\mathfrak{e} := \eta$, then $\mathfrak{e} : \mathbb{1}_{\mathfrak{X}} \overset{\bullet}{\to} \mathfrak{T}$ is a natural transformation. By multiplying the leftmost diagram in Theorem 1.105, part c from the right with $\mathfrak{F}$, and the diagram on the right hand side with $\mathfrak{G}$, upon gluing these diagrams together one obtains the diagram which gives the laws of a unit for a monad.

We have shown

PROPOSITION 1.107
Each adjunction defines a monad.

The converse of this holds as well: each monad defines an adjunction, as we will show below. We need for this Eilenberg-Moore algebras which will be defined and considered presently. For an extensive discussion of these algebras in the context of the monad induced by the subprobability functor, the reader is referred to Chapter 3.

Eilenberg-Moore Algebras. Given a monad $(\mathfrak{T}, \mathfrak{e}, \mathfrak{m})$ in a category $\mathfrak{C}$, a pair $\langle x, h \rangle$ consisting of an object x and a morphism $h : \mathfrak{T}(x) \to x$ in $\mathfrak{C}$ is called an *Eilenberg-Moore algebra* for the monad iff the following diagrams commute

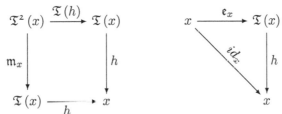

An *algebra morphism* $f : \langle x, h \rangle \to \langle x', h' \rangle$ between the algebras $\langle x, h \rangle$ and $\langle x', h' \rangle$ is a morphism $f : x \to x'$ in $\mathfrak{C}$ which makes the diagram

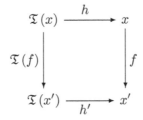

commute. Eilenberg-Moore algebras together with their morphisms form a category $\mathfrak{Alg}_{(\mathfrak{T}, \mathfrak{e}, \mathfrak{m})}$. We will usually omit the reference to the monad.

Fix for the moment $(\mathfrak{T}, \mathfrak{e}, \mathfrak{m})$ as a monad in category $\mathfrak{C}$, and let $\mathfrak{Alg} := \mathfrak{Alg}_{(\mathfrak{T}, \mathfrak{e}, \mathfrak{m})}$ be the associated category of Eilenberg-Moore algebras.

We state a sequence of auxiliary statements that will assist in showing that the monad defines an adjunction, and that the monad associated with this adjunction is the originally given one.

LEMMA 1.108
The pair $\langle \mathfrak{T}(x), \mathfrak{m}_x \rangle$ is a $\mathfrak{T}$-algebra for each x in $\mathfrak{C}$.

PROOF   This is immediate from the laws for $\mathfrak{e}$ and $\mathfrak{m}$ in a monad.   ☐

These algebras are usually called the *free algebras* for the monad. Morphisms in $\mathfrak{C}$ translate into morphisms in $\mathfrak{Alg}$ through functor $\mathfrak{T}$.

LEMMA 1.109
If $f : x \to y$ is a morphism in $\mathfrak{C}$, then $\mathfrak{T}(f) : \langle \mathfrak{T}(x), \mathfrak{m}_x \rangle \to \langle \mathfrak{T}(y), \mathfrak{m}_y \rangle$ is a morphism in $\mathfrak{Alg}$.

PROOF   Because $\mathfrak{m} : \mathfrak{T}^2 \overset{\bullet}{\to} \mathfrak{T}$, we see $\mathfrak{m}_y \circ \mathfrak{T}^2(f) = \mathfrak{T}(f) \circ \mathfrak{m}_x$. This is just the defining equation for a morphism in $\mathfrak{Alg}$.   ☐

This yields as an immediate consequence:

COROLLARY 1.110
Define $\mathfrak{F}(x) := \langle \mathfrak{T}(x), \mathfrak{m}_x \rangle$ for an object x in $\mathfrak{C}$, and $\mathfrak{F}(f) := \mathfrak{T}(f)$ for an arrow $f : x \to y$ in $\mathfrak{C}$. Then $\mathfrak{F} : \mathfrak{C} \to \mathfrak{Alg}$ is a functor.
Define $\mathfrak{G}(\langle x, h \rangle) := x$ for the $\mathfrak{T}$-algebra $\langle x, h \rangle$, and put $\mathfrak{G}(f) := f$ for the algebra morphism f, then $\mathfrak{G} : \mathfrak{Alg} \to \mathfrak{C}$ is a functor as well.

For this pair of functors we are about to define unit and counit for the adjunction. This is straightforward for the unit and quite immediate for the counit. To be specific:

LEMMA 1.111
Define $\eta_x := \mathfrak{T}(x)$ for the object x in $\mathfrak{C}$, and $\varepsilon_{\langle x, h \rangle} := \langle \mathfrak{T}(x), \mathfrak{m}_x \rangle$ for the $\mathfrak{T}$-algebra $\langle x, h \rangle$. Then $\eta : \mathbb{1}_{\mathfrak{C}} \overset{\bullet}{\to} \mathfrak{G} \circ \mathfrak{F}$ and $\varepsilon : \mathfrak{F} \circ \mathfrak{G} \overset{\bullet}{\to} \mathbb{1}_{\mathfrak{Alg}}$ are natural transformations.

PROOF   One first notes that $h : \langle \mathfrak{T}(x), \mathfrak{m}_x \rangle \to \langle x, h \rangle$ is a morphism of $\mathfrak{T}$-algebras, whenever $\langle x, h \rangle$ is a $\mathfrak{T}$-algebra. Then one notes that the diagram

$$
\begin{array}{ccc}
\langle \mathfrak{T}(x), \mathfrak{m}_x \rangle & \overset{h}{\longrightarrow} & \langle x, h \rangle \\
{\scriptstyle \mathfrak{T}(f)} \downarrow & & \downarrow {\scriptstyle f} \\
\langle \mathfrak{T}(y), \mathfrak{m}_y \rangle & \underset{h'}{\longrightarrow} & \langle y, h' \rangle
\end{array}
$$

commutes, whenever $f : \langle x, h \rangle \to \langle y, h' \rangle$ is an algebra morphism. Now the assertion follows from the observation that $(\mathfrak{F} \circ \mathfrak{G})(\langle x, h \rangle) = \langle \mathfrak{T}(x), \mathfrak{m}_x \rangle$ and $(\mathfrak{F} \circ \mathfrak{G})(f) = \mathfrak{T}(f)$ holds.   ☐

This yields the desired result: we did define an adjunction, and this adjunction yields the originally given monad as its monad according to Proposition 1.107.

PROPOSITION 1.112

If $(\mathfrak{T}, \mathfrak{e}, \mathfrak{m})$ is a monad in the category $\mathfrak{C}$ with $\mathfrak{Alg}$ as its category of $\mathfrak{T}$-algebras, then $\langle \mathfrak{F}, \mathfrak{G}, \eta, \varepsilon \rangle$ defines an adjunction. The monad given by this adjunction is $(\mathfrak{T}, \mathfrak{e}, \mathfrak{m})$.

PROOF It is straightforward to verify that $(\mathfrak{G}\varepsilon) \circ (\eta\mathfrak{G})$ is the identity of $\mathfrak{G}$, and $(\varepsilon\mathfrak{F}) \circ (\mathfrak{F}\eta)$ is the identity of $\mathfrak{F}$. Thus Proposition 1.106 says that we have in fact an adjunction. Because $(\mathfrak{G}\varepsilon\mathfrak{F})(x) = \mathfrak{G}\left(\varepsilon_{\mathfrak{F}(x)}\right) = \mathfrak{G}\left(\mathfrak{m}_x\right)$, the multiplication of the monad defined by the adjunction is the originally given multiplication; similarly one shows that the units coincide. □

The Kleisli Construction. If $(\mathfrak{T}, \mathfrak{e}, \mathfrak{m})$ is a monad in category $\mathfrak{C}$, then a *Kleisli morphism* $f : a \rightsquigarrow b$ between objects a and b is a morphism $f : a \rightarrow \mathfrak{T}(b)$. The composition $g * f$ of Kleisli morphisms $f : a \rightsquigarrow b$ and $g : b \rightsquigarrow c$ is defined through

$$g * f := \mathfrak{m}_c \circ \mathfrak{T}(g) \circ f,$$

where $\circ$ is the composition in $\mathfrak{C}$. The properties of a monad take care of associativity and the fact that the identity morphism in the original monad gives rise to an identity for the Kleisli composition. Let $f : a \rightarrow b$, $g : b \rightarrow c$ and $h : c \rightarrow d$ be morphisms. Then

$$\mathfrak{e}_b * f = f = f * \mathfrak{e}_a,$$

since

$$\mathfrak{e}_b * f = \mathfrak{m}_a \circ \mathfrak{T}(\mathfrak{e}_a) \circ f = \mathfrak{m}_a \circ \mathfrak{e}_{\mathfrak{T}(\mathfrak{e}_a)} \circ f = f =$$

$$id_b \circ f = \mathfrak{m}_b \circ \mathfrak{e}_{\mathfrak{T}(\mathfrak{e}_b)} \circ f = \mathfrak{m}_b \circ \mathfrak{T}(f) \circ_a \overset{(\sharp)}{=} f * \mathfrak{e}_a.$$

Equation ($\sharp$) uses the naturalness of $\mathfrak{e} : \mathbb{1}_{\mathfrak{C}} \overset{\bullet}{\rightarrow} \mathfrak{T}$, which implies that the diagram on the left hand side commutes.

Similarly, composition is associative, since

$$h*(g*f) = \mathfrak{m}_d \circ \mathfrak{T}(h) \circ \mathfrak{m}_c \circ \mathfrak{T}(g) \circ f =$$

$$\mathfrak{m}_d \circ (\mathfrak{T}(h) \circ \mathfrak{m}_c) \circ \mathfrak{T}(g) \circ f \overset{(\flat)}{=} \mathfrak{m}_d \circ \left(\mathfrak{m}_{\mathfrak{T}(a)} \circ \mathfrak{T}^2(h)\right) \circ \mathfrak{T}(g) \circ f =$$

$$\mathfrak{m}_a \circ \mathfrak{T}(\mathfrak{m}_a) \circ \mathfrak{T}^2(h) \circ \mathfrak{T}(g) \circ f = \mathfrak{m}_a \circ \mathfrak{T}(h*g) \circ f = (h*g)*f.$$

Again, naturalness is used in equation ($\flat$), this time of the multiplication $\mathfrak{m} : \mathfrak{T}^2 \overset{\bullet}{\to} \mathfrak{T}$, which makes sure that the above diagram on the right hand side is commutative.

The category so constructed is usually called the *Kleisli category* for the monad. This category can be used to build an adjunction which in turn has exactly the monad under consideration as its monad. To be specific, let $(\mathfrak{T}, \mathfrak{e}, \mathfrak{m})$ be a monad in category $\mathfrak{C}$, and denote the associated Kleisli category by $\mathfrak{K}$. If $f \in \mathfrak{X}(a, \mathfrak{T}(b))$ is a morphism in $\mathfrak{X}$, make $\kappa(f) \in \mathfrak{K}(a,b)$ the associated Kleisli morphism. Define $\mathfrak{F} : \mathfrak{X} \to \mathfrak{K}$ as the identity on objects, and put $\mathfrak{F}(f) := \kappa(\mathfrak{e}_b \circ f)$ for $f \in \mathfrak{X}(a, \mathfrak{T}(b))$. Define the functor $\mathfrak{G} : \mathfrak{K} \to \mathfrak{X}$ through $\mathfrak{G}(a) := \mathfrak{T}(a)$, and set $\mathfrak{G}(t) := \mathfrak{m}_b \circ \mathfrak{T}(t)$ for $t \in \mathfrak{K}(a,b)$.

PROPOSITION 1.113

With the notations from above,

a. $\mathfrak{F} : \mathfrak{X} \to \mathfrak{K}$ *and* $\mathfrak{G} : \mathfrak{K} \to \mathfrak{X}$ *are functors,*

b. *define* $\varphi = \varphi_{a,b}$ *as the bijection that is given through*

$$\mathfrak{K}(\mathfrak{F}(a), b) = \mathfrak{K}(a, b) \cong \mathfrak{X}(a, \mathfrak{T}(b)) = \mathfrak{X}(a, \mathfrak{G}(b)),$$

then φ *is natural in a and b,* $(\mathfrak{F}, \mathfrak{G}, \varphi)$ *is an adjunction,*

c. *the monad associated with this adjunction is* $(\mathfrak{T}, \mathfrak{e}, \mathfrak{m})$.

PROOF   1. From the discussion above it is clear that $\mathfrak{F}$ and $\mathfrak{G}$ are functors, and it is easily checked that φ is natural in a and b. For example, if $f : b \to c$ in $\mathfrak{X}$, then

$$\varphi_{a,c}(f_*(g)) = \varphi_{a,c}(f \circ g) =$$

$$\mathfrak{T}(f \circ g) = \mathfrak{T}(f) \circ \mathfrak{T}(g) = (\mathfrak{T}(f))_*(\mathfrak{T}(g)) = \mathfrak{T}(f)_* (\varphi_{a,b}(g)),$$

whenever $g : a \to b$ is an arrow in $\mathfrak{X}$. Thus $\varphi_{a,-}$ is natural.

Consequently, $(\mathfrak{F}, \mathfrak{G}, \varphi)$ is an adjunction. Computing the unit η for this adjunction, we see

$$\eta_a = \varphi(id_{\mathfrak{F}(a)}) = \varphi(\kappa(\eta_a)) = \mathfrak{e}_a,$$

and we have $\kappa(\varepsilon_a) = \kappa(id_{\mathfrak{T}(a)})$ for the counit, so that we obtain

$$(\mathfrak{G}\varepsilon\mathfrak{F})(a) = \mathfrak{G}\varepsilon_a = \mathfrak{G}\left(() \kappa(id_{\mathfrak{T}(a)})\right) = \mathfrak{m}_a \circ \mathfrak{T}\left(id_{\mathfrak{T}(a)}\right) = \mathfrak{m}_a$$

for the multiplication. This completes the proof. ⬚

Thus we see that the Kleisli construction can also be obtained from an adjunction. But this is not the end of the story, which should be told for the sake of completeness. Mac Lane (MacLane, 1997, Chapter VI.5) shows that for a given monad all those adjunctions which define the monad form a category, the initial object of which is the Kleisli construction, and the terminal object is given through the Eilenberg-Moore algebras. Thus the Kleisli construction and the Eilenberg-Moore algebras form the opposite and extreme ends of a whole spectrum of adjunctions.

1.7   Bibliographic Notes

Borel Sets, Analytic Spaces.   The classical reference to Borel sets, analytic sets and their interplay is of course Kuratowski's monograph (Kuratowski, 1966), albeit the notation is quite arcane and somewhat outdated. Most results in these introductory pages are folklore and can be found in many monographs; the exposition in Parthasarathy's classic on probability measures on metric spaces (Parthasarathy, 1967) is particularly noteworthy. I found the expositions in Srivastava's book (Srivastava, 1998) on Borel sets and in Kechris' book (Kechris, 1994) on descriptive set theory most helpful, so I used these books as guides both for the exposition and for most of the proofs. No claim for originality is being made here with this material. The discussion of analytic sets in Arveson's book (Arveson, 1976, Chapter 3) is particularly concise, so I tried to come close to his style of presenting this somewhat technical topic. The discussion on universal measurability and the proof for the existence of a universally measurable right inverse are taken from Arveson as well.

Measurable Selectors.   Measurable relations are a valuable tool in such diverse fields as stochastic dynamic programming (Wagner, 1977) and descriptive set theory (Kechris, 1994). Overviews are provided in (Srivastava, 1998, Chapter 5) and (Himmelberg, 1975; Wagner, 1977). The monograph (Castaing and Valadier, 1977) is a concise overview of the field by the end of the 1970s. The proofs for Proposition 1.57 are partly taken from there, partly from (Himmelberg and van Vleck, 1974) and from (Castaing, 1967).

Probability Measures   The weak topology on the probability measures on a Polish space is a standard topic in Probability Theory; see for example (Shiryaev, 1996; Loève, 1962; Billingsley, 1995). Billingsley (Billingsley, 1968) and Parthasarathy (Parthasarathy, 1967) did devote part of their monograph to it; (Billingsley, 1999) is an update that is much less technical than its predecessor. The proofs of the crucial statements are taken from (Parthasarathy, 1967) and from (Kellerer, 1972) with an occasional glance towards (Kechris, 1994).

Categories   The standard reference is Mac Lane's treatise (MacLane, 1997); (Borceux, 1994a; Borceux, 1994b) gives an encyclopaedic overview. (Barr and Wells, 1999) provides a representation the topics of which are oriented to the needs of computer science; (Fiadeiro, 2005) caters for the needs of software engineers. The present discussion of adjunctions and their relation to monads and to Eilenberg-Moore algebras follows rather closely the one given by Mac Lane (MacLane, 1997, Chapters IV, VI) and the one given by Barr and Wells in (Barr and Wells, 1985, 3.2). The reader is referred to Mac Lane's book for a discussion of the relationship between monads, algebras and adjunctions that is briefly mentioned at the end of Section 1.6.3. (Wadler, 1992) discusses the interaction between functional programming and monads in a graded series of examples.

Chapter 2

Stochastic Relations as Monads

2.1   Introduction

Consider a relation $R \subseteq X \times Y$ that relates an input $x \in X$ to a set $R(x) \subseteq Y$ of outputs. Each output carries the same weight, so each output has the same chance of being selected. But this is sometimes too coarse a description. It basically outlines all possibilities and selects — in the absence of other criteria — one of them. In a quantitatively oriented scenario one would attach some weight to each possible outcome and select the output with the highest weight. This can be done through assigning each outcome a probability, so that $K(x)$ is a probability distribution over Y. Looking at the alternative between the nondeterministic and the probabilistic approach, we see that in the nondeterministic case R is modeled as a map from X to the power set $\mathfrak{Pow}(Y)$ of Y, K in turn is modeled as a map from X to the set $\mathfrak{P}(Y)$ of all probabilities over Y. In the finite case one can convert between these models, albeit not without loss of information: given R, put $K(x) := \mathsf{card}(R(x))/\mathsf{card}(Y)$, given K, define $R(x) := \{y \in Y \mid K(x)(y) > 0\}$, so that K appears as a kind of refinement to K.

But this similarity is somewhat superficial, the relationship between relations and their stochastic cousins lies much deeper, and this is the story which this chapter is going to tell. We will find out that there is a common underlying structure for both of them by looking at the construction from a category theory point of view by showing that both arise through the same kind of construction from a monad. We show that the Kleisli construction provides an informal link between these kinds of relations by establishing that both kinds of relations arise through this construction, albeit over different monads. Selecting as a base category the category of sets and the monad related to the power set functor, we will obtain nondeterministic relations through the Kleisli construction. Selecting the category of measurable spaces and the monad for the subprobability functor, we will obtain stochastic relations through this construction. They will be defined formally here as a result of these discussions.

This is the rough picture, which will be refined somewhat. We will point out systematically some similarities. This is done through a discussion of the corresponding monads. When having a monad, one usually wants to know what the algebras for this monad looks like (because the algebras permit a reconstruction of an adjunction giving rise to that monad, just as the Kleisli construction does). We make this step as well for both monads, where the development for the monad based on the power set functor is well known, but not completely for one based on the subprobability functor. It will be shown in Chapter 3 what the algebras for the latter functor look like, and here we obtain also an explicit characterization.

We are dealing in this chapter mainly with categorical constructions, and we investigate the category of stochastic relations a bit more closely. We prepare for dealing with the problem of the existence of pullbacks, which then will be undertaken in Chapter 4 in detail. This question in turn will be later of some significance, when we discuss bisimulations, behavioral equivalence, and their relations to modal and temporal logic through logical equivalence. The problem of finding a pullback is quite trivial for nondeterministic relations (you basically write the pullback down explicitly), but it is far too strongly posed for stochastic relations. Even the request for weak pullbacks is not weak enough. We will show in Chapter 4 that semi-pullbacks exist in the category of stochastic relations over analytic spaces, and that this is the most we can expect: no weak pullbacks usually exist, as an example shows. Reflecting this on the background of similarities between both kinds of relations, we see that constructions that are easily carried out for the set-theoretic case are undertaken with difficulties for the probabilistic case (if at all). We will encounter this phenomenon later on again. It suggests that a construction like an abstract specification of relations that can be interpreted sensibly both over nondeterministic and over stochastic relations may work in special cases, but may be difficult to pursue in general.

Functorial Issues.   The present chapter introduces first the subprobability functor on the category of measurable spaces. It investigates this functor, shows that it gives rise to a monad, has a look at the Kleisli product and identifies for a special case the algebras for this functor.

To emphasize the similarities between nondeterministic and stochastic relations, we first look at the monad that is defined through the power set functor on $\mathfrak{Set}$. When modeling a software architecture in Section 2.4, we will require an additional argument to the indication of the system's work, and we assume for this purpose that this is modeled through a monoid H with 1 as an identity. Such a monoid could be a group, the free semigroup over an alphabet, or a $\vee$-semilattice with a smallest element. This additional argument will enter the constructions here at little additional cost but will provide an additional amount of flexibility and demonstrates the flexibility and adaptability of this construction.

Case Study. This chapter contains a case study as well, indicating the broad range of the concepts defined and investigated here. It deals with architectural modeling, demonstrating that architectural models for a very popular software architecture may be formulated relationally. Since the basic mathematical demands for a relational model are identical for nondeterministic and for stochastic relations, the model is formulated sufficiently general, so that both families of relations are covered (actually, the incorporation of the monoid permits tuning the functor a bit by incorporating additional information that may be used for bookkeeping and the like).

This chapter as a whole shows that the similarities between these families of relations considered are plentiful and interesting. They are translated from properties of the associated monads, in particular from the respective functors, which sit in their rôle as masterminds in the background and control the properties of their Kleisli products, sometimes remaining discreetly in the background, sometimes entering the bright sunlight through a direct argument.

2.2 The Manes Monad

The power set functor $\mathfrak{Pow}$ assigns to each set X its power set $\mathfrak{Pow}(X)$, and assigns to each map $f : X \to Y$ the map $\mathfrak{Pow}(f) : \mathfrak{Pow}(X) \to \mathfrak{Pow}(Y)$, mapping $A \subseteq X$ to

$$\mathfrak{Pow}(f)(A) := f[A] := \{f(x) \mid x \in A\}.$$

Define $\mathfrak{m} : \mathfrak{Pow}^2 \xrightarrow{\bullet} \mathfrak{Pow}$ through

$$\mathfrak{m}_X : \mathfrak{Pow}(\mathfrak{Pow}(X)) \ni A \mapsto \bigcup A \in \mathfrak{Pow}(X)$$

and $\mathfrak{e} : \mathfrak{I} \xrightarrow{\bullet} \mathfrak{Pow}(X)$ through

$$\mathfrak{e}_X : X \ni x \mapsto \{x\} \in \mathfrak{Pow}(X).$$

Elementary calculations show that both $\mathfrak{m}$ and $\mathfrak{e}$ form indeed natural transformations: Let $f : X \to Y$ be a map, then this diagram is commutative:

$$
\begin{array}{ccc}
\mathfrak{Pow}(\mathfrak{Pow}(X)) & \xrightarrow{\ \mathfrak{m}_X\ } & \mathfrak{Pow}(X) \\
{\scriptstyle \mathfrak{Pow}(\mathfrak{Pow}(f))}\Big\downarrow & & \Big\downarrow{\scriptstyle \mathfrak{Pow}(f)} \\
\mathfrak{Pow}(\mathfrak{Pow}(Y)) & \xrightarrow[\ \mathfrak{m}_Y\ ]{} & \mathfrak{Pow}(Y)
\end{array}
$$

In fact, if $A \in \mathfrak{Pow}\,(\mathfrak{Pow}\,(X))$, then

$$(\mathfrak{m}_Y \circ \mathfrak{Pow}\,(\mathfrak{Pow}\,(f)))\,(A) = \bigcup \{f\,[x] \mid x \in A\} = (\mathfrak{Pow}\,(f) \circ \mathfrak{m}_X)\,(A).$$

It is also not difficult to establish that $\langle \mathfrak{Pow}, \mathfrak{e}, \mathfrak{m} \rangle$ satisfies the laws of a monad; this monad will be referred to as the *Manes monad*.

Augmenting the construction by adding semigroup H, we define

$$\mathfrak{M}\,(X) := \mathfrak{Pow}\,(H \times X),$$

and if $f : X \to Y$ is a map, $A \subseteq H \times X$, then

$$\mathfrak{M}\,(f)\,(A) := \{\langle h, f(x) \rangle \mid \langle h, x \rangle \in A\}$$

defines the action of the functor on the morphisms of $\mathfrak{Set}$. Now define

$$\mathfrak{m}_X : \mathfrak{M}\,(\mathfrak{M}\,(X)) \to \mathfrak{M}\,(X)$$

upon setting

$$\mathfrak{m}_X(A) := \bigcup_{\langle h_1, b \rangle \in A} \{\langle h_1 h_2, x \rangle \mid \langle h_2, x \rangle \in b\},$$

then an easy but somewhat space consuming calculation reveals that $\mathfrak{m} : \mathfrak{M}^2 \overset{\bullet}{\to} \mathfrak{M}$ is a natural transformation. The natural transformation $\mathfrak{e} : \mathfrak{I}_{\mathfrak{Set}} \overset{\bullet}{\to} \mathfrak{M}$ is defined by $\mathfrak{e}_X : x \mapsto \{\langle 1, x \rangle\}$. A standard calculation shows then that $\langle \mathfrak{M}, \mathfrak{e}, \mathfrak{m} \rangle$ is a monad in $\mathfrak{Set}$.

A Kleisli morphism for the augmented monad between sets X and Y is a relation between X and $H \times Y$. This is well investigated for the case that the monoid H is trivial, cf. (Barr and Wells, 1999, 16.1.4); it generalizes to the present case. If $R : X \to \mathfrak{M}\,(Y)$ and $S : Y \to \mathfrak{M}\,(Z)$ are Kleisli morphisms, then we may either see R and S as maps to the corresponding power sets, or we interpret $R \subseteq X \times (H \times Y)$ and $S \subseteq Y \times (H \times Z)$ as relations; both views will be made use of interchangeably, depending on the convenience of use.

The (Kleisli-) product of R and S is identified in Proposition 2.1 which summarizes this example.

PROPOSITION 2.1

$\langle \mathfrak{M}, \mathfrak{e}, \mathfrak{m} \rangle$ *is a monad in the category* $\mathfrak{Set}$. *The Kleisli product for the relations* $R \subseteq X \times (H \times Y)$ *and* $S \subseteq Y \times (H \times Z)$ *is given through*

$$(S * R)(x) = \{\langle h_1 h_2, z \rangle \mid \exists y \in Y : \langle h_1, y \rangle \in R(x) \wedge \langle h_2, z \rangle \in S(y)\}.$$

Assume that the semigroup H is trivial, then functor $\mathfrak{M}$ equals the power set functor $\mathfrak{Pow}$; the natural transformations $\mathfrak{m}$ and $\mathfrak{e}$ are adjusted as well. We obtain as a consequence of Proposition 2.1 this well-known version:

COROLLARY 2.2

$\langle \mathfrak{Pow}, e, m \rangle$ *is a monad in the category* $\mathfrak{Set}$*. The Kleisli product for the relations* $R \subseteq X \times Y$ *and* $S \subseteq Y \times Z$ *is given through*

$$(S * R)(x) = \{z \mid \exists y \in Y : y \in R(x) \wedge z \in S(y)\}.$$

2.3 The Giry Monad

We will investigate now the probabilistic counterpart to the Manes monad. It assigns to each measurable space its subprobability measures. We know from Section 1.6 that $\mathfrak{G} : \mathfrak{Meas} \to \mathfrak{Meas}$ is a functor; before we investigate it further and closer, we will augment its work by a monoid as well.

2.3.1 Adding a Monoid

The Giry monad on $\mathfrak{Meas}$ assumes the monoid H being endowed with a σ-algebra $\mathcal{H}$ which makes multiplication measurable, when $H \times H$ carries the product σ-algebra $\mathcal{H} \otimes \mathcal{H}$.

Examples for measurable monoids are given by topological monoids; the Borel sets then form the canonical measurable structure. Topological groups are probably the most prominent examples. If H is a $\vee$-semilattice with a smallest element, then it is not difficult to see that $\vee$ is a continuous operation, when H is endowed with the interval topology (i.e., the topology which has open intervals as subbase). Taking again the Borel sets for this topology, we see that these semilattices yield measurable monoids, too.

We state as a preparation for the definition of the monad's multiplication. Assume that X and Y are measurable spaces.[1]

LEMMA 2.3

Let $f : X \to Y$ *be a measurable map, and assume that* C *is a measurable subset of* $H \times Y$*. Then* $\Gamma_C(\nu, s) := \nu(\{\langle t, x \rangle \mid \langle st, f(x) \rangle \in C\})$ *is a real-valued measurable map on* $\mathfrak{G}(H \times X) \times H$*.*

---

[1] Unless there is the danger of ambiguity, we will omit the notation of the σ-algebras from now on. They are assumed to be fixed. The following conventions will be adhered to: The space of sub-probabilities will be endowed with the corresponding weak-*-σ-algebra (see Section 1.5.2 for a definition), products will carry the product σ-algebra of their factors, and Polish or analytic spaces will always have their Borel sets, unless otherwise indicated. Deviations from these conventions will be made explicit.

PROOF   1. Consider the set $\mathcal{C}$ of all members C of $\mathcal{H} \otimes \mathcal{B}$ for which Γ_C has the desired property. We will analyze the properties of $\mathcal{C}$ now with the goal of showing that $\mathcal{C}$ equals $\mathcal{H} \otimes \mathcal{B}$.

2. Assume that $C = G \times D$ with $G \in \mathcal{H}$ and $D \in \mathcal{B}$. Since the semigroup multiplication is measurable, we know that $G' := \{\langle g, h \rangle \mid gh \in G\}$ is an element of $\mathcal{H} \otimes \mathcal{H}$, so that

$$\{\langle t, x \rangle \mid \langle st, f(x) \rangle \in G \times D\} = G'_s \times f^{-1}[D]$$

(recall that G'_s is the vertical cut of G' at s, see Section 1.5.4, and that $f^{-1}[D] \subseteq X$ is measurable). Hence

$$\Gamma_{G \times D}(\nu, s) = \nu(G'_s \times f^{-1}[D]).$$

An argument very similar to that establishing Lemma 1.86 shows that

$$\langle \nu, s \rangle \mapsto \nu(Q_s \times W)$$

is $(\mathcal{H} \otimes \mathcal{A})^{\bullet} \otimes \mathcal{H}$-measurable, whenever $Q \in \mathcal{H} \otimes \mathcal{H}$ and $W \in \mathcal{A}$. This implies that $\Gamma_{G \times D}$ constitutes a measurable map, so that all measurable rectangles are members of $\mathcal{C}$. The measurable rectangles form a family of sets which is closed under finite intersections.

3. Because subprobabilities are finitely and countably additive, we have

$$\Gamma_{(H \times X) \setminus C}(\nu, s) = \nu(H \times X) - \Gamma_C(\nu, s),$$

thus $\mathcal{C}$ is closed under complementation, and

$$\Gamma_{\bigcup_{n \in \mathbb{N}} C_n}(\nu, s) = \sum_{n \in \mathbb{N}} \Gamma_{C_n}(\nu, s),$$

whenever $(C_n)_{n \in \mathbb{N}}$ is a disjoint sequence, thus $\mathcal{C}$ is closed under disjoint countable unions. From the π-λ-Theorem 1.1 we infer now that $\mathcal{C}$ contains all sets $C \in \mathcal{H} \otimes \mathcal{B}$. This establishes the claim. $\Box$

Lemma 2.3 states that Γ_C is jointly measurable in both arguments. This entails that we may use Γ_C as an integrand, given a subprobability on its domain for integration. It has also as a consequence that upon fixing one argument, the arising partial map is measurable, so that Γ_C is measurable separately in each variable (it is well known that joint measurability is strictly stronger than separate measurability in each variable). We will make use of this observation as well.

Now define after these somewhat lengthy preparations for $x \in X$, the measurable subset $A \subseteq H \times X$ and the measure $\mu \in \mathfrak{S}(H \times \mathfrak{S}(H \times X))$ the functor $\mathfrak{S}$, unit $\mathfrak{e}$ and multiplication $\mathfrak{m}$ through

$$\mathfrak{G}(X) := \mathfrak{S}(H \times X, \mathcal{H} \otimes \mathcal{A})$$

$$\mathfrak{e}_X(x) := \delta_{\langle 1, x \rangle}$$

$$\mathfrak{m}_X(\mu)(A) := \int_{H \times \mathfrak{S}(H \times X)} p(\{\langle h, x \rangle \mid \langle gh, x \rangle \in A\})\, \mu(d\langle g, p \rangle).$$

Lemma 2.3 tells us that the integrand for the definition of $\mathfrak{m}_X$ is actually measurable.

If $f : X \to Y$ is measurable, we put

$$\mathfrak{G}(f)(\mu)(B) := \mu(\{\langle h, x \rangle \mid \langle h, f(x) \rangle \in B\})$$
$$= \mu\left((id_H \times f)^{-1}[B]\right)$$
$$= \mathfrak{S}(id_H \times f)(\mu)(B).$$

Consequently, $\mathfrak{G}(f) : \mathfrak{G}(X) \to \mathfrak{G}(Y)$ is measurable, and if $\psi \in \mathcal{F}(H \times Y)$, the Change of Variable formula (Proposition 1.95) implies that

$$\int_{H \times Y} \psi(h, y)\, \mathfrak{G}(f)(\mu)(d\langle h, y \rangle) = \int_{H \times X} \psi(h, f(x))\, \mu(d\langle h, x \rangle)$$

holds.

We are now in a position to show that $\langle \mathfrak{G}, \mathfrak{e}, \mathfrak{m} \rangle$ is a monad in $\mathfrak{Meas}$, adapting and extending Giry's proofs (Giry, 1981) to the situation at hand.

LEMMA 2.4

$\mathfrak{G}$ *is an endofunctor in* $\mathfrak{Meas}$, $\mathfrak{e} : \mathfrak{I}_{\mathfrak{Meas}} \overset{\bullet}{\to} \mathfrak{G}$ *and* $\mathfrak{m} : \mathfrak{G}^2 \overset{\bullet}{\to} \mathfrak{G}$ *are natural transformations.*

PROOF 1. It is immediate that $\mathfrak{G} : \mathfrak{Meas} \to \mathfrak{Meas}$ is a functor, and that $\mathfrak{e}$ is a natural transformation.

2. Let $f : X \to Y$ be a measurable map, then we know that for $\mu \in \mathfrak{G}(Y)$ and for the measurable subset $B \subseteq H \times Y$ these equations hold

$$\left(\mathfrak{m}_Y \circ \mathfrak{G}^2 f\right)(\mu)(B) =$$

$$\int_{H \times \mathfrak{S}(H \times Y)} (\mathfrak{G}f)(q)(\{\langle h, y \rangle \mid \langle gh, y \rangle \in B\})\, \mu(d\langle s, q \rangle) =$$

$$\int_{H \times \mathfrak{S}(H \times X)} q(\{\langle h, x \rangle \mid \langle gh, f(x) \rangle \in B\})\, \mu(d\langle s, q \rangle).$$

Again, an appeal to Lemma 2.3 makes sure that we are permitted to compute the integral, since the integrand constitutes a bounded function that is measurable jointly in both variables.

The latter expression coincides with $(\mathfrak{G}f \circ \mathfrak{m}_X)(\mu)(B)$. Thus we have established that

$$\left(\mathfrak{m}_Y \circ \mathfrak{G}^2 f\right) = \left(\mathfrak{G}f \circ \mathfrak{m}_X\right)$$

holds. Consequently, $\mu : \mathfrak{G}^2 \overset{\bullet}{\to} \mathfrak{G}$ is a natural transformation.   ☐

This is a preparation for establishing:

PROPOSITION 2.5
$\langle \mathfrak{G}, \mathfrak{e}, \mathfrak{m} \rangle$ *is a monad in* $\mathfrak{Meas}$.

PROOF   1. We need to demonstrate that both the associative and the unit laws hold. The Change of Variable formula implies that

$$\int_{G \times \mathfrak{G}(X)} \psi \, d\left(\mathfrak{G}\mathfrak{e}_X\right)(\mu) = \int_{H \times X} \psi(h, \mathfrak{e}_X)(x)) \; p(d\langle h, x \rangle),$$

whenever $\mu \in \mathfrak{G}(X)$, and $\psi \in \mathcal{F}(H \times \mathfrak{G}(X))$ is measurable and bounded. Consequently,

$$\begin{aligned}
(\mathfrak{m}_X \circ \mathfrak{G}\mathfrak{e}_X)(\mu)(B) &= \int_{H \times X} \mathfrak{e}_X(\{\langle g, x \rangle \mid \langle hg, x \rangle \in B\}) \; \mu(d\langle h, x \rangle) \\
&= \mu(B) \\
&= \left(\mathfrak{m}_X \circ \mathfrak{e}_{\mathfrak{G}(X)}\right)(\mu)(B)
\end{aligned}$$

is true for every $\mu \in \mathfrak{G}(X)$, and for every measurable subset B of $H \times X$. This establishes the unit laws.

2. As far as the associative law is concerned, fix $r \in \mathfrak{G}^3(X)$, and a measurable subset E of $H \times \mathfrak{G}(X)$. The Change of Variable formula implies that

$$\left(\mathfrak{m}_X \circ \mathfrak{m}_{\mathfrak{G}(X)}\right)(r)(E) =$$
$$\int_{H \times \mathfrak{G}(X)} q(\{\langle g, y \rangle \mid \langle hg, y \rangle \in E\}) \; \mathfrak{m}_{\mathfrak{G}(X)}(r)(d\langle h, q \rangle) =$$
$$\int_{H \times \mathfrak{G}^2(X)} \left(\int_{H \times \mathfrak{G}(X)} q(\{\langle j, y \rangle \mid \langle ghj, y \rangle \in E\}) \; p(d\langle h, q \rangle) \right) r(d\langle g, p \rangle).$$

On the other hand, expanding the definitions, and applying the Change of Variables formula suitably, it is seen that these transformations hold:

$$\left(\mathfrak{m}_X \circ \mathfrak{G}\mathfrak{m}_X\right)(r)(E) =$$
$$\int_{H \times \mathfrak{G}(X)} p(\{\langle h, y \rangle \mid \langle gh, y \rangle \in E\}) \; \mathfrak{G}\mathfrak{m}_X(r)(d\langle g, p \rangle) =$$
$$\int_{H \times \mathfrak{G}^2 X} \mathfrak{m}_X(q)(\{\langle h, y \rangle \mid \langle gh, y \rangle \in E\}) \; r(d\langle g, q \rangle) = \left(\mathfrak{m}_X \circ \mathfrak{m}_{\mathfrak{G}(X)}\right)(r)(E).$$

This shows that the associative law is valid. ⬜

We identify the product in the Kleisli category associated with this monad:

PROPOSITION 2.6

Let X, Y and Z be measurable spaces. Assume that $K : X \rightsquigarrow Y$ and $L : Y \rightsquigarrow Z$ are Kleisli morphisms for the monad $\langle \mathfrak{G}, \mathfrak{e}, \mathfrak{m} \rangle$. Then the Kleisli product $L * K$ for K and L is given through

$$(L * K)(x)(C) = \int_{H \times Y} L(y)(\{\langle h, x \rangle \mid \langle gh, x \rangle \in C\}) \; K(x)(d\langle g, y \rangle).$$

PROOF 1. Let C be a measurable subset of $H \times Z$, then the definition of the Kleisli product yields

$$
\begin{aligned}
(L * K)(x)(C) &= (\mathfrak{m}_Z \circ \mathfrak{G}(L) \circ K)(x)(C) \\
&= \mathfrak{m}_Z((\mathfrak{G}(L) \circ K)(x))(C) \\
&= \int_{H \times \mathfrak{G}(Z)} \mu(\{\langle t, z \rangle \mid \langle st, z \rangle \in C\}) \; (\mathfrak{G}(L) \circ K)(x)(d\langle s, \mu \rangle).
\end{aligned}
$$

2. If $\psi \in \mathcal{F}(H \times \mathfrak{G}(Z))$ and $\mu \in \mathfrak{G}(Y)$, the Change of Variables formula implies that

$$\int_{H \times \mathfrak{G}(Z, C)} \psi \; d\mathfrak{G}(L)(\mu) = \int_{H \times Y} \psi(t, L(y)) \; \mu(d\langle t, y \rangle).$$

Inserting this into the equation above, the result follows. ⬜

2.3.2 Stochastic Relations

Summarizing the discussion for a trivial monoid, we obtain as an extension to Proposition 1.94 the following Corollary. It is stated separately because we will use it over and over again.

COROLLARY 2.7

$\langle \mathfrak{G}, \mathfrak{e}, \mathfrak{m} \rangle$ is a monad in the category $\mathfrak{Meas}$ of measurable spaces with measurable maps as morphisms. Assume that $K : X \rightsquigarrow Y$ and $L : Y \rightsquigarrow Z$ are Kleisli morphisms for this monad. Then the Kleisli product $L * K$ for K and L is given through

$$(L * K)(x)(C) = \int_Y L(y)(C) \; K(x)(dy).$$

We will investigate these Kleisli morphisms in greater detail: they constitute just the stochastic relations. The name suggests the similarity with set-theoretic (or nondeterministic) relations.

DEFINITION 2.8   A stochastic relation $\mathsf{K} = (X, Y, K)$ *between the measurable spaces* X *and* Y *is a Kleisli morphism* $K : X \rightsquigarrow Y$ *for the monad* $\langle \mathfrak{S}, \mathfrak{e}, \mathfrak{m} \rangle$.

This is another, easier to handle characterization of stochastic relations which probably comes as a surprise: the Kleisli morphisms are the sub-Markov kernels introduced in Section 1.5.3.

PROPOSITION 2.9
  The following statements are equivalent for measurable spaces X *and* Y:

a. $K : X \rightsquigarrow Y$ *is a stochastic relation.*

b. $K : X \times \mathcal{B} \to [0, 1]$ *is a map such that*

  i. $x \mapsto K(x)(B)$ *is measurable for each* $B \in \mathcal{B}$,
  ii. $K(x) \in \mathfrak{S}(Y)$ *for each* $x \in X$.

  Here $\mathcal{B}$ *is the* σ-*algebra on* Y.

PROOF   1. Suppose that K is a stochastic relation, then $K : X \to \mathfrak{S}(Y)$ is a measurable map. The definition of the weak-*-σ-algebra implies that the evaluation map $x \mapsto K(x)(B)$ is measurable for each $B \in \mathcal{B}$.
  2. Assume conversely that K has the properties from part b. It is clear that K maps X to $\mathfrak{S}(Y)$, so measurability has to be established. Again, this follows readily from the definition of the weak-*-σ-algebra.   ∎

Stochastic relations were introduced as *sub-Markov kernels* (see Definition 1.81), and they have been studied already in the context of regular conditional probabilities and of disintegration. We want to emphasize the relational nature, so we rather stick to the name of stochastic relations from now on. It is closer to the Computer Science point of view.

COROLLARY 2.10
  Assume that $K : X \rightsquigarrow Y$ *is a stochastic relation, and* $D \in \mathcal{A} \otimes \mathcal{B}$, *where* $\mathcal{A}$ *and* $\mathcal{B}$ *are the respective* σ-*algebras. Then* $x \mapsto K(x)(D_x)$ *is a measurable map.*

PROOF   Consider

$$\mathcal{D} := \{ D \in \mathcal{A} \otimes \mathcal{B} \mid x \mapsto K(x)(D_x) \text{ is measurable} \}.$$

Evidently both $\emptyset$ and $X \times Y$ are members of $\mathcal{D}$, and since we can calculate D_x for $D = A \times B$ as

$$D_x = \begin{cases} B, x \in A, \\ \emptyset, x \notin A, \end{cases}$$

we see that all measurable rectangles are members of $\mathcal{D}$; this generator is closed under finite intersections. If $D \in \mathcal{D}$, then

$$((X \times Y) \setminus D)_x = Y \setminus D_x,$$

and

$$K(x)(Y \setminus D_x) = K(x)(Y) - K(x)(D_x)$$

(compare Section 1.3.2). Consequently, $\mathcal{D}$ is closed under complementation. Similarly, if $(D_n)_{n \in \mathbb{N}} \subseteq \mathcal{D}$ is a sequence of disjoint sets, then, since

$$\left(\bigcup_{n \in \mathbb{N}} D_n \right)_x = \bigcup_{n \in \mathbb{N}} (D_n)_x,$$

and since the infinite sum of a sequence of measurable functions is measurable again, provided it exists (which it does in this case), we may conclude that $\bigcup_{n \in \mathbb{N}} D_n \in \mathcal{D}$. Thus $\mathcal{D}$ is closed under complementation and disjoint unions, and it contains a generator that is closed under finite intersections. From the π-λ-Theorem 1.1 we see that

$$\mathcal{D} = \sigma \left(\{ A \times B \mid A \in \mathcal{A}, B \in \mathcal{B} \} \right),$$

hence the assertion is true for all product measurable sets. $\square$

Discussion. Proposition 2.9 supports the view that a stochastic relation models randomly changing phenomena. Assume first that $K : S \rightsquigarrow S$ is a stochastic relation on a state space S for some system. If the system is in state $s \in S$, then $K(s)(T)$ is interpreted as the probability that the system will change its state to a member of the measurable set $T \subseteq S$. Second, assume that X and Y are interpreted as the spaces of inputs and outputs of some randomly operating device. Then the value $K(x)(B)$ for a stochastic relation $K : X \rightsquigarrow Y$ is interpreted as the probability for an output to be a member of the measurable set $B \subseteq Y$ after the system has received input $x \in X$. Models like this are particularly attractive when outputs come from an uncountable set: here it is not always reasonable to assign to each individual $y \in Y$ a probability, because these individuals may be difficult to capture individually, or because they do not carry enough weight by themselves. On the other hand it appears sensible to assign *sets* of outputs the probability to be involved. It should be mentioned that methods of nonstandard analysis (Lindstrøm, 1988; Keisler, 1988) try to balance these seemingly irreconcilable points of view.

Another point worth mentioning is that there may exist inputs x for which the probability $K(x)(Y)$ that an output is delivered at all is not unity (for otherwise we would have postulated that K maps X to $\mathfrak{P}(Y)$ rather than to $\mathfrak{S}(Y)$). This permits modeling systems that may encounter situations in which no output at all will be given, e.g., because the computation leading to an output does not terminate; see (Morgan et al., 1996).

Suppose that the base spaces X and Y are identical, then a stochastic relation may be interpreted as a coalgebra for the subprobability functor $\mathfrak{S}$; see, e.g., (Rutten, 2000). This is of interest when modeling state transitions as hinted at above. The coalgebraic point of view appears quite attractive structurally, because it suggests to fit stochastic relations tightly under the roof of coalgebras, making tried and tested approaches available for investigating problems of stochastic relations. Unfortunately, this route can only be followed with partial success. There are two reasons for this: First, we will see that the subprobability functor has some idiosyncratic properties making work with it sometimes a little strenuous (for example, due to the lack of weak pullbacks, see Proposition 4.14). Second, a coalgebra $\langle x, c \rangle$ for functor $\mathfrak{F}$ is defined as a morphism $c : x \to \mathfrak{F}x$, so the codomain of morphism c is just the image of its domain under $\mathfrak{F}$. This is rather restrictive, both structurally and regarding applications. "Unfolding" domain and codomain into two independent objects provides much needed maneuverability, as we will experience almost everywhere.

Morphisms. Given two stochastic relations K_1 and K_2 with $\mathsf{K}_i : X_i \rightsquigarrow Y_i, i = 1, 2$ a morphism $\mathsf{f} : \mathsf{K}_1 \to \mathsf{K}_2$ is composed of two maps $\phi : X_1 \to X_2$ and $\psi : Y_1 \to Y_2$. Both maps should be measurable, and we will assume that both maps are onto. This is due to the observation that in the target system each element should be traced back to an element in the source system (so that there is no overabundance of elements in K_2 relative to K_1). We formulate as the compatibility condition relating the probabilistic structures K_1 and K_2 that

$$K_1(x_1)(\psi^{-1}[B_2]) = K_2(\phi(x_1))(B_2)$$

holds for each $x_1 \in X_1$ and each $B_2 \in \mathcal{B}_2$. Staying with the input-output model, we postulate that the probability of answering with an element of B_2 after input $\phi(x_1)$ equals the probability of answering after input x_1 with an element which will be mapped by ψ to B_2 (thus with an element of $\psi^{-1}[B_2]$). The equation above may be reformulated as

$$\mathfrak{S}(\psi) \circ K_1 = K_2 \circ \phi$$

(composition $\circ$ denoting composition of maps), and this leads to the following fundamental definition:

DEFINITION 2.11 *Given two stochastic relations $K_i : (X_i, \mathcal{A}_i) \rightsquigarrow (Y_i, \mathcal{B}_i)$ with $i = 1, 2$, a morphism $\mathsf{f} : \mathsf{K}_1 \to \mathsf{K}_2$ is a pair $\mathsf{f} = (\phi, \psi)$ of surjective maps such that* *a.* $\phi : X_1 \rightarrow X_2$ *is $\mathcal{A}_1$-$\mathcal{A}_2$-measurable,*

b. $\psi : Y_1 \rightarrow Y_2$ *is $\mathcal{B}_1$-$\mathcal{B}_2$-measurable,*

c. the diagram

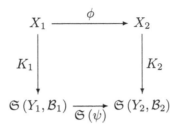

is commutative.

Listing Some Categories. We will usually not work in the very general category of all measurable spaces but rather restrict ourselves to some more specialized base categories like $\mathfrak{BPol}$, the category of **Pol**ish spaces with Borel measurable maps as morphisms or the category $\mathfrak{Anl}$ of **an**alytic spaces, also with Borel measurable maps as morphisms. The category $\mathfrak{Stoch}$ has **stoch**astic relations $\mathsf{K} = (X, Y, K)$ for measurable spaces X, Y as objects and pairs of surjective maps according to Definition 2.11 as morphisms. Stochastic relations will be investigated also over these more specialized categories, and morphisms between stochastic relations are available there as well. $\mathfrak{PolStoch}$ and $\mathfrak{anStoch}$ denote the category of all stochastic relations where the basic objects are taken from $\mathfrak{BPol}$ resp. $\mathfrak{Anl}$. The objects of $\mathfrak{PolStoch}$ will sometimes be called *Polish objects*, and, accordingly, *analytic objects* will be the objects in category $\mathfrak{anStoch}$.

2.4 Case Study: Architectural Modeling through Monads

Nondeterministic and stochastic relations are for some problems really instances of the same relational phenomenon: we need little beyond the corresponding monads, and the relevant aspects of the applications will be taken care of through the Kleisli morphisms for that monad. This basic observation lies at the heart of Moggi's λ_c-calculus, but in contrast to Moggi's work we are interested here in exploring the commonalities and the differences of two very specific monads. We will investigate the problem of modeling a popular, simple software architecture with this approach: given a monad with some additional features, we investigate modeling this architecture, and we show that both the Manes monad and the Giry monad are instances of it. It will be argued at the end that, since the architecture is so simple, this uniform approach of modeling is successful, and that a slightly more complicated software architecture will need more sophisticated categorial properties.

A pipeline is a popular architecture which connects computational components (filters) through connectors (pipes) so that computations are performed in a stream like fashion. The data are transported through the pipes between filters, gradually transforming inputs to outputs. This kind of stream processing has been made popular through UNIX pipes that serially connect independent components for performing a sequence of tasks. Because of its simplicity and its easy to grasp functionality it is a pet architecture for demonstrating ideas about formalizing the architectural design space (not unlike the data type `Stack` for algebraic specifications or abstract data types). We will show in this section how to formalize this architecture in terms of monads, hereby including specifications through set theoretic or probabilistic relations as special cases.

Software Architectures.   The structural aspects of a large programming system are captured through its (software) architecture. Initially, this term was used rather loosely. Work being done during the 1990s, in particular by M. Shaw and her associates, has established a body of knowledge in the software engineering community about methods for structuring large systems. This translates into practical tools like architectural design languages.

An architecture for a system separates computation from control on the system's level; while the former is represented by algorithms formulated in a programming language, the latter is formulated in terms of *components* (which carry out the computations) and *connectors* (which transport data from one component to another one). Connectors are elevated to first class rank making it possible to reason explicitly about connecting components. Considering an architecture then means identifying connectors and components and describing the interplay between them. Since the emphasis is on structure, formalizing an architecture helps in investigating its salient features; formalizations can be proposed on different levels.

The formalization of an architecture permits reasoning about it since it provides precise and abstract models that usually come with analytical techniques. This is in marked contrast to architectural techniques where the shape of an architecture and its architectural parameters are determined experimentally ((Doberkat et al., 2000) provides an example for constructing a substantial real life system). Shaw and Garlan (Shaw and Garlan, 1996, Sec.6) discuss architectural formalisms. They distinguish three levels of formalization:

The architecture of a specific system. This permits a precise characterization of the system-level functions that determine the overall product functionality.

The formalization of an architectural style. By describing architectural abstractions it becomes possible to analyze various static or dynamic properties of common architectural patterns or reference architectures which are used informally, e.g., as reference architectures. Essential ingredients in such a formalization are provided by connectors and by components.

A theory of software architecture. By classifying architectures with a mathematical machinery, a deductive basis for analyzing systems is provided.

We will focus on the intermediate level and investigate an architecture where the computational elements are represented through relations.

Relations. Nondeterministic and probabilistic constructions share, as we have seen in the previous sections, a common structure in representing the Kleisli construction for a monad. The case of nondeterministic relations is covered through the power set functor on the category of sets, and the stochastic case through the functor which assigns each measurable set the space of all subprobability measures, as we have seen in the discussion in Section 2.3. Thus monads (and their associated Kleisli categories) form the common abstraction for both cases, bringing us into the realm of Moggi's compelling argumentation and well known that monads form a suitable basis for modeling computations.

Consequently, the architectural modeling will be done on the basis of a monad.

Categories vs. Architectures. Categories with their emphasis on structure are a suitable formal tool for modeling software architectures. Focussing on structure implies the independence on any representation in a specification or programming language; technically this is achieved through the use of morphisms and functors. Synthesizing a design sometimes means formulating the components and combining them through a suitable colimit, cf. (Fiadeiro and Maibaum, 1996). Wermelinger and Fiadeiro (Wermelinger and Fiadeiro, 1998) discuss some salient features of an architectural modeling through categories in the context of their modeling mobile programs. Specifically they point out that this approach represents programs as objects. Morphisms show how programs can be composed; the explicit use of connectors facilitates the separation of computation and coordination. Moreover they point out that the mechanisms for interconnecting components yielding complex systems are formalized using universal constructs, in this way providing a stage for arguing about these mechanisms formally.

When modeling an architecture, one has to take care at least of the computational components and the connectors. Working in a category, the connectors are represented as objects while the computational components are modeled as morphisms between the objects. Since computations will be represented as monads, the most natural way is representing a component through

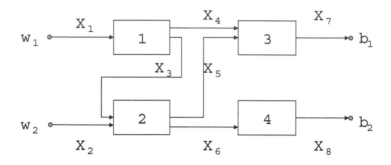

Figure 2.1:   A Pipeline as a System of Pipes and Filters

the work of the corresponding functor $\mathfrak{T}$. Here the Kleisli construction enters the game: suppose for simplicity that the input and the output for a component λ are modeled respectively through the objects x and y, then the computation performed by λ is represented through a Kleisli morphism $x \to \mathfrak{T}y$. These assignments are described when modeling a particular architecture, and the work of an instance of this architecture is described in terms of these assumptions. We will outline this point for a pipeline architecture, the architectural style being simple enough to be studied without being captivated in the intricacies of a discussion of an overwhelming number of technical and architecture specific issues. It is rather general, hence not tied to a particular domain or application, and it is semantically rich enough to illustrate the concepts proposed and investigated here.

Pipelines.   Filters transform streams of data functionally; each filter has input ports from which data are read, and output ports, to which results are written. Computation is performed incrementally and locally: a portion of the data available at the input ports is read, transformed, and written to the output ports which in turn serve as input ports for other components or as outputs for the system. The filters may be assumed to work concurrently. It is characteristic for this style that the data passing through a filter enters only through its input ports, and leaves only through its output ports; global data are not available. A pipe links an input port to an output port and transmits data from one component to another. Pipelines are in this taxonomy a substyle which performs the computations without cycles.

   Figure 2.1 shows an example for a simple pipeline. The system has two inputs w_1 and w_2 and two outputs b_1 and b_2. It has four independent components $1, \ldots, 4$. The edges are labeled with the types of the inputs the components accept, and produce, resp.: for example, component 1 accepts inputs of type X_1 and produces outputs of types X_3 and X_4, the former serving as an input to component 2 together with an input of type X_2, the latter serving as an input to component 3 together with an input of type X_5, which is produced by component 2. The entire system accepts two inputs of type X_1 and X_2 and produces two outputs of type X_7 and X_8.

We assume that the system forms a directed graph with filters as nodes and pipes as edges. The graph is assumed to be acyclic, so that loops among filters are not permitted; hence we will address the *linear* substyle of the architecture. The common pipelines are usually linear: think of UNIX pipes or of linear arrangements in which data are generated and collected from different sources.

Nevertheless, acyclicity is an assumption which from the mathematical point of view quite notably restricts a general model of pipes and filters. Suppose that we have a component C which has among its output ports the ports r and s, r being a "backward" port, s being a forward one. Here *backward* means that r's output is being fed into the input port of another component the output of which is pipelined directly or indirectly into one of the input ports of component C. With other words: C lies on a cycle in the graph constituting the system's topology. The functional character of the components implies that each input is associated with some output for all ports, so that an undefined value at one of these ports must not occur. Modeling this situation requires taking care of this iterative structure, and providing a neutral element of some sorts, indicating that the functional output of C flows through s only in certain situations. We will return to this point when discussing possible extensions in Section 2.4.7.1.

But before entering the discussion with formal machinery, we will first have a look at a simple example, so that the ideas will become more transparent.

2.4.1   A First Example

Consider the pipeline system of pipes and filters represented through the graph in Figure 2.1 again. It has the roots $\{w_1, w_2\}$, the leaves $\{b_1, b_2\}$ and the filters $\{1, 2, 3, 4\}$. This graph exhibits a little irregularity in that paths to a node from different roots may differ in length. Thus a signal from root w_1 to node 2 is routed through node 1 before arriving at node 2, whereas a signal originated from root w_2 arrives directly at node 2. In order to get a uniform treatment, we introduce **noops** which are to have the effect that all paths from a root to a node have the same length (we will call graphs with this property *stratified* later on, cf. Definition 2.19). Thus the graph is not stratified, but the version in Figure 2.2 is. Note that we have introduced two new artificial nodes Δ_2 and Δ_4.

We partition the set of nodes into classes S_j such that a node n is in S_j iff the length of a path from a root to n is exactly j. Hence we have these sets $S_0, \ldots, S_4$:

$$S_0 = \{w_1, w_2\}, S_1 = \{1, \Delta_2\}, S_2 = \{\Delta_4, 2\}, S_3 = \{3, 4\}, S_4 = \{b_1, b_2\}.$$

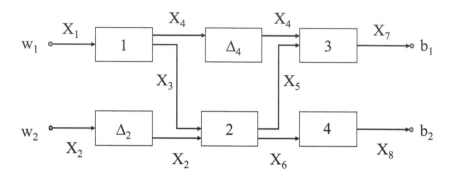

Figure 2.2:   Stratified Pipeline

We will associate sets with the edges, and relations between the corresponding sets with the nodes. Sets are meant to indicate the type of information flowing along that edge, relations associated with a node to indicate the processing being performed by this node. Hence we have for node 2 relation R_2 with $R_2 \subseteq (X_2 \times X_3) \times (X_5 \times X_6)$. The noop nodes are associated with no processing at all, thus the information is just passed unchanged through them, indicated by Δ, and we have

$$\Delta_2 := \{\langle x, x \rangle \mid x \in X_2\},$$
$$\Delta_4 := \{\langle x, x \rangle \mid x \in X_4\}.$$

Work in these partitions proper is characterized by independent processing of the partition's nodes, thus we take the Cartesian product of the sets labeling the incoming edges as the input sets for the partition, similarly for the output sets. The construction yields, e.g., for component S_2 this relation:

$$\{\langle x_2, x_3, x_4, x_4, x_5, x_6 \rangle \mid x_4 \in X_4, \langle x_2, x_3, x_5, x_6 \rangle \in R_2\}.$$

The work of the entire pipeline is then described through the product of the relations that represent the work of the individual partitions. This yields a relation which is a subset of $(X_1 \times X_2) \times (X_7 \times X_8)$. The result is, as expected:

$$\{\langle x_1, x_2, x_7, x_8 \rangle \mid \exists x_3 \in X_3, x_4 \in X_4, x_5 \in X_5, x_6 \in X_6 :$$
$$\langle x_1, x_3, x_4 \rangle \in R_1, \langle x_2, x_3, x_5, x_6 \rangle \in R_2, \langle x_4, x_5, x_7 \rangle \in R_3, \langle x_6, x_8 \rangle \in R_4\}.$$

This relation may be manipulated further. It may take part in horizontal and vertical operations. Horizontal operations form architectures by concatenating components, so that the resulting pipelines get longer and longer; vertical operations refine an architecture by replacing a component through an entire subsystem (we are a bit in conflict with the notion of horizonal and vertical composition in category theory, cp. (MacLane, 1997, II.4, II.5 and XII.3) where these terms are used for the composition of natural transformations for different functors; the present terminology is, however, quite graphically used in Software Engineering, too, so we stick to it here). These fundamental architectural operations will be discussed in detail in Section 2.4.6.

2.4.2 First Steps

This section serves as a preparation for things to come: we formulate a compatibility condition which relates the product in the category under consideration to the monad which is used for modeling the computations.

The category $\mathfrak{X}^{(n)}$ has as objects n-tuples of objects of $\mathfrak{X}$, and n-tuples of morphisms, the composition being defined componentwise. Define functors $\mathfrak{G}_{\mathfrak{X}}^{(n)}, \mathfrak{H}_{\mathfrak{X}}^{(n)} : \mathfrak{X}^{(n)} \to \mathfrak{X}$ upon setting $(n \geq 1)$

$$\mathfrak{G}_{\mathfrak{X}}^{(n)}\langle x_1, \ldots, x_n \rangle := \mathfrak{T}x_1 \times \cdots \times \mathfrak{T}x_n$$

$$\mathfrak{H}_{\mathfrak{X}}^{(n)}\langle x_1, \ldots, x_n \rangle := \mathfrak{T}(x_1 \times \cdots \times x_n),$$

and, if $\phi_i : x_i \to y_i$ are morphisms in $\mathfrak{X}$, then

$$\mathfrak{G}_{\mathfrak{X}}^{(n)}\langle \phi_1, \ldots, \phi_n \rangle := \mathfrak{T}\phi_1 \times \cdots \times \mathfrak{T}\phi_n$$

$$\mathfrak{H}_{\mathfrak{X}}^{(n)}\langle \phi_1, \ldots, \phi_n \rangle := \mathfrak{T}(\phi_1 \times \cdots \times \phi_n).$$

$\mathfrak{T}$ models the computations performed in the components, and which are partially done in parallel. This in turn will be modeled through finite products. Hence $\mathfrak{T}$ should be naturally related to the product in $\mathfrak{X}$; the present proposal assumes compatibility which mediates between $\mathfrak{T}(x) \times \mathfrak{T}(y) \times \mathfrak{T}(z)$ and $\mathfrak{T}(x \times y \times z)$ using the natural transformation $1_{\mathfrak{T}} : \mathfrak{T} \overset{\bullet}{\to} \mathfrak{T}$ and introducing another one between $\mathfrak{G}_{\mathfrak{X}}^{(2)}$ and $\mathfrak{H}_{\mathfrak{X}}^{(2)}$. To be specific:

DEFINITION 2.12 *Monad $\mathfrak{T}$ is compatible with the product in $\mathfrak{X}$ iff there exists a natural transformation $\theta : \mathfrak{G}_{\mathfrak{X}}^{(2)} \overset{\bullet}{\to} \mathfrak{H}_{\mathfrak{X}}^{(2)}$ which makes this diagram commutative:*

$$
\begin{array}{ccc}
\mathfrak{T}(x) \times \mathfrak{T}(y) \times \mathfrak{T}(z) & \xrightarrow{(1_{\mathfrak{T}} \times \theta)_{\langle x,y,z \rangle}} & \mathfrak{T}(x) \times \mathfrak{T}(y \times z) \\
{\scriptstyle (\theta \times 1_{\mathfrak{T}})_{\langle x,y,z \rangle}} \downarrow & & \downarrow {\scriptstyle \theta_{\langle x, y \times z \rangle}} \\
\mathfrak{T}(x \times y) \times \mathfrak{T}(z) & \xrightarrow{\theta_{\langle x \times y, z \rangle}} & \mathfrak{T}(x \times y \times z)
\end{array}
$$

θ *is called the* mediating *transformation.*

A mediating transformation θ spawns a sequence $(\theta^{(n)})_{n \geq 1}$ of natural transformations

$$\theta^{(n)} : \mathfrak{G}_{\mathfrak{X}}^{(n)} \overset{\bullet}{\to} \mathfrak{H}_{\mathfrak{X}}^{(n)}$$

in the following way:

$$\theta_x^{(1)} := 1_{\mathfrak{T}x}$$
$$\theta_{\langle x,y\rangle}^{(2)} := \theta_{\langle x,y\rangle}$$
$$\theta_{\langle x_1,\ldots,x_{n+1}\rangle}^{(n+1)} := \theta_{\langle x_1\times\cdots\times x_n,x_{n+1}\rangle}\circ\left(\theta^{(n)}\times 1_{\mathfrak{T}}\right)_{\langle x_1,\ldots,x_{n+1}\rangle}$$

LEMMA 2.13

Suppose that $\mathfrak{T}$ is compatible with the product in $\mathfrak{X}$, and let θ be the mediating transformation. Defining θ as above, the sequence $(\theta^{(n)})_{n\in\mathbb{N}}$ has the following properties:

a. $\theta^{(n)}$ is a natural transformation,

b. for all $k,\ell\in\mathbb{N}$ and for all objects x_i,y_i we have

$$\theta_{\langle x_1\times\cdots\times x_k,y_1\times\cdots\times y_\ell\rangle}\circ\left(\theta^{(k)}\times\theta^{(\ell)}\right)_{\langle x_1,\ldots,x_k,y_1,\ldots,y_\ell\rangle} = \theta_{\langle x_1,\ldots,x_k,y_1,\ldots,y_\ell\rangle}^{(k+\ell)}$$

PROOF   1. The first part is established by induction on n, since the composition of natural transformations is again a natural transformation.

2. The second part is proved by induction on ℓ, the start of the induction representing just the inductive definition from above. The induction step is established through the commutativity of this diagram:

with

$$\alpha := \left(\theta^{(k)} \times \theta^{(\ell)} \times 1_{\mathfrak{X}}\right)_{\langle a_1,\ldots,a_k,b_1,\ldots,b_\ell,b\rangle}$$

$$\beta := (1_{\mathfrak{X}} \times \theta)_{\langle \prod_{i=1}^k a_i, \prod_{i=1}^\ell b_i \times b\rangle}$$

$$\gamma := \left(\theta^{(k+\ell)} \times 1_{\mathfrak{X}}\right)_{\langle a_1,\ldots,a_k,b_1,\ldots,b_\ell,b\rangle}$$

$$\delta := \theta_{\langle \prod_{i=1}^k a_i \times \prod_{i=1}^\ell b_i, b\rangle}$$

$$\kappa := (\theta \times 1_{\mathfrak{X}})_{\langle \prod_{i=1}^k a_i, \prod_{i=1}^\ell b_i, b\rangle}$$

$$\lambda := \theta_{\langle \prod_{i=1}^k a_i, \prod_{i=1}^\ell b_i \times b\rangle}.$$

The upper triangle is commutative because of the induction hypothesis; the lower square is just the condition on θ from Definition 2.12. Then the assertion follows, since $\lambda \circ \beta \circ \alpha = \delta \circ \kappa \circ \alpha = \delta \circ \gamma$, and

$$\beta \circ \alpha = \left(\theta^{(k)} \times \theta^{(\ell+1)}\right)_{\langle a_1,\ldots,a_k,b_1,\ldots,b_\ell,b\rangle}$$

$$\delta \circ \gamma = \theta^{(k+\ell+1)}_{\langle a_1,\ldots,a_k,b_1,\ldots,b_\ell,b\rangle}.$$

$\square$

Our sample categories have mediating transformations.

LEMMA 2.14
The map

$$\begin{cases} \mathfrak{M}(X) \times \mathfrak{M}(Y) \to \mathfrak{M}(X \times Y) \\ \langle A, B\rangle \qquad\qquad \mapsto \{\langle h_1 h_2, x, y\rangle \mid \langle h_1, x\rangle \in A, \langle h_2, y\rangle \in B\} \end{cases}$$

defines a natural transformation that mediates between the functor and the product in $\mathfrak{Set}$.

Let us discuss the Giry monad. Define for the measurable subset C of $H \times X_1 \times X_2$ and for $\mu_i \in \mathfrak{G}(X_i)$ $(i = 1, 2)$ their H-product $\mu_1 \otimes_H \mu_2$ through

$$(\mu_1 \otimes_H \mu_2)(C) := \int_{H \times X_1} \mu_2(\{\langle h_2, y\rangle \mid \langle h_1 h_2, x, y\rangle \in C\})\ \mu_1(d\langle h_1, x\rangle),$$

then $\mu_1 \otimes_H \mu_2 \in \mathfrak{G}(X_1 \times X_2)$. In fact, we can say more:

LEMMA 2.15
The H-product is associative, it constitutes a natural mediating transformation $\mathfrak{G}^{(2)}_{\mathfrak{G}} \overset{\bullet}{\to} \mathfrak{H}^{(2)}_{\mathfrak{G}}$.

PROOF   1. Associativity of the H-product is an easy consequence of Fubini's Theorem on product integration.

2. Let $f : X \to X'$ and $g : Y \to Y'$ be measurable maps, then we have for the measures $\mu_1 \in \mathfrak{G}(X), \mu_2 \in \mathfrak{G}(Y)$ and for the measurable subset C' of $H \times X' \times Y'$

$$(\mathfrak{G}(f)(\mu_1) \otimes_H \mathfrak{G}(g)(\mu_2))(C') =$$

$$\int_{H \times X} \mu_2(\{\langle h_2, y \rangle \mid \langle h_1 h_2, f(x), g(y) \rangle \in C'\}) \, \mu_1(d\langle h_1, x \rangle) =$$

$$(\mu_1 \otimes_H \mu_2)(\{\langle h, x, y \rangle \mid \langle g, f(x), g(y) \rangle \in C'\}) =$$

$$\mathfrak{G}(f \times g)(\langle \mu_1, \mu_2 \rangle \mapsto \mu_1 \otimes_H \mu_2)(C').$$

But this means that the H-product is natural for $\mathfrak{G}_{\mathfrak{G}}^{(2)}$ and $\mathfrak{H}_{\mathfrak{G}}^{(2)}$. It is easy to see that this transformation is mediating.   ⬚

The product in $\mathfrak{X}$ defines together with $\mathfrak{T}$ an associative operation:

DEFINITION 2.16   *Let* $\tau : a \to \mathfrak{T}b$ *and* $\tau' : a' \to \mathfrak{T}b'$ *be morphisms, then define* $\tau \times_{\mathfrak{T}} \tau' : a \times a' \to \mathfrak{T}(b \times b')$ *upon setting* $\tau \times_{\mathfrak{T}} \tau' := \theta_{\langle b, b' \rangle} \circ \tau \times \tau'$.

Example 2.17 will show that $\times_{\mathfrak{T}}$ does not exhibit the universal properties which would be necessary to form a product (thus the category $\mathfrak{X}_{\mathfrak{T}}$ does not necessarily have finite products, even if $\mathfrak{X}$ has them).

EXAMPLE 2.17

Let, for simplicity, H be the trivial monoid $\{1\}$, which we omit from the notation. Suppose that $\times_{\mathfrak{G}}$ is a product in $\mathfrak{M}_{\mathfrak{G}}$, and fix two nonempty measurable spaces X_1 and X_2. There exists a measurable space X and the two projections $p_i : X \to \mathfrak{G}(X_i)$ such that, whenever $K_i : S \to \mathfrak{G}(X_i)$ $(i = 1, 2)$ is a morphism, we can find a morphism $K : S \to \mathfrak{G}(X)$ such that $K_i = p_i * K$ holds for $i = 1, 2$. This means that

$$K_i(s)(B_i) = \int_X p_i(x)(B_i) \; K(s)(dx)$$

always holds. Now let always $K_i(s)(X_i)$ equal 1. This implies that $p_i(x)(X_i)$ equals 1 $K(s)$-almost everywhere for each s, and for each K which can be so constructed. Note that p_1, p_2 do not depend on the specific choice of K_1, K_2. But then we have for any $L_i : S \to \mathfrak{G}(X_i)$ with product L:

$$L_1(s)(X_1) = \int_X p_1(x)(X_1) \; L(s)(dx)$$

$$= \int_X p_2(x)(X_2) \; L(s)(dx)$$

$$= L_2(s)(X_2).$$

Since we cannot always maintain $L_1(s)(X_1) = L_2(s)(X_2)$ it follows that $\mathfrak{M}_{\mathfrak{G}}$ does not have finite products.

We have, however:

COROLLARY 2.18

Let $\tau : a \rightarrow \mathfrak{T}b, \tau : a' \rightarrow \mathfrak{T}b', \tau : a'' \rightarrow \mathfrak{T}b''$ *be morphisms in* $\mathfrak{X}$, *then*
$(\tau \times_{\mathfrak{T}} \tau') \times_{\mathfrak{T}} \tau'' = \tau \times_{\mathfrak{T}} (\tau' \times_{\mathfrak{T}} \tau'')$.

PROOF Lemma 2.13 shows that both sides of this equation equal

$$\theta^{(3)}_{\langle b, b', b'' \rangle} \circ (\tau \times \tau' \times \tau'').$$

$\square$

2.4.3 The Basic Construction

We will associate a computation to a pipeline by composing computations performed in its components. This construction will be first carried out due to technical reasons for graphs that exhibit a certain regularity: the nodes are partitioned into layers so that the information flows strictly from one layer to the next one. This restriction is introduced for reasons of synchronization: the inputs at each port of a component in a layer are uniformly available at the same time, and so are the outputs. It makes modeling somewhat easier, but it is really only a technical device. We remove it in Section 2.4.5, after we have shown in Section 2.4.4 how to manipulate a *dag* (a *directed acyclic graph*) so that it is satisfied.

Fix in this section a finite dag $\mathcal{G} = (V, E)$ with roots W and leaves B; for convenience we assume the set V of nodes to be somehow linearly ordered. Put for node n

$$\bullet n := \{m \in V \mid \langle m, n \rangle \in E\}$$
$$n \bullet := \{m \in V \mid \langle n, m \rangle \in E\}$$

as the sets of nodes which have an edge into that node or out of it, resp. $\mathcal{G}$ is not supposed to have any isolated nodes, i.e., nodes n with $\bullet n \cup n \bullet = \emptyset$.

We define sets $(S_j)_{0 \leq j \leq k}$ through

$$S_0 := W,$$
$$S_{j+1} := \{n \in V \mid \bullet n \subseteq S_j\} \ (j \geq 0).$$

DEFINITION 2.19   *The dag $\mathcal{G} = (V, E)$ is called* stratified *iff the sets* $(S_j)_{0 \leq j \leq k}$ *form a partition of V for some k. The maximal index k such that $S_k \neq \emptyset$ is denoted by $\Lambda(\mathcal{G})$.*

Let for the rest of this section $\mathcal{G}$ be a stratified graph.

LEMMA 2.20

The set of inputs $\{\langle m, n \rangle \mid \langle m, n \rangle \in E, n \in S_j\}$ into the set S_j of nodes equals the set of outputs $\{\langle m, \ell \rangle \mid \langle m, \ell \rangle \in E, m \in S_{j-1}\}$ from the sets S_{j-1} for $j \geq 1$.

PROOF   This follows directly from the fact that $\mathcal{G}$ is stratified.   ⬜

This observation shows that each node n is in some uniquely determined set S_j. If n is an inner node (thus if $j > 0$ and $j < k$), then $\langle m, n \rangle \in E$ implies $m \in S_{j-1}$, and $\langle n, m \rangle \in E$ implies $m \in S_{j+1}$. Depicting $S_0, \ldots, S_k$ as blocks from left to right, information flows into n only from nodes in S_{j-1}, thus from nodes on the left, and flows from n only into nodes in S_{j+1}, hence into nodes on the right.

We associate now objects from category $\mathfrak{X}$ with edges, and nodes with morphisms in $\mathfrak{X}_{\mathfrak{T}}$. To be specific, each edge $\langle k, n \rangle \in E$ is assigned an object $\gamma_{\langle k, n \rangle}$ in $\mathfrak{X}$. If $\mathfrak{T}$ is the Manes functor $\mathfrak{M}$, this means that an edge $\langle k, n \rangle$ is assigned a set which represents the flow from node k to node n. For $\mathfrak{T}$ as the Giry functor $\mathfrak{G}$, the edge is assigned a measurable space which also represents the flow along this edge: if it is used as an input, then it is the sample space of all inputs for a probabilistic relation; if it is used as an output, then it represents the space of all probability measures over this space, cf. Example 2.21.

The input to node n and the output from this node are then reflected respectively through the respective products

$$\mathsf{i}(\gamma, n) := \prod \{\gamma_{\langle k, n \rangle} \mid k \in \bullet n\} \ (n \notin W)$$

$$\mathsf{o}(\gamma, n) := \prod \{\gamma_{\langle n, k \rangle} \mid k \in n\bullet\} \ (n \notin B)$$

Each inner node n is labeled with a Kleisli morphism

$$\mathsf{a}(\gamma, n) : \mathsf{i}(\gamma, n) \to \mathfrak{T}(\mathsf{o}(\gamma, n)),$$

so that $\mathsf{a}(\gamma, n)$ models the work being performed by node n.

EXAMPLE 2.21
Suppose that $\bullet n = \{m_1, \ldots, m_r\}$ and $n\bullet = \{\ell_1, \ldots, \ell_s\}$.

1. For the Manes monad we assign the edges $\langle m_1, n \rangle, \ldots, \langle m_r, n \rangle$ to the sets $X_1, \ldots, X_r$, and sets $Y_1, \ldots, Y_s$ to the edges $\langle n, \ell_1 \rangle, \ldots, \langle n, \ell_s \rangle$. The node n itself is assigned a relation

$$\mathsf{a}(\gamma, n) \subseteq (X_1 \times \cdots \times X_r) \times (H \times Y_1 \times \cdots \times Y_s).$$

Suppose that $\langle x_1, \ldots, x_r \rangle \in X_1 \times \cdots \times X_r$ is an input to node n which is related to output $\langle h, y_1, \ldots, y_s \rangle$. That tuple represents the node's work, and we have two kinds of results: the tuple $\langle y_1, \ldots, y_s \rangle$ which in turn is being communicated to other nodes in the pipeline, and $h \in H$ which may be interpreted as an immediate result which could be read off this processing element. Proposition 2.1 indicates that all these results, which will not be communicated as input to other filters, will be accumulated as control percolates through the system.

2. For the Giry monad, X_i and Y_j are measurable spaces, and

$$\mathsf{a}(\gamma, n) : X_1 \times \cdots \times X_r \rightsquigarrow H \times Y_1 \times \cdots \times Y_s$$

is a stochastic relation. Thus for an input $\langle x_1, \ldots, x_r \rangle \in X_1 \times \cdots \times X_r$, and for a measurable $B \subseteq H \times Y_1 \times \cdots \times Y_s$ we get $\mathsf{a}(\gamma, n)(x_1, \ldots, x_r)(B)$ as the probability that the computation in node n terminates, and that $\langle h, y_1, \ldots, y_s \rangle$ will be a member of B; the interpretation of the components for this tuple is the same as above.

This indicates that a relational environment for modeling the basic scenario for a pipeline architecture is provided, capturing both the nondeterministic and the probabilistic case.

DEFINITION 2.22   *Call $\langle \mathcal{G}, \gamma \rangle$ a pipeline system (abbreviated as PF-system) over the monad $\langle \mathfrak{T}, \mathfrak{e}, \mathfrak{m} \rangle$ iff the following conditions hold:*

- *$\mathcal{G} = \langle V, E \rangle$ is a directed graph with W and B as the sets of roots, and leaves, resp.*

- *$\forall \langle n, m \rangle \in E : \gamma_{\langle n, m \rangle}$ is an object in $\mathfrak{X}$,*

- *$\forall n \in V \setminus (W \cup B) : \mathsf{a}(\gamma, n) : \mathsf{i}(\gamma, n) \to \mathfrak{T}(\mathsf{o}(\gamma, n))$ is a morphism in $\mathfrak{X}$.*

The system $\langle \mathcal{G}, \gamma \rangle$ is called stratified *iff $\mathcal{G}$ is stratified.*

Since the monad will be fixed in the sequel, we will not mention it explicitly when talking about PF-systems; unless explicitly mentioned, PF-systems will be stratified in this section.

Now define for $0 < j \leq k$ the object

$$\mathsf{g}(\gamma, S_j) := \prod \{ \mathsf{i}(\gamma, n) \mid n \in S_j \},$$

then $\mathbf{g}\,(\gamma, S_j)$ indicates the kind of flow into S_j (which is, because of Observation 2.20, the flow out of S_{j-1}); hence the component S_j has what could be called the input signature $\mathbf{g}\,(\gamma, S_{j-1})$ and the output signature $\mathbf{g}\,(\gamma, S_j)$.

The work being done in S_j can be represented through the Kleisli morphism

$$\mathsf{A}\,(\gamma, S_j) : \mathbf{g}\,(\gamma, S_{j-1}) \to \mathfrak{T}\,(\mathbf{g}\,(\gamma, S_j)),$$

with

$$\mathsf{A}\,(\gamma, S_j) := \theta^{(\#S_j)}_{\langle n|n\in S_j\rangle} \circ \prod \{\mathsf{a}\,(\gamma, n) \mid n \in S_j\},$$

where θ is the natural transformation which mediates between $\mathfrak{T}$ and the product in $\mathfrak{X}$, cf. Definition 2.12. The linear order on V makes $\theta^{(\#S_j)}_{\langle n|n\in S_j\rangle}$ uniquely determined. The work of the entire system is then represented through

$$\mathsf{P}\,(\mathcal{G}, \gamma) := \mathsf{A}\,(\gamma, S_{k-1}) * \ldots * \mathsf{A}\,(\gamma, S_1).$$

The construction shows that

$$\mathsf{P}\,(\mathcal{G}, \gamma) : \mathbf{g}\,(\gamma, S_1) \to \mathfrak{T}\,(\mathbf{g}\,(\gamma, S_k))$$

is a Kleisli morphism between the inputs to the system and the outputs from it, thus representing the system's work.

The example that follows discusses a particular pipeline system, stratifies the graph and exercises the construction proposed here for the Giry monad. The set theoretic case has already been dealt with in Section 2.4.1.

EXAMPLE 2.23

We discuss the example outlined in Section 2.4.1 (Figure 2.2) again, this time for the stochastic case. We assume that the monoid carries a measurable structure which makes multiplication measurable, cf. Section 2.3.1. The Δ_i $(i = 2, 4)$ are Dirac kernels: we put

$$\Delta_i(x) := \delta_{\langle 1, x\rangle} \in \mathfrak{G}\,(X_i).$$

Node n is this time represented through a stochastic relation K_n between the appropriate sets, e.g.,

$$K_2 : X_2 \times X_3 \rightsquigarrow H \times X_5 \times X_6.$$

The construction gives then, e.g., for component S_2:

$$\mathsf{A}\,(\gamma, S_2)\,(x_2, x_3, x_4) = K_2(x_2, x_3) \otimes_H \Delta_4(x_4);$$

thus the probability that the computation in components S_2 will give an element of the measurable set $D \subseteq H \times X_4 \times X_5 \times X_6$ after input of $\langle x_1, x_2, x_3\rangle \in$

$X_1 \times X_2 \times X_3$ is computed as

$$\mathsf{A}\,(\gamma, S_2)\,(x_2, x_3, x_4)\,(D) =$$

$$\int_{H \times X_5 \times X_6} \Delta_4(x_4)\,(\{\langle h_4, x_4' \rangle \mid \langle h_2 h_4, x_4', x_5, x_6 \rangle \in D\}) \times$$
$$\times K_2(x_2, x_3)(d\langle h_2, x_5, x_6 \rangle) =$$
$$K_2(x_2, x_3)\,(\{\langle h_2, x_5, x_6 \rangle \mid \langle h_2, x_4, x_5, x_6 \rangle \in D\})\,.$$

Let $f : H \times X_4 \times X_5 \times X_6 \to \mathbb{R}$ be a bounded and measurable function, then a computation of the Kleisli product according to Proposition 2.6 shows that $(x_1 \in X_1, x_2 \in X_2)$

$$\int_{H \times X_4 \times X_5 \times X_6} f\,d((\mathsf{A}\,(\gamma, S_2) * \mathsf{A}\,(\gamma, S_1)))\,(x_1, x_2) =$$

$$\int_{H \times X_3 \times X_4} \int_{H \times X_5 \times X_6} f(gh, x_4, x_5, x_6)\,K_2(x_2, x_3)(d\langle h, x_5, x_6 \rangle) \times$$
$$K_1(x_1)(d\langle g, x_3, x_4 \rangle).$$

We get in this way for $\langle x_1, x_2 \rangle \in X_1 \times X_2$ and the measurable subset $F \subseteq H \times X_7 \times X_8$

$$\mathsf{P}\,(\mathcal{G}, \gamma)\,(x_1, x_2)(F)$$

$$= \int_{H \times X_4 \times X_5 \times X_6} \mathsf{A}\,(\gamma, S_3)\,(x_4, x_5, x_6)\,(\{\langle h, x_7, x_8 \rangle \mid \langle gh, x_7, x_8 \in F\}) \times$$
$$\times d\,(\mathsf{A}\,(\gamma, S_2) * \mathsf{A}\,(\gamma, S_1))\,(x_1, x_2)(d\langle g, x_4, x_5, x_6 \rangle)$$

$$= \int_{H \times X_3 \times X_4} \int_{H \times X_5 \times X_6} \int_{H \times X_7} K_4(x_6)(\{\langle h_1, x_8 \rangle \mid \langle g_1 hgh_1, x_7, x_8 \rangle \in F\}) \times$$
$$\times K_3(x_4, x_5)(d\langle g, x_7 \rangle) K_2(x_2, x_3)(d\langle h, x_5, x_6 \rangle) K_1(x_1)(d\langle g_1, x_3, x_4 \rangle)$$

as the work of the entire pipeline.

We will now prepare for removing the condition that a PF-system should be stratified.

2.4.4 Stratifying Graphs

The assumption in carrying out the basic construction in Section 2.4.3 has been that the graph underlying the PF-system is stratified. But graphs rarely are, so it becomes necessary to make provisions for generalizing the construction to general directed graphs. The strategy is to devise a way of stratifying a graph, to perform the construction on the new graph, and to make sure that all graphs that are stratified versions of the given one perform the same work. The present section is auxiliary in character and provides an algorithm for stratifying. Section 2.4.5 will do the generalization.

ALGORITHM 2.24

$Edges := E; \ Nodes := V; \ zeta := 0;$
while $Edges \neq \emptyset$ **do**
      forall $n \in range \ Edges$ **do**
         $H(n) := \{a \mid \langle a, n \rangle \in Edges, \alpha(a) = 0\};$
      od;
      forall $a \in domain \ H$ **do**
         $d(a) := zeta;$
      od;
      $Edges := Edges \setminus \{\langle a, n \rangle \mid \langle a, n \rangle \in Edges, \alpha(a) = 0\};$
      forall $n \in Nodes$ **do**
         $r := \#\{k \mid \langle k, n \rangle \in Edges\};$
         if $r = 0$ **then**
               $\alpha(n) := 0;$
         else
               for $j := r + 1$ **to** $\alpha(n)$ **do**
                   choose m **from** $H(n);$
                   $H(n) := H(n) \setminus \{m\};$
                   $q := \texttt{newq()};$
                   $\alpha(q) := 0; \ V := V \cup \{q\};$
                   $E := (E \setminus \{\langle m, n \rangle\}) \cup \{\langle m, q \rangle, \langle q, n \rangle\};$
                   $Edges := Edges \cup \{\langle q, n \rangle\};$
               od; -- **forall**
         fi;
      od; -- **forall**
      $zeta := zeta + 1;$
od; -- **while** ♣

Figure 2.3:   Algorithm `Stratify`

Algorithm 2.24 produces from $\mathcal{G} = (V, E)$ a stratified graph $\mathcal{G}' = (V', E')$ with $V \subseteq V'$ and $E' \cap (V \times V) \subseteq E$. It assumes that $\mathcal{G}$ does not have any isolated nodes, and that each node lies on a path from a root to a leaf. We assume that we have a source Q of fresh nodes which is disjoint from $V \cup E$; invoking the function `newq()` will produce a fresh node. The map α initially gives the in-degree of a node; we use some auxiliary values which will be needed and discussed in the sequel.

Thus we iterate over all edges, removing roots as we go; for a node n we use $H(n)$ for recording which nodes have edges leading into n that will be removed. When we see that a node n has no longer any edges having n as a target, this node will be promoted to a root (and removed in due course); promotion to a root means changing the in-degree $\alpha(n)$ to 0. If it turns out, however, that there are still edges going into that node (note that in this case

$\#H(n)$ equals $\alpha(n) - r$), we replace each edge $\langle m, n \rangle$ by a pair of edges $\langle m, q \rangle$ and $\langle q, n \rangle$, where q is a fresh node which is put into the set V of nodes.

Since each dag has roots, and since $\mathcal{G}$ is assumed to have no isolated nodes, it is not difficult to see that Algorithm 2.24 terminates. It is also evident that the new graph $\mathcal{G}' = (V', E')$ has the given one as a subgraph in the sense that $V \subseteq V'$ and $E' \cap (V \times V) \subseteq E$ both hold.

LEMMA 2.25

Let $n \in E'$ be a node in $\mathcal{G}'$, and assume that there exists a path from a root of $\mathcal{G}'$ to n. Then this path has length $d(n)$.

PROOF   1. We proceed by induction on the value of *zeta*. The beginning is trivial, since exactly the roots of $\mathcal{G}$ are removed, and no new roots are introduced.

2. Now let *zeta* $= k$, and assume that $d(n)$ equals $k + 1$. This means that $\alpha(n)$ is set to 0 when *zeta* has the value k. We distinguish the case that n is a new node introduced in this step from the case that $\alpha(n)$ is set to 0 because $r = 0$ holds.

- If n is a new node, we can find an edge $\langle m_1, n_1 \rangle$ which gave rise to this creation, hence that edge is replaced by the pair of edges $\langle m_1, n \rangle$ and $\langle n, n_1 \rangle$. Edge $\langle m_1, n_1 \rangle$ is a member of the set *Edges* before control enters the body of the actual loop, thus will be removed. The induction hypothesis makes sure that each path from a root to m in the graph constructed so far has length k, thus $d(n) = k + 1$.

- If n is no new node, the assumption that there is a path in the new graph to n implies that, since there is no node m with $\langle m, n \rangle \in Edges$, there are edges $\langle m, n \rangle$ which have been deleted in the step before. For all these m we have $d(m) = k$. In the new graph all these nodes m have the property that each path from a root to them has length k.

This implies the assertion.   ⬜

An immediate consequence of Lemma 2.25 is

PROPOSITION 2.26

Algorithm 2.24 produces a stratified graph.

PROOF   Using the notation from above, put $S_j := \{n \in V' \mid d(n) = j\}$. Then S_0 is the set of roots for $\mathcal{G}'$ as well as for $\mathcal{G}$, and if node n is in S_{j+1}, then all its predecessors (w. r. t. $\mathcal{G}'$) are in S_j. These sets are mutually disjoint, and $S_{k'} = \emptyset$ for all $k' \geq k$ for some minimal index k. Since $n \in S_{d(n)}$ holds for each node $n \in V'$, we see that $(S_j)_{0 \leq j \leq k}$ forms a partition of V'. ⬜

Armed with this tool, we now enter the discussion of the general case.

2.4.5   The General Case

We will demonstrate that all the stratified PF-systems which can be constructed from a given one will do the same work, so that this morphism is an invariant, and that it is sensible to assign it to a nonstratified PF-system as its work. This has as a remarkable consequence that two constructions can be carried out that help in composing larger systems from smaller ones: we show in Section 2.4.6 how two PF-systems can be glued together (as a horizontal extension), and that hierarchical refinement is available as construction technique, permitting the expansion of a node by an entire subsystem. This is a vertical extension.

Both the PF-system $\langle \mathcal{G}, \gamma \rangle$, and $\mathcal{G} = \langle V, E \rangle$ as the graph underlying it are fixed. The sets W and B denote the roots, and the leaves of $\mathcal{G}$, resp. We fix also the set Q which serves as a reservoir of fresh nodes for stratification.

We begin with an adaptation of Algorithm 2.24 to PF-systems by taking the labels for edges and nodes coming with such a system into account. To be specific, suppose we replace an edge $\langle m, n \rangle$ from the set of edges by the pair $\langle m, q \rangle$ and $\langle q, n \rangle$ with the fresh node $q \in Q$. Then we put

$$\gamma_{\langle m,q \rangle} := \gamma_{\langle m,n \rangle},$$
$$\gamma_{\langle q,n \rangle} := \gamma_{\langle m,n \rangle},$$
$$\mathsf{a}\,(\gamma, q) := \mathfrak{e}_{\gamma_{\langle q,n \rangle}}$$

(remember that $\mathfrak{e}$ denotes the unit for the corresponding monad). Thus if the edge $\langle m, n \rangle$ carries type a, where a is an object in $\mathfrak{X}$, then the new edges carry this type, and the node inserted is assigned the Kleisli morphism $\mathfrak{e}_a$; note that the natural transformation $\mathfrak{e}$ provides the identities in the Kleisli category $\mathfrak{X}_{\mathfrak{T}}$. In terms of pipelines, by inserting $\mathfrak{e}_{\gamma_{\langle q,n \rangle}}$ we insert a **noop** into the system, since the filter introduced in this way evidently does not do any other work than transporting inputs unchanged to outputs. In this way we obtain from $\langle \mathcal{G}, \gamma \rangle$ a stratified PF-system $\langle \mathcal{G}_1, \gamma \rangle$, reusing γ for simplicity.

The graph constructed by Algorithm 2.24 is an extension of the given graph. This is made precise now.

DEFINITION 2.27   *The graph $\mathcal{G}' = (V', E')$ is called a Q-extension to $\mathcal{G}$ iff*

a. *$\mathcal{G}'$ is stratified with $E' \cap (V \times V) \subseteq E$, and $\mathcal{G}'$ has the same roots as $\mathcal{G}$,*

b. *$V \subseteq V'$, and $V' \setminus V \subseteq Q$,*

c. *if $\langle n, m \rangle \in E \setminus E'$, then there exists a unique path $n = q_0, \ldots, q_k = m$ from n to m in $\mathcal{G}'$ with $\langle q_i, q_{i+1} \rangle \in E'$ for $0 \leq i < k$,* *d. for all $q \in V' \setminus V$, $\#(\bullet q) = \#(q \bullet) = 1$.*

Thus a Q-extension has new nodes from the fountain Q of nodes only, an edge in E is either an edge in E', or its endpoints are connected through a unique path that runs entirely through Q (apart from the endpoints, of course). The new nodes in $\mathcal{G}'$ do not have a rich social life by being neighbor to only two other nodes, thus such a node receives inputs from exactly one node and propagates it to a unique other node.

LEMMA 2.28
The graph constructed from Algorithm 2.24 is an Q-extension to $\mathcal{G}$.

PROOF If $\mathcal{G}'$ is the graph constructed from $\mathcal{G}$, then $\mathcal{G}'$ has been shown to be stratified in Proposition 2.26. The construction makes sure that the other conditions from Definition 2.27 are satisfied. ▯

Any Q-extension can be decorated as indicated above: the nodes from Q receive $\mathfrak{e}_x$ as their function, where x is an appropriate object which labels the edges leading into that node, and out of it, resp. This leads to the notion of an Q-extension to a PF-system which will not be formally defined since the definition is obvious (the reader is invited to formulate it).

We want to establish that the work of a PF-system is an invariant for all Q-extensions to a given PF-system. For this we should make sure that the composition of Kleisli morphisms and the operation $\times_{\mathfrak{T}}$ which resembles a product so closely relate to each other like composition and product:

DEFINITION 2.29 *The monad $\langle \mathfrak{T}, \mathfrak{e}, \mathfrak{m} \rangle$ satisfies the $\sharp$-condition iff*

a. $\mathfrak{e}_{a \times b} = \mathfrak{e}_a \times_{\mathfrak{T}} \mathfrak{e}_b$ for all objects a and b in $\mathfrak{X}$,

b. for the morphisms $f_i : a_i \to \mathfrak{T} b_i, g_i : b_i \to \mathfrak{T} c_i$ $(i = 1, 2)$ the equality

$$(g_1 \times_{\mathfrak{T}} g_2) * (f_1 \times_{\mathfrak{T}} f_2) = (g_1 * f_1) \times_{\mathfrak{T}} (g_2 * f_2)$$

holds.

Thus the identity on $a \times b$ in $\mathfrak{X}_{\mathfrak{T}}$ is obtained from the respective identities on a and b by performing the $\times_{\mathfrak{T}}$-operation. In terms of computation, combining the identities on a and on b independently to a component yields the identity in $a \times b$. The second condition explains the name: viewing the Kleisli composition $*$ as a horizontal operation along the flow of information which indicates piping, and $\times_{\mathfrak{T}}$ as a vertical operation modeling independent composition, the equation is visualized in Figure 2.4. Hence piping of composed computations is tantamount to composing piped computations.

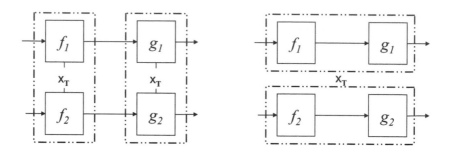

Figure 2.4: $(g_1 \times_{\mathfrak{T}} g_2) * (f_1 \times_{\mathfrak{T}} f_2)$ vs. $(g_1 * f_1) \times_{\mathfrak{T}} (g_2 * f_2)$

Let us investigate our reference categories:

PROPOSITION 2.30

Both the Manes and the Giry category satisfy the ♯-condition, provided the monoid H which comes with the respective monads is commutative.

PROOF   1. The first condition is readily established for both monads.

2. Let $R_i : A_i \to \mathfrak{M}(B_i), S_i : B_i \to \mathfrak{M}(C_i) \, (i = 1, 2)$ be morphisms with $\mathfrak{M}$ as the functor underlying the Manes monad. Then these equalities hold for $\langle a_1, a_2 \rangle \in A_1 \times A_2$:

$$((S_1 \times_{\mathfrak{M}} S_2) * (R_1 \times_{\mathfrak{M}} R_2)) (a_1, a_2) =$$
$$\{\langle h_1 h_2 h_3 h_4, c_1, c_2 \rangle \mid \exists b_1, b_2 : \langle h_1, b_1 \rangle \in R_1(a_1), \langle h_2, b_2 \rangle \in R_2(a_2),$$
$$\langle h_3, c_1 \rangle \in S_1(b_1), \langle h_4, c_2 \rangle \in S_2(b_2)\},$$

and

$$((S_1 * R_1) \times_{\mathfrak{M}} (S_2 * R_2)) (a_1, a_2) =$$
$$\{\langle h_1 h_2 h_3 h_4, c_1, c_2 \rangle \mid \exists b_1, b_2 : \langle h_1, b_1 \rangle \in R_1(a_1), \langle h_2, c_2 \rangle \in S_1(b_1),$$
$$\langle h_3, b_2 \rangle \in R_2(a_2), \langle h_4, c_2 \rangle \in S_2(b_2)\}$$

3. Let $K_i : A_i \to \mathfrak{G}(B_i), L_i : B_i \to \mathfrak{G}(C_i) \, (i = 1, 2)$ be morphisms with $\mathfrak{G}$ as the functor underlying the Manes monad. Here A_i, B_i, C_i are measurable spaces, and the monoid H is assumed to be measurable as well. Now

$$((L_1 \times_{\mathfrak{G}} L_2) * (K_1 \times_{\mathfrak{G}} K_2)) (a_1, a_2)$$

is a finite measure on $H \times C_1 \times C_2$, and so is

$$((L_1 * K_1) \times_{\mathfrak{M}} (L_2 * K_2)) (a_1, a_2),$$

hence it is sufficient for establishing equality to show that the integrals for an arbitrary measurable and bounded function $\psi : H \times C_1 \times C_2 \to \mathbb{R}$ coincide. A calculation using Fubini's Theorem on product integration establishes that

$$\int_{H \times C_1 \times C_2} \psi \, d\left((L_1 \times_\mathfrak{G} L_2) * (K_1 \times_\mathfrak{G} K_2)\right)(a_1, a_2) =$$

$$\int_{H \times B_1} \int_{H \times B_2} \int_{H \times C_1} \int_{H \times C_2} \psi(h_1 h_2 h_3 h_4, c_1, c_2) \, L_2(b_2)(d\langle h_4, c_2 \rangle) \times$$
$$L_1(b_1)(d\langle h_3, c_1 \rangle) \times \, K_2(a_2)(d\langle h_2, b_2 \rangle) \, K_1(a_1)(d\langle h_1, b_1 \rangle)$$

and

$$\int_{H \times C_1 \times C_2} \psi \, d\left((L_1 \times_\mathfrak{G} L_2) * (K_1 \times_\mathfrak{G} K_2)\right)(a_1, a_2) =$$

$$\int_{H \times B_1} \int_{H \times C_1} \int_{H \times B_2} \int_{H \times C_2} \psi(h_1 h_2 h_3 h_4, c_1, c_2) \, L_2(b_2)(d\langle h_4, c_2 \rangle) \times$$
$$K_2(a_2)(d\langle h_3, b_2 \rangle) \times \, L_1(b_2)(d\langle h_2, c_1 \rangle) \, K_1(a_1)(d\langle h_1, b_1 \rangle).$$

4. These equalities establish the claim. It is interesting to observe in which way in both cases the roles of h_2 and h_3 get interchanged, reflecting the way in which morphisms change positions. ⬚

An easy induction using the second assertion in Lemma 2.13 establishes that the Kleisli identity on $a_1 \times \cdots \times a_n$ can be calculated through the identities on the components. The $\sharp$-condition makes also sure that we may shift computations between products (the easy inductive proof is left to the reader):

LEMMA 2.31
Assume that the $\sharp$-condition holds. Then

a. *The equality*

$$\mathfrak{e}_{a_1 \times \cdots \times a_n} = \mathfrak{e}_{a_1} \times_\mathfrak{X} \ldots \times_\mathfrak{X} \mathfrak{e}_{a_n}$$

holds for all objects $a_1, \ldots, a_n$ in $\mathfrak{X}$,

b. *If $\sigma_i : a_i \to \mathfrak{T} b_i$ and $\tau_i : b_i \to \mathfrak{T} c_i$ are morphisms in $\mathfrak{X}$, then*

$$(\tau_1 \times_\mathfrak{X} \ldots \times_\mathfrak{X} \tau_n) * (\sigma_1 \times_\mathfrak{X} \ldots \times_\mathfrak{X} \sigma_n) =$$
$$(\tau_1 \times_\mathfrak{X} \ldots \times_\mathfrak{X} \tau_{j-1} \times_\mathfrak{X} (\tau_j * \sigma_j) \times_\mathfrak{X} \tau_{j+1} \times_\mathfrak{X} \ldots \times_\mathfrak{X} \tau_n) *$$
$$* (\sigma_1 \times_\mathfrak{X} \ldots \times_\mathfrak{X} \sigma_{j-1} \times_\mathfrak{X} \mathfrak{e}_{a_j} \times_\mathfrak{X} \sigma_{j+1} \times_\mathfrak{X} \ldots \times_\mathfrak{X} \sigma_n) =$$
$$(\tau_1 \times_\mathfrak{X} \ldots \times_\mathfrak{X} \tau_{j-1} \times_\mathfrak{X} \mathfrak{e}_{b_j} \times_\mathfrak{X} \tau_{j+1} \times_\mathfrak{X} \ldots \times_\mathfrak{X} \tau_n) *$$
$$* (\sigma_1 \times_\mathfrak{X} \ldots \times_\mathfrak{X} \sigma_{j-1} \times_\mathfrak{X} (\tau_j * \sigma_j) \times_\mathfrak{X} \sigma_{j+1} \times_\mathfrak{X} \ldots \times_\mathfrak{X} \sigma_n).$$

The equations in part b of Lemma 2.31 are useful in our context: $\sigma_1 \times_{\mathfrak{T}}$ $\ldots \times_{\mathfrak{T}} \sigma_n$ and $\tau_1 \times_{\mathfrak{T}} \ldots \times_{\mathfrak{T}} \tau_n$ represent the computations in consecutive blocks of a PF-system. Then we may shift the computation of a component out of a block into the next or the previous one without changing the result; shifting means among others replacing the morphism by the appropriate identity. We will use this observation in the proof of Proposition 2.34 for establishing the invariance result.

From now on we assume that the $\sharp$-condition is satisfied.

A Little Digression.   In fact, we can say more about representing $\times_{\mathfrak{T}}$-products of morphisms: they can be written as Kleisli-products of a very special kind. The discerning reader will no doubt observe that the kind of representation derived from the discussion that follows will not be needed for the present constructions of PF-systems. It appears to be interesting, nevertheless.

DEFINITION 2.32   *Assume $n > 1$, let $\tau_i : a_i \to \mathfrak{T}b_i$ be morphisms for $1 \leq i \leq n$, and let ξ be a permutation of $\{1, \ldots, n\}$. Then $\langle \sigma_1, \ldots, \sigma_n \rangle$ is the ξ-expansion of $\langle \tau_1, \ldots, \tau_n \rangle$ iff σ_j can be written as $\zeta_{j,1} \times_{\mathfrak{T}} \ldots \times_{\mathfrak{T}} \zeta_{j,n}$ such that*

a. each $\zeta_{j,i}$ is either $\mathfrak{e}_{a_i}, \mathfrak{e}_{b_i}$ or one of $\tau_1, \ldots, \tau_n$,

b. $\zeta_{j,k} \in \{\tau_1, \ldots, \tau_n\}$ iff $\xi(j) = k$,

c. if $\zeta_{j,k} = \tau_i$, then

$$\zeta_{\ell,k} = \begin{cases} \mathfrak{e}_{a_i}, & \ell > j \\ \mathfrak{e}_{b_i}, & \ell < j. \end{cases}$$

For example, the permutation $(13)(2)$ of $\{1, 2, 3\}$ corresponds to

$$\begin{pmatrix} \zeta_{1,1} & \zeta_{1,2} & \zeta_{1,3} \\ \zeta_{2,1} & \zeta_{2,2} & \zeta_{2,3} \\ \zeta_{3,1} & \zeta_{3,2} & \zeta_{3,3} \end{pmatrix} = \begin{pmatrix} \mathfrak{e}_{b_1} & \mathfrak{e}_{b_2} & \tau_3 \\ \mathfrak{e}_{b_1} & \tau_2 & \mathfrak{e}_{a_3} \\ \tau_1 & \mathfrak{e}_{a_2} & \mathfrak{e}_{a_3} \end{pmatrix}.$$

Thus if $\langle \sigma_1, \ldots, \sigma_n \rangle$ is a ξ-expansion of $\langle \tau_1, \ldots, \tau_n \rangle$, then assuming $\xi(j) = i$, σ_j can be written as

$$\mathfrak{e}_{b_1} \times_{\mathfrak{T}} \ldots \times_{\mathfrak{T}} \mathfrak{e}_{b_{i-1}} \times_{\mathfrak{T}} \tau_i \times_{\mathfrak{T}} \mathfrak{e}_{a_{i+1}} \times_{\mathfrak{T}} \ldots \times_{\mathfrak{T}} \mathfrak{e}_{a_n}$$

indicating that τ_i is doing its work, whereas $\tau_1, \ldots, \tau_{i-1}$ did do their work already (thus the identity on the range is incorporated) and that $\tau_{i+1}, \ldots, \tau_n$ will still have to do their work (hence the identity of the respective domains are incorporated into the $\times_{\mathfrak{T}}$-product).

LEMMA 2.33

Let, under the assumptions of Definition 2.32, $\langle \sigma_1, \ldots, \sigma_n \rangle$ be an ξ-expansion of $\langle \tau_1, \ldots, \tau_n \rangle$, then

$$\tau_1 \times_{\mathfrak{T}} \ldots \times_{\mathfrak{T}} \tau_n = \sigma_1 * \ldots * \sigma_n$$

holds.

PROOF 1. The proof proceeds by induction on n. For $n = 2$ the only ξ-expansions of $\langle \tau_1, \tau_2 \rangle$ are $\langle \mathfrak{e}_{b_1} \times_{\mathfrak{T}} \tau_2, \tau_1 \times_{\mathfrak{T}} \mathfrak{e}_{a_2} \rangle$ and $\langle \tau_1 \times_{\mathfrak{T}} \mathfrak{e}_{b_2} \mathfrak{e}_{a_1} \times_{\mathfrak{T}} \tau_2 \rangle$. The $\sharp$-condition then permits directly establishing the claim.

2. The inductive step considers the ξ-expansion $\langle \sigma_1, \ldots, \sigma_{n+1} \rangle$ of the $n+1$-tuple $\langle \tau_1, \ldots, \tau_{n+1} \rangle$. Then $\xi(j_{n+1}) = n+1$, and we can write

$$\sigma_i = \begin{cases} \sigma_i' \times_{\mathfrak{T}} \mathfrak{e}_{b_{n+1}}, & i < j_{n+1} \\ \sigma_i' \times_{\mathfrak{T}} \mathfrak{e}_{a_{n+1}}, & i > j_{n+1} \end{cases}$$

for some σ_i'. It is easy to see that

$$\langle \sigma_1', \ldots, \sigma_{j_{n+1}-1}', \sigma_{j_{n+1}+1}', \ldots, \sigma_{n+1}' \rangle$$

is an ξ'-expansion for $\langle \tau_1, \ldots, \tau_n \rangle$, where ξ' is the permutation of $\{1, \ldots, n\}$ derived from ξ. Now write $\sigma_{j_{n+1}} = \mathfrak{e}_{c_1} \times_{\mathfrak{T}} \ldots \times_{\mathfrak{T}} \mathfrak{e}_{c_n}$ where $c_i \in \{a_i, b_i\}$ is suitably chosen according to the definition of the expansion, then we have by the induction hypothesis, by the $\sharp$-condition, and by Lemma 2.31

$$\sigma_1 * \ldots * \sigma_{n+1} = \left(\sigma_1' \times_{\mathfrak{T}} \mathfrak{e}_{b_{n+1}} \right) * \ldots * \left(\sigma_{j_{n+1}-1}' \times_{\mathfrak{T}} \mathfrak{e}_{b_{n+1}} \right) *$$

$$* \, \sigma_{j_{n+1}} * \left(\sigma_{j_{n+1}+1}' \times_{\mathfrak{T}} \mathfrak{e}_{a_{n+1}} \right) * \ldots * \left(\sigma_{n+1}' \times_{\mathfrak{T}} \mathfrak{e}_{a_{n+1}} \right)$$

$$= \left(\sigma_1' * \ldots * \sigma_{j_{n+1}-1}' * \left(\mathfrak{e}_{c_1} \times_{\mathfrak{T}} \ldots \times_{\mathfrak{T}} \mathfrak{e}_{c_n} \right) \right) \times_{\mathfrak{T}} \tau_{n+1} *$$

$$* \left(\sigma_{j_{n+1}+1}' * \ldots * \sigma_{n+1}' \right) \times_{\mathfrak{T}} \mathfrak{e}_{a_{n+1}}$$

$$= \left(\sigma_1' * \ldots * \sigma_{j_{n+1}-1}' * \mathfrak{e}_{c_1 \times \cdots \times c_n} * \sigma_{j_{n+1}+1}' * \ldots * \sigma_{n+1}' \right) \times_{\mathfrak{T}} \tau_{n+1}$$

$$= \left(\sigma_1' * \ldots * \sigma_{j_{n+1}-1}' * \sigma_{j_{n+1}+1}' * \ldots \sigma_{n+1}' \right) \times_{\mathfrak{T}} \tau_{n+1}$$

$$= \tau_1 \times_{\mathfrak{T}} \ldots \times_{\mathfrak{T}} \tau_{n+1}.$$

This establishes the claim. □

Returning To The Discussion. Assume that n is an inner node in $\mathcal{G}$ with

$$\mathsf{a}\,(\gamma, n) : \prod \{ \gamma_{\langle k, n \rangle} \mid k \in \bullet n \} \to \mathfrak{T} \left(\prod \{ \gamma_{\langle n, k \rangle} \mid k \in n \bullet \} \right)$$

as its label, and assume that the edge $\langle k, n \rangle$ is replaced by the edges $\langle k, q \rangle$ and $\langle q, n \rangle$ for some $q \in Q$. The new edges are labeled through the object $\gamma_{\langle k,n \rangle}$, and the new node n carries the label $e_{\gamma_{\langle k,n \rangle}}$. Other edges leading into node n are also replaced. The net effect of inserting a node just in front of node n is replacing a (γ, n) by

$$\mathsf{a}\,(\gamma, n) * e_{\prod \{\gamma_{\langle k,n \rangle} \mid k \in \bullet n\}}$$

which equals of course a (γ, n). Similarly, replacing an edge $\langle n, k \rangle$ by edges $\langle n, q \rangle, \langle q, k \rangle$ and introducing labels on edges and on $q \in Q$ accordingly has the effect of replacing a (γ, n) by

$$e_{\prod \{\gamma_{\langle n,k \rangle} \mid k \in n \bullet\}} * \mathsf{a}\,(\gamma, n)\,,$$

equalling a (γ, n), too. This is a translation of the idea of inserting "neutral" nodes into the graph in order to render it stratified. In fact, two Q-extensions to $\mathcal{G}$ differ only by such neutral nodes on paths between nodes taken from $\mathcal{G}$.

PROPOSITION 2.34
Suppose $\langle \mathcal{G}, \gamma \rangle$ is a PF-system with $\langle \mathcal{G}_1, \gamma \rangle$ and $\langle \mathcal{G}_2, \gamma \rangle$ as Q-extensions. Then $\mathsf{P}\,(\gamma, \mathcal{G}_1) = \mathsf{P}\,(\gamma, \mathcal{G}_2)\,.$

PROOF   1. The proof proceeds by induction on

$$N := \max\{\Lambda(\mathcal{G}_1), \Lambda(\mathcal{G}_2)\}.$$

The jth partition element of graph $\mathcal{G}_i$ will denoted by $S_j^{(i)}$.

2. The induction starts at $N = 2$. This step inspects each node n of $\mathcal{G}$ in turn. Suppose $n \in \left(S_1^{(1)} \setminus S_1^{(2)} \right) \cap V$, then, since graph $\mathcal{G}_2$ is an Q-extension to $\mathcal{G}$, for each predecessor w of n in $\mathcal{G}_1$ there exists a node $q_w \in Q$ such that $\langle w, q_w \rangle, \langle q_w, n \rangle$ are edges in $\mathcal{G}_2$ which are labeled by the object $\gamma_{\langle w,n \rangle}$; node q_w itself carries the label $e_{\gamma_{\langle w,n \rangle}}$. Lemma 2.31 implies that the morphism a (γ, n) which participates in defining $\mathsf{P}\,(\gamma, \mathcal{G}_2)$ has a factor

$$e_{\gamma_{\langle w_1,n \rangle}} \times \cdots \times \gamma_{\langle w_r,n \rangle}$$

to the right, where $w_1, \ldots, w_r$ are in that order all predecessors of n in $\mathcal{G}_1$. A similar argument applies to $n \in \left(S_1^{(2)} \setminus S_1^{(1)} \right) \cap V$, so that $\mathsf{P}\,(\gamma, \mathcal{G}_1)$ differs only by factors from $\mathsf{P}\,(\gamma, \mathcal{G}_2)$ which are identity Kleisli morphisms. Hence the assertion holds for $N = 2$.

3. Let $\max\{\Lambda(\mathcal{G}_1), \Lambda(\mathcal{G}_2)\} = N + 1$. We may and do assume w.l.g. that $V \cap \left(S_1^{(2)} \cup S_1^{(1)} \right) \neq \emptyset$, for, otherwise, no node of V is directly connected to a root in either extension, so we may construct new graphs by eliminating the respective sets S_1 without changing the work of either graph.

We construct from the PF-system $\langle \mathcal{G}_1, \gamma \rangle$ a PF-system $\langle \mathcal{G}_3, \gamma \rangle$ which is an Q-extension to $\langle \mathcal{G}, \gamma \rangle$ such that

$$S_1^{(3)} = \{n \in V \mid \exists w \in W : w \to_Q^* n \text{ in } \mathcal{G}_1\}$$
$$\cup \{n \in V \mid \exists w \in W : w \to_Q^* n \text{ in } \mathcal{G}_2\},$$

where $\to_Q^*$ indicates that there exists a (unique) path of nonnegative length that runs — with the exception of the endpoints — entirely through Q. Moreover, $\mathsf{P}(\gamma, \mathcal{G}_1) = \mathsf{P}(\gamma, \mathcal{G}_3)$ will hold.

Initially, $\langle \mathcal{G}_3, \gamma \rangle := \langle \mathcal{G}_1, \gamma \rangle$. Assume $n \in V \cap S_1^{(1)}$ such that $n \in S_t^{(2)}$ for some $t > 1$. Let $w_1, \ldots, w_r$ be all predecessors to n in $\mathcal{G}_1$. Since $\mathcal{G}_2$ is an Q-extension, there exist nodes $q_{1,2}, \ldots, q_{1,t-1}, \ldots, q_{r,2}, \ldots, q_{r,t-1}$ in Q such that

$$w_1 = q_{1,1} \cdots q_{1,t} = n$$
$$\vdots \qquad \vdots$$
$$w_r = q_{r,1} \cdots q_{r,t} = n$$

form paths that run with the exception of their endpoints entirely through Q. The edges on the ith path are labeled with the object $\gamma_{\langle w_i, n \rangle}$, and a $(\gamma, q_{i,j}) = e_{\gamma_{\langle w_i, n \rangle}}$.

Let $k_1, \ldots, k_s$ be all successors to n in $\mathcal{G}_2$. Remove the nodes $\{q_{i,j} \mid 1 \le i \le r, 2 \le i \le t-1\}$ and the edges, including $\langle w_i, q_{i,2} \rangle$ and $\langle q_{i,t-1}, n \rangle$ for $1 \le i \le r$ from $\mathcal{G}_3$, and add nodes $q'_{1,2}, \ldots, q'_{1,t-1}, \ldots, q'_{s,2}, \ldots, q'_{s,t-1}$ as well as edges so that we have the paths

$$n = q'_{1,1} \cdots q'_{1,t} = k_1$$
$$\vdots \qquad \vdots$$
$$n = q'_{s,1} \cdots q'_{s,t} = k_s.$$

Put $\gamma_{\langle q'_{i,j}, q'_{i,j+1} \rangle} := \gamma_{\langle n, k_i \rangle}$, and set $\gamma(q'_{i,j}) := e_{\gamma_{\langle n, k_i \rangle}}$ $(i > 1, j < t)$.

Consequently, graph $\mathcal{G}_3$ remains an Q-extension to $\mathcal{G}$, and from Lemma 2.31, part b, we see that $\mathsf{P}(\gamma, \mathcal{G}_1) = \mathsf{P}(\gamma, \mathcal{G}_3)$ holds. Working in this way through $V \cap \left(S_1^{(1)} \cup S_1^{(2)} \right)$ will eventually produce the desired graph.

In the same manner we construct a PF-system $\langle \mathcal{G}_4, \gamma \rangle$ with $S_1^{(4)} = S_1^{(2)}$ such that $\mathsf{P}(\gamma, \mathcal{G}_2) = \mathsf{P}(\gamma, \mathcal{G}_4)$ holds.

Remove the roots from $\mathcal{G}_3$; this yields the graph $\widetilde{\mathcal{G}_3}$ which is an Q-extension to

$$\widetilde{\mathcal{G}} := (V \setminus W, E \cap (V \setminus W \times V \setminus W)).$$

Similarly, remove the roots from $\mathcal{G}_4$ yielding $\widetilde{\mathcal{G}_4}$, which is also an Q-extension to $\widetilde{\mathcal{G}}$. Since $\max\{\Lambda(\widetilde{\mathcal{G}_3}), \Lambda(\widetilde{\mathcal{G}_4})\} \leq N$, the induction hypothesis applies, so that

$$
\begin{aligned}
\mathsf{P}\left(\gamma, \mathcal{G}_1\right) &= \mathsf{P}\left(\gamma, \mathcal{G}_3\right) \\
&= \mathsf{P}\left(\gamma, \widetilde{\mathcal{G}_3}\right) * \mathsf{A}\left(S_1^{(3)}, \gamma\right) \\
&= \mathsf{P}\left(\gamma, \widetilde{\mathcal{G}_4}\right) * \mathsf{A}\left(S_1^{(4)}, \gamma\right) \\
&= \mathsf{P}\left(\gamma, \mathcal{G}_4\right) \\
&= \mathsf{P}\left(\gamma, \mathcal{G}_2\right)
\end{aligned}
$$

holds. ⬜

This proposition shows that the work described by an Q-extension of a PF-system does only depend on the underlying PF-system, so that we are now in a position to define the work of such a system — which need not be stratified — through its stratified step-twins.

DEFINITION 2.35　*We define the work* $\mathsf{P}\left(\gamma, \mathcal{G}\right)$ *being done by the PF-system* $\langle \mathcal{G}, \gamma \rangle$ *as the work* $\mathsf{P}\left(\gamma, \mathcal{G}_1\right)$ *of one of its Q-extensions* $\langle \mathcal{G}_1, \gamma \rangle$.

Consequently the work of a PF-system may be conveniently computed through one of its Q-extensions.

2.4.6　System Evolution

Our constructions support system evolution in a quite general sense. A PF-system may evolve horizontally or vertically. Horizontal evolution concatenates pipelines, with data transformations possibly serving as glue between the parts. Vertical evolution refines a pipeline by substituting a component through an entire subsystem. Both operations are vital in composing systems from smaller ones, so that larger systems can be built up through a suitable sequence of them.

2.4.6.1　Concatenation

Let $\langle \mathcal{G}_1, \gamma \rangle$ and $\langle \mathcal{G}_2, \chi \rangle$ be two PF-systems, $\mathcal{G}_i = \langle V_i, E_i \rangle$. The idea in concatenating both is to pipe the output from the first system to the input of the second one, hence

$$
V_1 \cap V_2 = B_1 = W_2
$$

should hold: the output nodes from the first system should coincide with the input nodes for the second one; otherwise, these systems do not share nodes. Neither an input node nor an output node carries any functionality in our model, but by lumping them together, we may wish to perform some work (combining pipes often requires some transformation, e.g., of formats, between input and output). Hence we assume for each node $n \in B_1 = W_2$ the existence of a Kleisli morphism

$$\mathsf{a}\,(\tau, n) : \prod\{\gamma_{\langle k,n\rangle} \mid \langle k,n\rangle \in E_1\} \to \mathfrak{T}\left(\prod\{\chi_{\langle n,j\rangle} \mid \langle n,j\rangle \in E_2\}\right).$$

This permits defining the τ-*concatenation* $\langle \mathcal{G}_1, \gamma\rangle +_\tau \langle \mathcal{G}_2, \chi\rangle$ as $\langle \mathcal{H}, \kappa\rangle$ with

$$\mathcal{H} := \langle V_1 \cup V_2, E_1 \cup E_2\rangle,$$

$$\kappa_{\langle k,n\rangle} := \begin{cases} \gamma_{\langle k,n\rangle}, & \langle k,n\rangle \in E_1, \\ \chi_{\langle k,n\rangle}, & \text{otherwise}, \end{cases}$$

$$\mathsf{a}\,(\kappa, n) := \begin{cases} \mathsf{a}\,(\gamma, n), & n \in V_1 \setminus (W_1 \cup B_1), \\ \mathsf{a}\,(\tau, n), & n \in B_1, \\ \mathsf{a}\,(\chi, n), & n \in V_2 \setminus (W_2 \cup B_2). \end{cases}$$

We get as a consequence of Proposition 2.34:

PROPOSITION 2.36
Under the conditions above, $\langle \mathcal{H}, \kappa\rangle := \langle \mathcal{G}_1, \gamma\rangle +_\tau \langle \mathcal{G}_2, \chi\rangle$ is a PF-system, and

$$\mathsf{P}\,(\kappa, \mathcal{H}) = \mathsf{P}\,(\chi, \mathcal{G}_2) * (\mathsf{a}\,(\tau, n_1) \times_{\mathfrak{T}} \ldots \times_{\mathfrak{T}} \mathsf{a}\,(\tau, n_k)) * \mathsf{P}\,(\gamma, \mathcal{G}_1),$$

where $B_1 = W_2 = \{n_1, \ldots, n_k\}$.

Thus τ provides the glue for composing the PF-systems, and the work being done exhibits the work performed when combining both systems. The glue alluded at here is different from but similar in function to the glue introduced in (Wermelinger and Fiadeiro, 1998).

2.4.6.2 Substitution

Systems are often built through successive stages of refinements, where a part of a system is first represented as a node, and this node is then replaced in subsequent steps by an entire subsystem. This may graphically be described as *glass-box refinement*.

Let $\mathcal{G}_i = \langle V_i, E_i\rangle$ be dags with respective roots W_i and leaves B_i, and let $n \in V_1$ be a node such that (dots taken in $\mathcal{G}_1$)

$$\bullet n = W_2, n\bullet = B_2.$$

Thus an incoming edge for n comes from a root in $\mathcal{G}_2$, and an outgoing edge goes to a leaf in $\mathcal{G}_2$. For technically simplifying the representation, we assume that only the nodes in $W_2 \cup B_2$ are common to V_1 and V_2. We assume further that we have a selection map

$$\psi : W_2 \cup B_2 \to V_2 \setminus (W_2 \cup B_2)$$

which will help constructing new edges when absorbing $\mathcal{G}_2$ into $\mathcal{G}_1$ by associating with each root or leaf an inner node as source or target of an edge, as we will see. We require that $\psi[W_2] \cap \psi[B_2] = \emptyset$, since otherwise cycles in the replacement graph would result. Define the ψ-*replacement*

$$\mathcal{G}_1[\mathcal{G}_2 \backslash_\psi n]$$

of node n through graph $\mathcal{G}_2$ as the graph $\langle U, D \rangle$ by

$$U := (V_1 \setminus \{n\}) \cup V_2,$$
$$D := \left(E_1 \cap (V_1 \setminus \{n\})^2\right) \cup E_2 \cup$$
$$\{\langle w, \psi(w)\rangle \mid w \in W_2\} \cup \{\langle \psi(b), b\rangle \mid b \in B_2\}.$$

Thus we build the new graph by combining all nodes with the exception of n, the node to be replaced. All edges leading into n or out of it are removed, and replaced by edges into $\mathcal{G}_2$: if $\langle w, n\rangle \in E_1$ is an edge in $\mathcal{G}_1$, the node w must be a root in $\mathcal{G}_2$, then this edge will be replaced in the replacement graph by the edge $\langle w, \psi(w)\rangle$, similarly for edges $\langle n, b\rangle \in E_1$. Since the graphs $\mathcal{G}_1$ and $\mathcal{G}_2$ do not have cycles, and since ψ assigns by assumption different nodes to roots and to leaves, $\langle U, D\rangle$ does not have any cycles either.

We apply this construction to PF-systems now. Suppose that in addition to the assumptions made so far $\langle \mathcal{G}_2, \gamma\rangle$ and $\langle \mathcal{G}_2, \chi\rangle$ are PF-systems. For getting our machinery going, the new edges need labels from $\mathfrak{X}$; these edges should not violate the typing constraints imposed on node n. Call the selection map ψ *viable* iff

$$\forall w \in W_2 : \gamma_{\langle w,n\rangle} = \chi_{\langle w, \psi(w)\rangle} \wedge \forall b \in B_2 : \gamma_{\langle n,b\rangle} = \chi_{\langle \psi(b),b\rangle}$$

holds. This entails that the Kleisli morphisms $\mathsf{a}\,(\gamma, n)$ and $\mathsf{P}\,(\chi, \mathcal{G}_2)$ have the same signatures.

We define the PF-system

$$\langle \mathcal{G}_1[\mathcal{G}_2\backslash_\psi n], \gamma[\chi\backslash_\psi n]\rangle$$

in the obvious way by taking the values γ, and χ for edges in E_1 or in E_2, resp., depending on where they come from, and by setting for the new edges

$$\gamma[\chi\backslash_\psi n]_{\langle w, \psi(w)\rangle} := \gamma_{\langle w,n\rangle},$$

similarly for $\gamma[\chi\backslash_\psi n]_{\langle \psi(b),b\rangle}$. The labels for the nodes are left unchanged, coming either from γ or from χ. Then Proposition 2.34 implies that we may compute the work for the composed system in these steps:

- compute $\mathsf{P}\,(\chi, \mathcal{G}_2)$, hence the work of the system which is to refine node n,

- substitute for $\mathsf{a}\,(\gamma, n)$ the morphism $\mathsf{P}\,(\chi, \mathcal{G}_2)$, leaving the rest of γ alone; technically: form $\gamma[\mathsf{P}\,(\chi, \mathcal{G}_2) \setminus n]$,

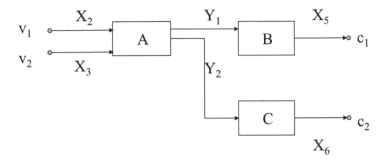

Figure 2.5: Replacing Subsystem

- compute the work done by the PF-system based on graph $\mathcal{G}_1$ with the modified value for γ.

Formally:

PROPOSITION 2.37

$P\left(\gamma[\chi\backslash_\psi n], \mathcal{G}_1[\mathcal{G}_2\backslash_\psi n]\right) = P\left(\gamma[P\left(\chi, \mathcal{G}_2\right)\backslash n], \mathcal{G}_1\right)$, *provided the selection map ψ is viable.*

Reading this equation from left to right, we see what happens when a node is substituted by an entire subsystem. Reading it from right to left it permits us to state the effect of shrinking a subsystem into a single node — this may be helpful when system evolution goes both ways, expanding nodes to subsystems, and replacing a subsystem by another one.

To illustrate: The PF-system in Figure 2.1 shall be refined as an extension to and a continuation of the discussion in Section 2.4.1. Node 2 will be replaced by the subsystem depicted in Figure 2.5.

Assign relations with the following signatures to the inner nodes:

$$R_A \subseteq (X_2 \times X_3) \times (Y_1 \times Y_2),$$
$$R_B \subseteq Y_1 \times X_5,$$
$$R_C \subseteq Y_2 \times X_6.$$

The work being done by this subsystem is then given by

$$\{\langle x_2, x_3, x_5, x_6\rangle \mid \exists y_1 \in Y_1, y_2 \in Y_2 :$$
$$\langle x_2, x_3, y_1, y_2\rangle \in R_A, \langle y_1, x_5\rangle \in R_B, \langle y_2, x_6\rangle \in R_C\}.$$

The reader is invited to formulate $\langle \mathcal{G}_2, \chi\rangle$ and a viable map ψ. Proposition 2.37 then gives the work of the entire system, when the replacement has been done,

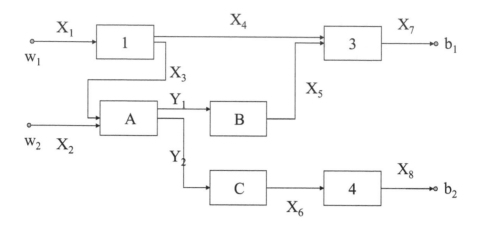

Figure 2.6:   System after Replacement

as

$$\{\langle x_1, x_2, x_7, x_8 \rangle \mid \exists x_3 \in X_3, x_4 \in X_4, x_5 \in X_5, x_6 \in X_6 \exists y_1 \in Y_1, y_2 \in Y_2 :$$
$$\langle x_1, x_3, x_4 \rangle \in R_1, \langle x_2, x_3, y_1, y_2 \rangle \in R_A, \langle y_1, x_5 \rangle \in R_B, \langle y_2, x_6 \rangle \in R_C$$
$$\langle x_4, x_5, x_7 \rangle \in R_3, \langle x_6, x_8 \rangle \in R_4\}.$$

The resulting system is given in Figure 2.6.

For the stochastic case one assumes similarly that stochastic relations

$$K_A : X_2 \times X_3 \rightsquigarrow H \times Y_1 \times Y_2,$$
$$K_B : Y_1 \rightsquigarrow H \times X_5,$$
$$K_C : Y_2 \rightsquigarrow H \times X_6$$

are assigned to the subsystem's nodes. In view of Proposition 2.30 the monoid H should be commutative. The work is computed through the Kleisli product, yielding an expression for the measure $K_2(x_2, x_3)$ in terms of K_A, K_B and K_C which is then substituted into the integral elaborated in Example 2.23. Specifically, the subsystem's work is described by the stochastic relation $K_s :$ $X_2 \times X_3 \rightsquigarrow H \times X_5 \times X_6$ such that for $Q \subseteq H \times X_5 \times X_6$

$$K_s(x_2, x_3)(Q) = \int_{H \times Y_1 \times Y_2} (K_B(y_1) \otimes_H K_C(y_2)) \times$$
$$\times (\{\langle h, x_5, x_6 \rangle \mid \langle h_a h, x_5, x_6 \rangle \in Q\}) K_A(x_1, x_2)(d\langle h_a, y_1, y_2 \rangle)$$
$$= \int_{H \times Y_1 \times Y_2} \int_{H \times X_6} K_B(y_1) (\{\langle h_b, x_5 \rangle \mid h_a h_c h_b, x_5, x_6 \rangle \in Q\}) \times$$
$$\times K_C(y_2)(d\langle h_c, x_6 \rangle) \, K_A(x_1, x_2)(d\langle h_a, y_1, y_2 \rangle).$$

Consequently, we have for $f : H \times X_5 \times X_6 \to \mathbb{R}$ measurable and bounded

$$\int_{H \times X_5 \times X_6} f(h, x_5, x_6) \, dK_s(x_2, x_3)(d\langle h, x_5, x_6 \rangle)$$

$$= \int_{H \times Y_1 \times Y_2} \int_{H \times X_6} \int_{H \times X_5} f(h_a h_b h_c, x_5, x_6) \, K_B(y_1)(d\langle h_b, x_5 \rangle) \times$$

$$\times K_C(y_2)(d\langle h_c, x_6 \rangle) \, K_A(x_1, x_2)(d\langle h_a, y_1, y_2 \rangle).$$

Note the accumulating behavior of the monoid's elements.

This integral is needed, because the final equation describing the system's work before replacement in Example 2.23 requires us to describe integration with respect to $K_2(x_1, x_3)$. Substituting, we get from that equation

$$\int_{H \times X_3 \times X_4} \int_{H \times Y_1 \times Y_2}$$

$$\int_{H \times X_6} \int_{H \times X_5} \int_{H \times X_7} K_4(x_6)(\{\langle h_1, x_8 \rangle \mid \langle g_1(h_a h_c h_b) g h_1, x_7, x_8 \rangle \in F\}) \times$$

$$K_3(x_4, x_5)(d\langle g, x_7 \rangle) K_B(y_1)(d\langle h_b, x_5 \rangle) K_C(y_2)(d\langle h_c, x_6 \rangle) \times$$

$$K_A(x_1, x_2)(d\langle h_a, y_1, y_2 \rangle) K_1(x_1)(d\langle g_1, x_3, x_4 \rangle).$$

This formidable expression describes the probability that input $\langle x_1, x_2 \rangle$ produces an element of the Borel set $F \subseteq H \times X_7 \times X_8$. It is interesting to see how each component leaves its trace in the result's monoid compartment.

2.4.7 Related Approaches

Modeling of pipelines through the specification language Z summarized and discussed, e.g., in (Shaw and Garlan, 1996; Shaw, 2001; Abowd et al., 1993) is evidently much closer to an implementation than the approach proposed in the present paper. Thus a person intending to implement a system with such an architecture is probably better off looking at a Z-specification, using well-known refinement techniques like the ones discussed by Spivey (Spivey, 1989, Ch. 5) for coming even closer to a realization as a running system.

The difference to the present approach, however, lies deeper: Shaw at al. emphasize the first class rank of architectural connectors (Shaw and Garlan, 1996; Abowd et al., 1993; Shaw, 2001; Shaw et al., 1995; Shaw and Garlan, 1995). This implies that filters and pipes are treated on the same eye-level. The scenario here marks a contrast: connectors are represented through objects in a category, components through morphisms of a rather special kind, putting these two kinds of entities on different levels. It may much be said in favor of dealing uniformly with connectors and with components, but it seems that an asymmetric treatment helps the intuition: computations are conceptually different from data transport, however complex the latter may be. The present approach reflects an approach like "Tell me, what your data are, then we will talk about computations on them", so useful in object-oriented software construction.

The approach proposed by Fiadeiro et al., see, e.g., (Wermelinger and Fiadeiro, 1998; Fiadeiro and Maibaum, 1996), using categories for modeling architectures, shows how different kinds of functors, in particular interface functors, may be put to use for constructing systems. This is illustrated in (Wermelinger and Fiadeiro, 1998) where a diagram is "compiled" through computing its colimit, leading to the early version of a program. Moreover, fundamental kinds of interactions of program components are studied using the patterns constructed in that paper. The focus lies on modeling just the interactions for a particular class of mobile programs, emphasizing the importance of connectors: "Software Architecture has put forward the concept of connector to express complex relationships between system components, thus facilitating the separation of coordination from computation" is the very first sentence in the paper's abstract. The computation proper, however, has not been addressed, and this is what we propose in the present paper. Reflecting the mobile nature of the programs discussed in (Wermelinger and Fiadeiro, 1998), and taking into account that no fixed topology is available for the computing nodes in such a scenario, another difference becomes visible: the topology of the communication and the direction of data flow remains fixed here but may be subject to change dynamically in a mobile context. But this is a completely different story, since PF-systems exhibit a fixed structure by their very nature.

Program evolution is supported by concatenating, and by hierarchically composing PF-systems. While the first operation is easily modeled in the Z-approach, only a hint at supporting the hierarchical composition is given in (Abowd et al., 1993), making it difficult to compare both approaches in this respect.

The *Reo* calculus introduced by Arbab and Rutten is based on timed data streams; it supports a topology of connections that is inherently dynamic, and it is based on a distinctive separation of data and time. Examples exhibit the intriguing yet powerful simplicity of this calculus (which also may accommodate mobility). Because we focus in the present work on a fixed communication topology, a comparison between *Reo* and the constructions proposed here shows that *Reo* is both more general and more specific: *Reo* focusses on set-theoretic relations only, monads are not mentioned. Timing plays a crucial role, so timed sequences are used. On the other hand, arbitrary relations are allowed, so are arbitrary compositions, which permits coinduction as the guiding principle for proofs. Finally, the calculus has all kinds of channels, not only pipes. The central difference is, however, that coordination forms a first class concept in this approach.

Barbosa (Barbosa, 2001) considers components, i.e., state-based dynamical systems. He models them as coalgebras for a class of endofunctors in the category of sets and shows how to obtain a behavioral model through parameterizations by a strong monad. This permits him to capture some important behavioral features like partiality or nondeterminism. The guiding proof principle is a variant of coinduction. Probabilistic settings are outside this rich thesis.

The FOCUS approach due to M. Broy and K. Stølen outlined in (Broy and Stølen, 2001) provides specification, refinement and verification techniques for the development of interactive systems, thus is not tied to one particular architecture. The basic construction is that of a timed stream of data, refinement is the basic architectural operation, where different kinds of refinements are investigated. For example, Broy and Stølen discuss glass-box refinements which is the exact counterpart to the substitution operation investigated in Section 2.4.6, and they discuss composing systems through an operator style (Broy and Stølen, 2001, 14.5 resp. 5.3.3). FOCUS has a much broader spectrum of application than the present approach. This is due to the fact that no particular architecture is aimed at; there is, however, no explicit notion of computation there, while the present approach encapsulates computations through monads.

2.4.7.1 Possible Extensions

The present discussion excludes those PF-systems that are cyclic by discussing linear pipelines only; further work should admit systems that contain cycles, and, more generally, by admitting decisions which component to invoke next. Here the *extensive categories* investigated, e.g., in (Carboni et al., 1993; Taylor, 1999) are helpful. Extensive categories stress the use of pullbacks somewhat; on the other hand, Corollary 4.15 will show that pullbacks in the category of stochastic relations with surjective measurable maps as morphisms do not exist. A first and very encouraging step towards investigating layered architectures within the framework of extensive categories can be found in (Lajios, 2006).

Introducing timing and explicit synchronization is another area that needs further consideration — the present model is abstract enough not to unduly constrain the modeler, but on the other hand some support could be offered, even at the price of restricting the model. As far as properties of the models are concerned, proof rules which permit stating properties for systems that are evolving according to Section 2.4.6 are of interest. For a different probabilistic approach to synchronization using automata see (D'Argenio, 1999; Bryans et al., 2003).

It is challenging to see how other architectural styles are tackled, and how to model dynamically changing communication topologies. On the relational side, we have narrowed down monads which represent the two major kinds of relations. The natural transformations θ and the $\sharp$-condition seem to be ingredients to a monad which models relations.

2.5   Bibliographic Notes

Categorial Aspects Of Probabilities.   M. Giry's paper (Giry, 1981) is
the first systematic investigation of categorial aspects of probability spaces
on Polish and general measurable spaces (while at that time there was a
substantial body of results on the connection between measure spaces on
compact spaces and convex sets (Semadeni, 1973)); the main line of development pursued here is taken from that paper. P. Panangaden (Panangaden,
1998; Panangaden, 1997) used Giry's construction to elaborate on the analogy
between set-theoretic and probabilistic relations; this line of argumentation
has also been emphasized in the present exposition. Panangaden's construction in (Panangaden, 1998) is slightly different from the one discussed here:
he takes two measurable spaces X and Y and uses a stochastic transformation as a morphism between X and Y; composition of morphisms is then
given by the Kleisli product (in analogy to policies and randomized policies in
stochastic dynamic optimization (Schäl, 1974), one might call them *randomized morphisms*). In later papers, e.g., (Desharnais et al., 2002; Desharnais
et al., 2000), however, the model presented here has been used, albeit usually in a coalgebraic context, for modeling state transitions. One important
field of investigations has been labeled Markov transition systems. They are
constituted of a state space S, a set A of actions, and for each action a probabilistic relation $k_a : S \rightsquigarrow S$, so that $k_a(s)(T)$ indicates the probability that
the system's state will be a member of Borel set $T \subseteq S$ upon action $a \in A$
in state $s \in S$. Such a transition system serves as a probabilistic Kripke
model for a very simple modal logic without negation that, as K. Larsen and
A. Skou (Larsen and Skou, 1991) have forcefully demonstrated, is useful in
testing. Bisimilarity of systems with its intimate connections to testing is
illustrated through the Hennessy-Milner Theorem and its ramifications. We
will discuss this in Chapter 5.

   This treatise concentrates on systems that are usually based on Polish or
analytic spaces, hence work in a nonfinite scenario. But the finite case is interesting, nevertheless. Some of its categorical and coalgebraic aspects have
been investigated in depth in A. Sokolova's thesis (Sokolova, 2005), in particular questions pertaining to composition and different forms of bisimulations.
That the finite case is not really a peculiar special case but may be used for
(hyperfinite) approximations is shown in (Doberkat, 2006a)

   Other constructions in the literature that relate the product in a category
with a monad might be of interest. Mac Lane (MacLane, 1997, Ch. XI.2)
defines a *monoidal functor* between monoidal categories which comes close to
the compatibility definition proposed here for an endofunctor, where the rôle
of the tensor product there is played by the product here. The present definition does not require any conditions on the terminal elements and its image
under the functor, thus it is weaker. Mac Lane formulates a transformation

quite similar to the one given in Lemma 2.13. The proof in (MacLane, 1997) refers, however, to a coherence theorem and appears a bit inaccessible by not making the construction transparent. Consequently, a direct proof is given here. Moggi (Moggi, 1991, Def. 3.2) on the other hand defines a *strong monad* in a category which is closed under finite products by postulating the existence of a natural transformation $t_{a,b} : a \times \mathfrak{T}b \to \mathfrak{T}(a \times b)$ having some properties which relate $\mathfrak{T}$ to the product in the category (t called a *tensorial strength*). In (Moggi, 1989, 3.2.3) it is shown how the tensorial strength induces a natural transformation $\mathfrak{G}_{\mathfrak{T}}^{(2)} \overset{\bullet}{\to} \mathfrak{H}_{\mathfrak{T}}^{(2)}$ in the terminology used here. Barbosa (Barbosa, 2001, 3.51 – 3.55) discusses a strength catalogue for distributive categories.

Architectural Issues. Since software architecture is a lively field of research in software engineering that has already entered most curricula in this area (Pleumann, 2004; Doberkat et al., 2005), many different approaches on diverse levels can be found in the literature. Apart from formal approaches, architectural description languages formulate abstractions closer to implementations than, say, an approach resting on category theory; see (Shaw and Garlan, 1995) for a discussion and assessment of these languages. Other formalisms are used as well. The paper (Medvidovics et al., 2002) by Medvidovics et al. investigates the suitability of the Unified Modeling Language for architectural descriptions. Probably more important, it discusses some desiderata for the language to be usable for architectural descriptions. Pipes and filters are targeted in the work reported in (Abowd et al., 1993; Shaw and Garlan, 1996). The discussion centers around a formulation of this architecture through a denotational framework for developing formal models of architectural styles. It is based on the specification language Z. Arbab and Rutten present *Reo*, a calculus of component connectors based on coinduction (Arbab and Rutten, 2002); coalgebras play a leading role in Barbosa's work on components (Barbosa, 2001) as well. A formal calculus of connectors is given in (Bruni et al., 2005), assigning connectors the rôle as mediators for the interaction between other computational components and connectors; formally, this model is an attempt to conciliate between categorical and the algebraic approaches to interaction. The FOCUS calculus developed by Broy and Stølen (Broy and Stølen, 2001) should be mentioned here as well. Category theory is used in formalizations, e.g., of architectures for mobile programs based on UNITY (Wermelinger and Fiadeiro, 1998; Fiadeiro and Maibaum, 1996). Finally, Lajios (Lajios, 2006) shows that additional assumptions are needed when decisions are to be modeled in an architecture (a decision might involve the selection for the next component to be visited); he models layered architectures using lextensive categories (Carboni et al., 1993) and shows that tools gleaned from graph transformations are a most welcome addition to modeling architectural transformations.

Chapter 3

Eilenberg-Moore Algebras for Stochastic Relations

3.1 Introduction

It is shown at the end of Section 1.6.3 that the adjunction constructed from the Eilenberg-Moore algebras and the one constructed through the Kleisli category form in some sense the extreme points in a category of all adjunctions from which the given monad can be recovered. From this, the algebraic interest to identify these algebras is derived. The algebras for the power set monad (dubbed here the *Manes monad*) are well known, and briefly discussed in Section 3.1. We will identify the algebras for the subprobability functor through smooth equivalence relations and through positive convex structures in this chapter, first through the equivalence relations they induce on the set of subprobability measures. This will be a vehicle for an identification of these algebras without having to refer to the underlying probabilistic structure. It is done initially for the subprobability functor and, with some small adjustments, for the probability functor as well. We provide some examples to illustrate the algebras. Finally, the left adjoint of the forgetful functor that assigns each algebra the underlying Polish space is identified; it is just the functor that maps each Polish space to all its subprobabilities (with the monad's multiplication as the associated algebra). We work in this chapter in the category c$\mathfrak{Pol}$ of **Pol**ish spaces with **c**ontinuous maps as morphisms. A possible and desirable extension to the discussion here would be identification of Eilenberg-Moore algebras for the subprobability functor on analytic spaces with Borel measurable maps as morphisms.

An Exercise: Algebras for the Manes Monad. The Kleisli construction that was introduced and briefly investigated in Section 1.6.3 helps in constructing an adjunction from which the monad can be recovered. Technically this is done by constructing a functor from the category into the Kleisli category, from which an adjoined pair of functors is easily determined, as we have seen. This observation was historically the original motivation for having a closer look at the Kleisli construction. Eilenberg-Moore algebras are another way of doing this, as we have seen in Section 1.6.3 as well. We will discuss and identify the Eilenberg-Moore algebras for the Giry monad.

Before doing this, and in order to indicate what we aim at with this identification, we mention as an illustration the algebras for the Manes monad $\langle \mathfrak{Pow}, \mathfrak{e}, \mathfrak{m} \rangle$ in the category $\mathfrak{Set}$ of sets that has been mentioned very briefly in Section 2.2. It is well known that the algebras for this monad may be identified with the complete sup-semi lattices (MacLane, 1997, Exercise VI.2.1). Doing this exercise is instructive for observing the components of a monad at work.

Assume first that $\leq$ is a partial order on a set X that is sup-complete, so that $\sup a$ exists for each $a \subseteq X$. Define $h(a) := \sup a$, then we have for each $A \in \mathfrak{Pow}(\mathfrak{Pow}(X))$ from the familiar properties of the supremum

$$\sup(\bigcup A) = \sup \{\sup a \mid a \in A\}.$$

This translates into $(h \circ \mathfrak{m}_X)(A) = (h \circ \mathfrak{Pow}(h))(A)$. Because $x = \sup\{x\}$ holds for each $x \in X$, we see that $\langle X, h \rangle$ defines an algebra.

Assume on the other hand that $\langle X, h \rangle$ is an algebra, and put for $x, x' \in X$ $x \leq x'$ iff $h(\{x, x'\}) = x'$. This defines a partial order: reflexivity and antisymmetry are obvious. Transitivity is seen as follows: assume $x \leq x'$ and $x' \leq x''$, then

$$
\begin{aligned}
h(\{x, x''\}) &= h(\{h(\{x\}), h(\{x', x''\})\}) \\
&= (h \circ \mathfrak{Pow}(h))(\{\{x\}, \{x', x''\}\}) \\
&= (h \circ \mathfrak{m}_X)(\{\{x\}, \{x', x''\}\}) \\
&= h(\{x, x', x''\}) \\
&= (h \circ \mathfrak{m}_X)(\{\{x, x'\}, \{x', x''\}\}) \\
&= (h \circ \mathfrak{Pow}(h))(\{\{x, x'\}, \{x', x''\}\}) \\
&= h(\{x', x''\}) \\
&= x''.
\end{aligned}
$$

It is clear from $\{x\} \cup \emptyset = \{x\}$ for every $x \in X$ that $h(\emptyset)$ is the smallest element. Finally, it has to be shown that $h(a)$ is the smallest upper bound for $a \subseteq X$ in the order $\leq$. We may assume that $a \neq \emptyset$. Suppose that $x \leq t$ holds for all

$x \in a$, then

$$
\begin{aligned}
h(a \cup \{t\}) &= h\left(\bigcup_{x \in a} \{x,t\}\right) \\
&= (h \circ \mathfrak{m}_X)\,(\{\{x,t\} \mid x \in a\}) \\
&= (h \circ \mathfrak{Pow}\,(h))\,(\{\{x,t\} \mid x \in a\}) \\
&= h\,(\{h(\{x,t\})\ x \in a\}) \\
&= h(\{t\}) \\
&= t.
\end{aligned}
$$

Thus, if $x \leq t$ for all $x \in a$, hence $h(a) \leq t$, thus $h(a)$ is an upper bound to a, and similarly, $h(a)$ is the smallest upper bound.

We will turn now to a characterization of the Eilenberg-Moore algebras for the Giry monad over some Polish space X that will be fixed throughout.

3.2 Characterization through Equivalence Relations

We show in this section that an algebra may be characterized in the way its fibres, i.e., the inverse images of points, partition the domain $\mathfrak{S}\,(X)$. The aspect that interests here is that these partitions are positive convex and take closed values. They observe an additional constraint due to continuity. This yields necessary and sufficient conditions for the characterization of partitions spawned by these algebras; a characterization of the morphisms in the category of all algebras is also derived.

3.2.1 Preparations

We need some elementary properties for later reference. They are collected in the next Lemma. First, however, we define

$$
\Omega := \{\langle \alpha_1, \ldots, \alpha_k \rangle \mid k \in \mathbb{N}, \alpha_i \geq 0, \sum_{i=1}^{k} \alpha_i \leq 1\}
$$

for the rest of the chapter, the elements of Ω being called *positive convex tuples* or simply *positive convex*.

LEMMA 3.1

a. Let $f : A \to B$ be a map between the Polish spaces A and B, and let

$$
\mu = \alpha_1 \cdot \delta_{a_1} + \ldots + \alpha_n \cdot \delta_{a_n}
$$

be the linear combination of Dirac measures for $a_1, \ldots, a_n \in A$ with positive convex $\langle \alpha_1, \ldots, \alpha_n \rangle \in \Omega$. Then $\mathfrak{S}(f)(\mu) = \alpha_1 \cdot \delta_{f(a_1)} + \ldots + \alpha_n \cdot \delta_{f(a_n)}$.

b. *Let $\mu_1, \ldots, \mu_n$ be subprobability measures on X, and let*

$$M = \alpha_1 \cdot \delta_{\mu_1} + \ldots + \alpha_n \cdot \delta_{\mu_n}$$

be the linear combination of the corresponding Dirac measures in $\mathfrak{S}(\mathfrak{S}(X))$ with positive convex coefficients $\langle \alpha_1, \ldots, \alpha_n \rangle \in \Omega$. Then $\mathfrak{m}_X(M) = \alpha_1 \cdot \mu_1 + \ldots + \alpha_n \cdot \mu_n$.

PROOF   The first part follows directly from the observation $\delta_x(f^{-1}[D]) = \delta_{f(x)}(D)$, and the second one is easily inferred from

$$\mathfrak{m}_X(\delta_\mu)(Q) = \int_{\mathfrak{S}(X)} \rho(Q)\, \delta_\mu(d\rho)$$
$$= \mu(Q)$$

for each Borel subset $Q \subseteq X$, and from the linearity of the integral.   ⬚

Both $\mathfrak{e}_X$ and $\mathfrak{m}_X$ are morphisms in $\mathfrak{cPol}$ for Polish X, as the following Lemma shows.

LEMMA 3.2
$\mathfrak{e}_X : X \to \mathfrak{S}(X)$ *and* $\mathfrak{m}_X : \mathfrak{S}(\mathfrak{S}(X)) \to \mathfrak{S}(X)$ *are continuous.*

PROOF   1. Continuity of $\mathfrak{e}_X$ is clear, since $x_n \to x$ implies

$$\int_X f\, d\mathfrak{e}_X(x_n) = \int_X f\, d\delta_{x_n} = f(x_n) \to f(x) = \int_X f\, d\mathfrak{e}_X(x),$$

whenever $f \in C(X)$ is continuous and bounded. Thus $\mathfrak{e}_X(x_n) \to_w \mathfrak{e}_X(x)$.

2. Let $(M_n)_{n \in \mathbb{N}}$ be a sequence in $\mathfrak{S}(\mathfrak{S}(X))$ with $M_n \to_w M_0$, then we get for $f \in C(X)$ through the Change of Variable formula, and because

$$\mu \mapsto \int_X f\, d\mu$$

is a member of $C(\mathfrak{S}(X))$, this chain

$$\int_{\mathfrak{S}(X)} f\, d\mathfrak{m}_X(M_n) = \int_{\mathfrak{S}(X)} \left(\int_X f\, d\mu \right) M_n(d\mu) \to$$
$$\int_{\mathfrak{S}(X)} \left(\int_X f\, d\mu \right) M_0(d\mu) = \int_{\mathfrak{S}(X)} f\, d\mathfrak{m}_X(M_0).$$

Thus $\mathfrak{m}_X(M_n) \to_w \mathfrak{m}_X(M_0)$ is established, as desired.   ⬚

3.2.2 Positive Convex Partitions

The natural approach is to think of these algebras in terms of an equivalence relation which may be thought to identify probability distributions, and to investigate either these relations or the partitions associated with them. These characterizations lead to the identification of the algebras as exactly the positive convex structures on their base space.

Assume that the pair $\langle X, h \rangle$ is an algebra, and define for each $x \in X$

$$G_h(x) := \{\mu \in \mathfrak{S}(X) \mid h(\mu) = x\} \left(= h^{-1}[\{x\}]\right).$$

Then $G_h(x) \neq \emptyset$ for all $x \in X$ due to h being onto. The algebra h will be characterized through properties of the sct-valued map G_h. We will need the *weak inverse* $\exists R$ for a set-valued map $R : X \to \mathfrak{Pow}(Y) \setminus \{\emptyset\}$; see Section 1.4. If Y is a topological space, if R takes closed values, and if $\exists R(W)$ is compact in X whenever $W \subseteq Y$ is compact, then R is called *k-upper-semicontinuous* (abbreviated as k.u.s.c.). If Y is compact, this is the usual notion of upper-semicontinuity known from topology.

The importance of being k.u.s.c. becomes clear at once from

LEMMA 3.3
Let $f : A \to B$ be a surjective map between the Polish spaces A and B, and put $G_f(b) := f^{-1}[\{b\}]$ for $b \in B$. Then f is continuous iff G_f is k.u.s.c.

PROOF   A direct calculation for the weak inverse shows $\exists G_f(A_0) = f[A_0]$ for each subset $A_0 \subseteq A$. The assertion now follows from the well-known fact that a map between metric spaces is continuous iff it maps compact sets to compact sets. □

Applying this observation to the set-valued map G_h, we obtain:

PROPOSITION 3.4
The set-valued map $x \mapsto G_h(x)$ has the following properties:

a. $\delta_x \in G_h(x)$ *holds for each $x \in X$.*

b. $\mathcal{G}_h := \{G_h(x) \mid x \in X\}$ *is a partition of $\mathfrak{S}(X)$ into closed and positive convex sets.*

c. $x \mapsto G_h(x)$ *is k.u.s.c.*

d. *Let $\sim_h$ be the equivalence relation on $\mathfrak{S}(X)$ induced by the partition $\mathcal{G}_h$. If $\mu_i \sim_h \mu_i'$ $(1 \leq i \leq n)$, then*

$$(\alpha_1 \cdot \mu_1 + \ldots + \alpha_n \cdot \mu_n) \sim_h (\alpha_1 \cdot \mu_1' + \ldots + \alpha_n \cdot \mu_n')$$

for the positive convex coefficients $\langle \alpha_1, \ldots, \alpha_n \rangle \in \Omega$.

PROOF   Because $\{x\}$ is closed, and h is continuous, $G_h(x) = h^{-1}[\{x\}]$ is a closed subset of $\mathfrak{S}(X)$. Because h is onto, every G_h takes nonempty values; it is clear that $\{G_h(x) \mid x \in X\}$ forms a partition of $\mathfrak{S}(X)$. Because h is continuous, G_h is k.u.s.c. by Lemma 3.3. Positive convexity will follow immediately from part d.

Assume that $h(\mu_i) = h(\mu_i') = x_i$ $(1 \leq i \leq n)$, and observe that $h(\delta_x) = x$ holds for all $x \in X$. Using Lemma 3.1, we get:

$$
\begin{aligned}
h(\alpha_1 \cdot \mu_1 + \ldots + \alpha_n \cdot \mu_n) &= (h \circ \mathfrak{m}_X)(\alpha_1 \cdot \delta_{\mu_1} + \ldots + \alpha_n \cdot \delta_{\mu_n}) \\
&= (h \circ \mathfrak{S}(h))(\alpha_1 \cdot \delta_{\mu_1} + \ldots + \alpha_n \cdot \delta_{\mu_n}) \\
&= h(\alpha_1 \cdot \delta_{h(\mu_1)} + \ldots + \alpha_n \cdot \delta_{h(\mu_n)}) \\
&= h(\alpha_1 \cdot \delta_{x_1} + \ldots + \alpha_n \cdot \delta_{x_n})
\end{aligned}
$$

In a similar way, $h(\alpha_1 \cdot \mu_1' + \ldots + \alpha_n \cdot \mu_n') = h(\alpha_1 \cdot \delta_{x_1} + \ldots + \alpha_n \cdot \delta_{x_n})$ is obtained. This implies the assertion.   □

Thus $\mathcal{G}_h$ is invariant under taking positive convex combinations. It is a positive convex partition in the sense of the following definition.

DEFINITION 3.5   *An equivalence relation ρ on $\mathfrak{S}(X)$ is said to be* positive convex *iff $\mu_i \, \rho \, \mu_i'$ for $1 \leq i \leq n$ and $\langle \alpha_1, \ldots, \alpha_n \rangle \in \Omega$ together imply*

$$
(\alpha_1 \cdot \mu_1 + \ldots + \alpha_n \cdot \mu_n) \, \rho \, (\alpha_1 \cdot \mu_1' + \ldots + \alpha_n \cdot \mu_n')
$$

for each $n \in \mathbb{N}$. A partition of $\mathfrak{S}(X)$ is called positive convex *iff its associated equivalence relation is.*

Note that the elements of a positive convex partition form positive convex sets. The converse to Proposition 3.4 characterizes algebras:

PROPOSITION 3.6

Assume $\mathcal{G} = \{G(x) \mid x \in X\}$ is a positive convex partition of $\mathfrak{S}(X)$ into closed sets which is indexed by X such that $\delta_x \in G(x)$ for each $x \in X$, and such that $x \mapsto G(x)$ is k.u.s.c. Define $h : \mathfrak{S}(X) \to X$ through $h(\mu) = x$ iff $\mu \in G(x)$. Then $\langle X, h \rangle$ is an algebra for the Giry monad.

PROOF   1. It is clear that h is well defined and surjective, and that

$$
\exists G(F) = h[F]
$$

holds for each subset $F \subseteq \mathfrak{S}(X)$. Thus $h[K]$ is compact whenever K is compact, because G is k.u.s.c. Hence h is continuous by Lemma 3.3.

2. An easy induction establishes that h respects positive convex combinations: if $h(\mu_i) = h(\mu_i')$ for $i = 1, \ldots, n$, and if $\alpha_1, \ldots, \alpha_n$ are positive convex

coefficients, then

$$h(\sum_{i=1}^{n} \alpha_i \cdot \mu_i) = h(\sum_{i=1}^{n} \alpha_i \cdot \mu_i').$$

We claim that

$$(h \circ \mathfrak{m}_X)(M) = (h \circ \mathfrak{S}(h))(M)$$

holds for each *discrete* $M \in \mathfrak{S}(\mathfrak{S}(X))$. In fact, let

$$M = \sum_{i=1}^{n} \alpha_i \cdot \delta_{\mu_i}$$

be such a discrete measure, then Lemma 3.1 implies that

$$\mathfrak{m}_X(M) = \sum_{i=1}^{n} \alpha_i \cdot \mu_i,$$

thus

$$(h \circ \mathfrak{m}_X)(M) = h\left(\sum_{i=1}^{n} \alpha_i \cdot \mu_i\right) = h\left(\sum_{i=1}^{n} \alpha_i \cdot \delta_{h(\mu_i)}\right) = (h \circ \mathfrak{S}(h))(M),$$

because we know also from Lemma 3.1 that

$$\mathfrak{S}(h)(M) = \sum_{i=1}^{n} \alpha_i \cdot \delta_{h(\mu_i)}$$

holds.

3. Since the discrete measures are dense in the weak topology (see Section 1.5.2), we find for $M_0 \in \mathfrak{S}(\mathfrak{S}(X))$ a sequence $(M_n)_{n \in \mathbb{N}}$ of discrete measures M_n with $M_n \to_w M_0$. Consequently, we get from the continuity of both h and $\mathfrak{m}_X$ (Lemma 3.2) together with the continuity of $\mathfrak{S}(h)$

$$(h \circ \mathfrak{m}_X)(M_0) = \lim_{n \to \infty} (h \circ \mathfrak{m}_X)(M_n) = \lim_{n \to \infty} (h \circ \mathfrak{S}(h))(M_n) = (h \circ \mathfrak{S}(h))(M_0).$$

This proves the claim.   ⬚

We have established

PROPOSITION 3.7
The algebras $\langle X, h \rangle$ for the Giry monad for Polish spaces X are exactly the positive convex k.u.s.c. partitions $\{G(x) \mid x \in X\}$ into closed subsets of $\mathfrak{S}(X)$ such that $\delta_x \in G(x)$ for all $x \in X$.

For characterizing the category $\mathfrak{Alg}$ of all Eilenberg-Moore algebras for the Giry monad we package the properties of partitions representing algebras into the notion of a G-partition. They will form the objects of category $\mathfrak{GPart}$.

DEFINITION 3.8 $\mathcal{G}$ *is called a* G-partition for X *iff*

a. $\mathcal{G} = \{G(x) \mid x \in X\}$ *is a positive convex partition for* $\mathfrak{S}(X)$ *into closed sets indexed by* X,

b. $\delta_x \in G(x)$ *holds for all* $x \in X$,

c. *the set-valued map* $x \mapsto G(x)$ *is k.u.s.c.*

Define the objects of category $\mathfrak{GPart}$ as pairs $\langle X, \mathcal{G} \rangle$ where X is a Polish space, and $\mathcal{G}$ is a G-partition for X. A morphism f between $\mathcal{G}$ and $\mathcal{G}'$ will map elements of $G(x)$ to $G'(f(x))$ through its associated map $\mathfrak{S}(f)$. Thus an element $\mu \in G(x)$ will correspond to an element $\mathfrak{S}(f)(\mu) \in G'(f(x))$.

DEFINITION 3.9   *A morphism for* $\mathfrak{GPart}$ $f : \langle X, \mathcal{G} \rangle \to \langle X', \mathcal{G}' \rangle$ *is a continuous map* $f : X \to X'$ *such that*

$$G(x) \subseteq \mathfrak{S}(f)^{-1}[G'(f(x))]$$

holds for each $x \in X$.

Define the functor $F : \mathfrak{Alg} \to \mathfrak{GPart}$ by associating with each algebra $\langle X, h \rangle$ its Giry partition $F(X, h)$ according to Proposition 3.7. Assume that $f : \langle X, h \rangle \to \langle X', h' \rangle$ is a morphism in $\mathfrak{Alg}$, and let $\mathcal{G} = \{G(x) \mid x \in X\}$ resp. $\mathcal{G}' = \{G'(x') \mid x' \in X'\}$ be the corresponding partitions. Then the properties of an algebra morphism yield

$$
\begin{aligned}
\mu \in \mathfrak{S}(f)^{-1}[G'(f(x))] &\Leftrightarrow \mathfrak{S}(f)(\mu) \in G'(f(x)) \\
&\Leftrightarrow (h' \circ \mathfrak{S}(f))(\mu) = f(x) \\
&\Leftrightarrow (f \circ h)(\mu) = f(x).
\end{aligned}
$$

Thus $\mu \in \mathfrak{S}(f)^{-1}[G'(f(x))]$, provided $\mu \in G(x)$. Hence f is a morphism in $\mathfrak{GPart}$ between $F(X, h)$ and $F(X', h')$. Conversely, let $f : \langle X, \mathcal{G} \rangle \to \langle X', \mathcal{G}' \rangle$ be a morphism in $\mathfrak{GPart}$ with $\langle X, \mathcal{G} \rangle = F(X, h)$ and $\langle X', \mathcal{G}' \rangle = F(X', h')$. Then

$$
\begin{aligned}
h(\mu) = x &\Leftrightarrow \mu \in G(x) \\
&\Rightarrow \mathfrak{S}(f)(\mu) \in G'(f(x)) \\
&\Leftrightarrow h'(\mathfrak{S}(f)(\mu)) = f(x),
\end{aligned}
$$

thus $h' \circ \mathfrak{S}(f) = f \circ h$ is inferred. Hence f constitutes a morphism in category $\mathfrak{Alg}$.

Summarizing, we have shown

PROPOSITION 3.10

The category 𝔄𝔩𝔤 *of Eilenberg-Moore algebras for the Giry monad is isomorphic to the category* 𝔊𝔓𝔞𝔯𝔱 *of G-partitions.*

3.2.3 Smooth Relations

The characterization of algebras so far encoded the crucial properties into a partition of $\mathfrak{S}(X)$, thus indirectly into an equivalence relation on that space. We can move directly to a particular class of these relations when looking at an alternative characterization of the algebras through smooth equivalence relations. In contrast to the characterization in 3.2 that started from the fibres $h^{-1}[\{x\}]$ we study here the kernel of h, i.e., the set

$$\mathsf{ker}(h) = \{\langle \mu, \mu' \rangle \mid h(\mu) = h(\mu')\}.$$

We characterize then algebras in terms of the kernel for the associated map, and we indicate how an algebra may be constructed from such a partition. The characterization is interesting in its own right and permits another characterization of morphisms for algebras, but it will also help in giving an intrinsic characterization of algebras in terms of positive convex structures.

Recall that an equivalence relation ρ on a Polish space A is called *smooth* iff there exists a Polish space B and a Borel measurable map $f : A \to B$ such that $\rho = \mathsf{ker}(f)$. A brief first characterization of smooth relations is given in Lemma 1.52. Proposition 1.53 gives a rather important property of smooth relations, viz., that their factor space is smooth again. These relations will turn out to be most interesting in the investigation of stochastic relations, so we will return to them later in Chapter 5 and study them in greater detail there. Despite the later and more systematic treatment of the topic, we will state some properties that will be needed here immediately.

For the Polish space A with topology $\mathcal{T}$ let $\mathcal{T}/\rho$ be the final topology on A/ρ with respect to the given topology $\mathcal{T}$ and η_ρ, i.e., the largest topology $\mathcal{T}'$ on A/ρ which makes η_ρ $\mathcal{T}$-$\mathcal{T}'$-continuous. Clearly a map $g : A/\rho \to B$ for a topological space $(B, \mathcal{S})$ is $\mathcal{T}/\rho$-$\mathcal{S}$-continuous iff $g \circ \eta_\rho : A \to B$ is $\mathcal{T}$-$\mathcal{S}$-continuous; compare Lemma 1.3. We will need this observation in the proof of Proposition 3.11.

Now let $\langle X, h \rangle$ be an algebra for the Giry monad. Obviously $\rho_h := \mathsf{ker}(h)$ defines a smooth equivalence relation ρ_h on the Polish space $\mathfrak{S}(X)$.

PROPOSITION 3.11

The equivalence relation ρ_h is positive convex, each equivalence class $[\mu]_{\rho_h}$ is closed and positive convex, and the factor space $\mathfrak{S}(X)/\rho_h$ is homeomorphic to

X when the former is endowed with the topology $\mathcal{W}/\rho_h$, $\mathcal{W}$ being the topology of weak convergence on $\mathfrak{S}(X)$.

PROOF   1. Positive convexity of ρ_h follows from the properties of h exactly as in the proof of Proposition 3.4. Positive convexity of the classes is inferred from this as well. Continuity of h implies that the classes are closed sets.

2. Define $\chi_h([\mu]_{\rho_h}) := h(\mu)$ for $\mu \in \mathfrak{S}(X)$. Then $\chi_h : X/\rho_h \to X$ is well defined and a bijection. Let $G \subseteq X$ be an open set, then

$$\eta_{\rho_h}^{-1} \left[\chi_h^{-1} [G] \right] = h^{-1} [G].$$

Because $\mathcal{W}/\rho_h$ is the largest topology on $\mathfrak{S}(X)/\rho_h$ that renders η_{ρ_h} continuous, and because $h^{-1}[G] \subseteq \mathfrak{S}(X)$ is open by assumption, we infer that $\chi_h^{-1}[G]$ is $\mathcal{W}/\rho_h$-open. Thus χ_h is continuous. On the other hand, if $(x_n)_{n\in\mathbb{N}}$ is a sequence in X converging to $x_0 \in X$, then $\delta_{x_n} \rightharpoonup_w \delta_{x_0}$ in $\mathfrak{S}(X)$, thus $[\delta_{x_n}]_{\rho_h} \to [\delta_{x_0}]_{\rho_h}$ in $\mathcal{W}/\rho_h$ by construction. Consequently χ_h^{-1} is also continuous. $\qquad\Box$

Thus each algebra induces a G-triplet consisting of its kernel and a homeomorphism:

DEFINITION 3.12   *A G-triplet $\langle X, \rho, \chi \rangle$ is a Polish space X with a smooth and positive convex equivalence relation ρ on $\mathfrak{S}(X)$ such that $\chi : \mathfrak{S}(X)/\rho \to X$ is a homeomorphism with $\chi([\delta_x]_\rho) = x$ for all $x \in X$. Here $\mathfrak{S}(X)/\rho$ carries the final topology with respect to the weak topology on $\mathfrak{S}(X)$ and η_ρ.*

Now assume that a G-triplet $\langle X, \rho, \chi \rangle$ is given. Define $h(\mu) := \chi([\mu]_\rho)$ for $\mu \in \mathfrak{S}(X)$. Then $\langle X, h \rangle$ is an algebra for the Giry monad: $h(\delta_x) = x$ follows from the assumption, and because $h = \chi \circ \eta_\rho$, the map h is continuous. An argument very similar to that used in the proof of Proposition 3.4 shows that $h \circ m_X = h \circ \mathfrak{S}(h)$ holds; this is so since ρ is assumed to be positive convex.

DEFINITION 3.13   *The continuous map $f : X \to X'$ between the Polish spaces X and X' constitutes a G-triplet morphism $f : \langle X, \rho, \chi \rangle \to \langle X', \rho', \chi' \rangle$ iff these conditions hold:*

a. $\mu \rho \mu'$ implies $\mathfrak{S}(f)(\mu) \rho' \mathfrak{S}(f)\mu'$,

b. the diagram

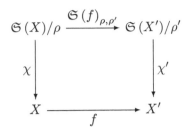

commutes, where

$$\mathfrak{S}\left(f\right)_{\rho,\rho'}\left([\mu]_{\rho}\right) := [\mathfrak{S}\left(f\right)\left(\mu\right)]_{\rho'}.$$

G-triplets with their morphisms form a category $\mathfrak{GTrip}$.

LEMMA 3.14

Each algebra morphism $f : \langle X, h \rangle \to \langle X', h' \rangle$ induces a G-triplet morphism $f : \langle X, \rho_h, \chi_h \rangle \to \langle X', \rho_{h'}, \chi_{h'} \rangle$.

PROOF   1. It is an easy calculation to show that $\mathfrak{S}\left(f\right)\left(\mu\right)\rho_{h'}\,\mathfrak{S}\left(f\right)\left(\mu\right)$ holds, provided $\mu\,\rho_h\,\mu'$. This is so because f is a morphism for the algebras.

2. Since for each $\mu \in \mathfrak{S}\left(X\right)$ there exists $x \in X$ such that $[\mu]_{\rho_h} = [\delta_x]_{\rho_h}$ (in fact, $h(\mu)$ would do, because $h(\mu) = h\left(\delta_{h(\mu)}\right)$, as shown above), it is enough to demonstrate that

$$\chi_{h'}\left(\mathfrak{S}\left(f\right)_{\rho_h,\rho'_{h'}}\left([\delta_x]_{\rho_h}\right)\right) = f(\chi_h([\delta_x]_{\rho_h}))$$

is true for each $x \in X$. Because $\mathfrak{S}\left(f\right)\left(\delta_x\right) = \delta_{f(x)}$, a little computation shows that both sides of the above equation boil down to $f(x)$. $\square$

The morphisms between G-triplets are just the morphisms between algebras (when we forget that these games play in different categories).

PROPOSITION 3.15

Let $f : \langle X, \rho, \chi \rangle \to \langle X', \rho', \chi' \rangle$ be a morphism between G-triplets, and let $\langle X, h \rangle$ resp. $\langle X', h' \rangle$ be the associated algebras. Then $f : \langle X, h \rangle \to \langle X', h' \rangle$ is an algebra morphism.

PROOF   Given $\mu \in \mathfrak{S}\left(X\right)$ we have to show that $(f \circ h)(\mu)$ equals $(h' \circ$

$\mathfrak{S}(f))(\mu)$. Since $h(\mu) = \chi([\mu]_\rho)$, we obtain

$$
\begin{aligned}
(f \circ h)(\mu) &= f\left(\chi([\mu]_\rho)\right) \\
&= \chi'\left(\mathfrak{S}(f)_{\rho,\rho'}([\mu]_\rho)\right) \\
&= \chi'\left([\mathfrak{S}(f)(\mu)]_{\rho'}\right) \\
&= (h' \circ \mathfrak{S}(f))(\mu).
\end{aligned}
$$

$\square$

Putting all these constructions with their properties together, we obtain as a second characterization

PROPOSITION 3.16

The category $\mathfrak{Alg}$ of algebras for the Giry monad is isomorphic to the category $\mathfrak{GTrip}$ of G-triplets.

Albeit being treated similarly, the probabilistic case requires a separate discussion. We define an equivalence relation ρ on $\mathfrak{P}(X)$ to be convex iff for each $n \in \mathbb{N}$ the conditions $\mu_i \ \rho \ \mu_i'$ for $1 \le i \le n$ and $\langle \alpha_1, \ldots, \alpha_n \rangle \in \Omega_{\mathbf{c}}$ together imply $(\sum_{i=1}^n \alpha_i \cdot \mu_i) \ \rho \ (\sum_{i=1}^n \alpha_i \cdot \mu_i')$, where

$$
\Omega_{\mathbf{c}} := \{\langle \alpha_1, \ldots, \alpha_k \rangle \mid \alpha_i \ge 0, \alpha_1 + \cdots + \alpha_k = 1\}
$$

are all convex coefficients. Then the ρ-classes form convex subsets of $\mathfrak{P}(X)$. We introduce *PG-triplets* $\langle X, \rho, \chi \rangle$ for a Polish space X, a smooth convex equivalence relation ρ and a homeomorphism $\chi : \mathfrak{P}(X)/\rho \to X$ such that $\chi\left([\delta_x]_\rho\right) = x$ for all $x \in X$. A continuous map $f : X \to X'$ then is a PG-triplet morphism $\langle X, \rho, \chi \rangle \to \langle X', \rho', \chi' \rangle$ iff

a. $\mu \ \rho \ \mu' \Rightarrow \mathfrak{P}(f)(\mu) = \mathfrak{P}(f)(\mu')$,

b. $\chi' \circ \mathfrak{P}(f)_{\rho,\rho'} = f \circ \chi$.

Here $\mathfrak{P}(f)_{\rho,\rho'}$ is defined in analogy to $\mathfrak{S}(f)_{\rho,\rho'}$ in Definition 3.13 as

$$
\mathfrak{P}(f)_{\rho,\rho'}\left([\mu]_\rho\right) := [\mathfrak{P}(f)(\mu)]_{\rho'}.
$$

We see then that each algebra morphism $f : \langle X, h \rangle \to \langle X', h' \rangle$ induces a PG-triplet morphism $f : \langle X, \rho_h, \chi_h \rangle \to \langle X', \rho_{h'}, \chi_{h'} \rangle$, and vice versa. The reader is invited to fill in the details.

Summarizing, this yields:

PROPOSITION 3.17

The category of algebras for the Giry monad for the probability functor is isomorphic to the full subcategory of G-triplets $\langle X, \rho, \chi \rangle$ with a smooth and convex equivalence relation such that $\chi : \mathfrak{P}(X)/\rho \to X$ is a homeomorphism.

This prepares us for another and more self-contained characterization of algebras. We will show now that $\mathfrak{StrConv}$ is isomorphic to $\mathfrak{Alg}$.

3.3   Positive Convex Structures

Suppose the Polish space X is embedded as a positive convex set into a linear space V over the reals as a positive convex structure. This means that, if $x_1, \ldots, x_k \in X$, $\langle \alpha_1, \ldots, \alpha_k \rangle \in \Omega$, then $\sum_{i=1}^{k} \alpha_i \cdot x_i \in X$. In addition, forming positive convex combinations should be compatible with the topological structure on X, so it should be continuous. This entails of course that $x_{i,n} \to x_{i,0}$ and $\alpha_n \to \alpha_0$ with $\alpha_0, \alpha_n \in \Omega$ together imply $\sum_{i=1}^{k} \alpha_{i,n} \cdot x_{i,n} \to \sum_{i=1}^{k} \alpha_{i,0} \cdot x_{i,0}$. These requirements are quite comparable to those for a topological vector space, postulating continuity of addition and scalar multiplication.

These observations meet the intuition about positive convexity, but it has the drawback that we have to look for a linear space V into which X to embed. It has the additional shortcoming that once we did identify V, the positive convex structure on X is fixed through the vector space, but we will see soon that we need some flexibility. Consequently, we propose an abstract description of positive convexity, much in the spirit of Pumplün's approach (Pumplün, 2003). Thus the essential properties (for us, that is) of positive convexity are described intrinsically for X without having to resort to a vector space. This leads to the definition of a positive convex structure.

DEFINITION 3.18   *A positive convex structure $\mathcal{P}$ on the Polish space X has for each $\alpha = \langle \alpha_1, \ldots, \alpha_n \rangle \in \Omega$ a continuous map $\alpha_{\mathcal{P}} : X^n \to X$ which we write as*

$$\alpha_{\mathcal{P}}(x_1, \ldots, x_n) = \sum_{1 \leq i \leq n}^{\mathcal{P}} \alpha_i \cdot x_i,$$

such that

a. $\sum_{1 \leq i \leq n}^{\mathcal{P}} \delta_{i,k} \cdot x_i = x_k$, *where $\delta_{i,j}$ is Kronecker's δ (thus $\delta_{i,j} = 1$ if $i = j$, and $\delta_{i,j} = 0$, otherwise),* *b. the identity*

$$\sum_{1 \leq i \leq n}^{\mathcal{P}} \alpha_i \cdot \left(\sum_{1 \leq k \leq m}^{\mathcal{P}} \beta_{i,k} \cdot x_k \right) = \sum_{1 \leq k \leq m}^{\mathcal{P}} \left(\sum_{1 \leq i \leq n}^{\mathcal{P}} \alpha_i \beta_{i,k} \right) \cdot x_k$$

holds whenever $\langle \alpha_1, \ldots, \alpha_n \rangle, \langle \beta_{i,1}, \ldots, \beta_{i,m} \rangle \in \Omega, 1 \leq i \leq n.$

Property a looks quite trivial, when written down this way. Rephrasing it states that the map

$$\langle \delta_{1,k}, \ldots, \delta_{n,k} \rangle_{\mathcal{P}} : T^n \to T,$$

which is assigned to the n-tuple $\langle \delta_{1,k}, \ldots, \delta_{n,k} \rangle$ through $\mathcal{P}$ acts as the projection to the kth component for $1 \leq k \leq n$. Similarly, property b may be re-coded in a formal but less concise way. Thus we will use freely the notation from vector spaces, omitting in particular the explicit reference to the structure whenever possible. Hence simple addition $\alpha_1 \cdot x_1 + \alpha_2 \cdot x_2$ will be written rather than $\sum_{1 \leq i \leq 2}^{\mathcal{P}} \alpha_i \cdot x_i$, with the understanding that it refers to a fixed positive convex structure $\mathcal{P}$ on X.

It is an easy exercise that for a positive convex structure the usual rules for manipulating sums in vector spaces apply, e.g., $1 \cdot x = x, \sum_{i=1}^{n} \alpha_i \cdot x_i = \sum_{i=1, \alpha_i \neq 0}^{n} \alpha_i \cdot x_i$, or the law of associativity, $(\alpha_1 \cdot x_1 + \alpha_2 \cdot x_2) + \alpha_3 \cdot x_3 = \alpha_1 \cdot x_1 + (\alpha_2 \cdot x_2 + \alpha_3 \cdot x_3)$. Nevertheless, care should be observed, for of course not all rules apply: we cannot in general conclude $x = x'$ from $\alpha \cdot x = \alpha \cdot x'$, even if $\alpha \neq 0$.

A morphism $\theta : \langle X_1, \mathcal{P}_1 \rangle \to \langle X_2, \mathcal{P}_2 \rangle$ between continuous positive convex structures is a continuous map $\theta : X_1 \to X_2$ such that

$$\theta \left(\sum_{1 \leq i \leq n}^{\mathcal{P}_1} \alpha_i \cdot x_i \right) = \sum_{1 \leq i \leq n}^{\mathcal{P}_2} \alpha_i \cdot \theta(x_i)$$

holds for $x_1, \ldots, x_n \in X$ and $\langle \alpha_1, \ldots, \alpha_n \rangle \in \Omega$. In analogy to linear algebra, θ will be called an *affine* map. Positive convex structures with their morphisms form a category $\mathfrak{StrConv}$.

3.4   Algebras through Positive Convex Structures

The algebras are also described without having to resort to $\mathfrak{S}(X)$. This is done through an intrinsic characterization using positive convex structures with affine maps. This characterization is comparable to the one given by Manes for the power set monad (which also does not resort explicitly to the underlying monad or its functor); see page 132.

LEMMA 3.19

Given an algebra $\langle X, h \rangle$, define for $x_1, \ldots, x_n \in X$ and the positive convex coefficients $\langle \alpha_1, \ldots, \alpha_n \rangle \in \Omega$

$$\sum_{i=1}^{n} \alpha_i \cdot x_i := h(\sum_{i=1}^{n} \alpha_i \cdot \delta_{x_i}).$$

This defines a positive convex structure on X.

PROOF   1. Because

$$h\left(\sum_{i=1}^{n} \delta_{i,j} \cdot \delta_{x_i}\right) = h(\delta_{x_j}) = x_j,$$

property a in Definition 3.18 is satisfied.

2. Proving property b, we resort to the properties of an algebra and a monad:

$$\sum_{i=1}^{n} \alpha_i \cdot \left(\sum_{k=1}^{m} \beta_{i,k} \cdot x_k\right) = h\left(\sum_{i=1}^{n} \alpha_i \cdot \delta_{\sum_{h=1}^{m} \beta_{i,k} \cdot x_k}\right) \tag{3.1}$$

$$= h\left(\sum_{,i-1}^{n} \alpha_i \cdot \delta_{h\left(\sum_{k=1}^{m} \beta_{i,k} \cdot \delta_{x_k}\right)}\right) \tag{3.2}$$

$$= h\left(\sum_{i=1}^{n} \alpha_i \cdot \mathfrak{S}(h)\left(\delta_{\sum_{k=1}^{m} \beta_{i,k} \cdot \delta_{x_k}}\right)\right) \tag{3.3}$$

$$= (h \circ \mathfrak{S}(h))\left(\sum_{i=1}^{n} \alpha_i \cdot \delta_{\sum_{k=1}^{m} \beta_{i,k} \cdot \delta_{x_k}}\right) \tag{3.4}$$

$$= (h \circ \mathfrak{m}_X)\left(\sum_{i=1}^{n} \alpha_i \cdot \delta_{\sum_{k=1}^{m} \beta_{i,k} \cdot \delta_{x_k}}\right) \tag{3.5}$$

$$= h\left(\sum_{i=1}^{n} \alpha_i \cdot \mathfrak{m}_X\left(\delta_{\sum_{k=1}^{m} \beta_{i,k} \cdot \delta_{x_k}}\right)\right) \tag{3.6}$$

$$= h\left(\sum_{i=1}^{n} \alpha_i \cdot \left(\sum_{k=1}^{m} \beta_{i,k} \cdot \delta_{x_k}\right)\right) \tag{3.7}$$

$$= h\left(\sum_{k=1}^{m} \left(\sum_{i=1}^{n} \alpha_i \cdot \beta_{i,k}\right) \delta_{x_k}\right) \tag{3.8}$$

$$= \sum_{k=1}^{m} \left(\sum_{i=1}^{n} \alpha_i \cdot \beta_{i,k}\right) x_k. \tag{3.9}$$

The equations (3.1) and (3.2) reflect the definition of the structure, equation (3.3) applies $\delta_{h(\tau)} = \mathfrak{S}(h)(\delta_\tau)$, equation (3.4) uses the linearity of $\mathfrak{S}(h)$

according to Lemma 3.1, equation (3.5) is due to h being an algebra. Winding down, equation (3.6) uses Lemma 3.1 again, this time for $\mathfrak{m}_X$, equation (3.7) uses that $\mathfrak{m}_X \circ \delta_\tau = \tau$, equation (3.8) is just rearranging terms, and equation (3.9) is the definition again. $\qquad\square$

Let conversely such a positive convex structure be given. We show that we can define a G-triplet from it. Let

$$\mathcal{T}_X := \{\sum_{i=1}^n \alpha_i \cdot \delta_{x_i} \mid n \in \mathbb{N}, x_1, \ldots, x_n \in X, \langle \alpha_1, \ldots, \alpha_n \rangle \in \Omega\},$$

then $\mathcal{T}_X$ is dense in $\mathfrak{S}(X)$. Put

$$h_0\left(\sum_{i=1}^n \alpha_i \cdot \delta_{x_i}\right) := \sum_{i=1}^n \alpha_i \cdot x_i,$$

then $h_0 : \mathcal{T}_X \to X$ is well defined. This is so since

$$\sum_{i=1}^n \alpha_i \cdot \delta_{x_i} = \sum_{j=1}^m \alpha'_j \cdot \delta_{x'_j}$$

implies that

$$\sum_{i=1, \alpha_i \neq 0}^n \alpha_i \cdot \delta_{x_i} = \sum_{j=1, \alpha'_j \neq 0}^m \alpha'_j \cdot \delta_{x'_j},$$

hence given i with $\alpha_i \neq 0$ there exists j with $\alpha'_j \neq 0$ such that $x_i = x'_j$ and vice versa. Consequently,

$$\sum_{i=1}^n \alpha_i \cdot x_i = \sum_{i=1, \alpha_i \neq 0}^n \alpha_i \cdot x_i = \sum_{j=1, \alpha'_j \neq 0}^n \alpha'_j \cdot x'_j = \sum_{j=1}^n \alpha'_j \cdot x'_j$$

is inferred from the properties of positive convex structures.

The map h_0 is uniformly continuous, because

$$d\left(h_0(\sum_{i=1}^n \alpha_i \cdot \delta_{x_i}), h_0(\sum_{j=1}^m d_j \cdot \delta_{y_j})\right) \leq \mathbf{d}_P\left(\sum_{i=1}^n \alpha_i \cdot \delta_{x_i}, \sum_{j=1}^m d_j \cdot \delta_{y_j}\right),$$

$\mathbf{d}_P$ denoting the Prohorov metric; see Section 1.5. We need uniform continuity here, because it is well known that otherwise a unique, continuous extension from the dense subset of discrete measures to the set of all measures cannot be guaranteed.

Define ρ_0 as the kernel of h_0, then ρ_0 is a smooth equivalence relation on $\mathcal{T}_X$, and it is not difficult to see that the set of topological closures

$$\{\mathsf{cl}\left([t]_{\rho_0}\right) \mid t \in \mathcal{T}_X\}$$

forms a partition of $\mathfrak{S}(X)$ through the following arguments:

a. the closures of different equivalence classes are disjoint,

b. given $\mu \in \mathfrak{S}(X)$, one can find a sequence $(t_n)_{n \in \mathbb{N}}$ in $\mathcal{T}_X$ with $t_n \rightharpoonup_w \mu$. Since X is Polish, in particular complete, the sequence $(h_0(t_n))_{n \in \mathbb{N}}$ converges to some t_0, and because h_0 is uniformly continuous, one concludes that $\mu \in \mathsf{cl}\left([t_0]_{\rho_0}\right)$. Thus each member of $\mathfrak{S}(X)$ is in some class.

This yields an equivalence relation ρ on $\mathfrak{S}(X)$. Uniform continuity of h_0 gives a unique continuous extension h of h_0 to $\mathfrak{S}(X)$, thus ρ equals the kernel of h, hence ρ is a smooth equivalence relation, and it is evidently positive convex. Defining on $\mathfrak{S}(X)/\rho$ the metric

$$D([\mu_1]_\rho, [\mu_2]_\rho) := d(h(\mu_1), h(\mu_2)),$$

it is rather immediate that the metric space $(\mathfrak{S}(X)/\rho, D)$ is homeomorphic to X with metric d, and that the topology induced by the metric is just the final topology with respect to the weak topology on $\mathfrak{S}(X)$ and η_ρ.

It is clear that each affine and continuous map between positive convex structures gives rise to a morphism between the corresponding G-triplets, and vice versa. Thus we have established:

PROPOSITION 3.20

The category of 𝔄𝔩𝔤 *of algebras for the Giry monad is isomorphic to the category* 𝔖𝔱𝔯𝔠𝔬𝔫𝔳 *of positive convex structures with continuous affine maps as morphisms.*

For the probability functor we again mirror the development, but this time we need not go into details. We obtain eventually this characterization for the category 𝔭𝔄𝔩𝔤 of algebras for the Giry monad, when restricted to the probability functor (with the obvious necessary adjustments made for morphisms):

PROPOSITION 3.21

The category of algebras for the Giry monad for the probability functor is isomorphic to the full subcategory of continuous convex structures.

For a partial history of this result see the Bibliographic Notes at the end of this chapter.

3.5 Examples

We illustrate the concept and propose some examples by looking at some well-known situations. Most of this section is not really new, probably apart from the proposed point of view. We first show that the monad carries for each Polish space an instance of an algebra with it. Then we prove that in the finite case an algebra exists only in the case of a singleton set. Finally a geometrically oriented example is discussed by investigating the barycenter of a probability on a compact and convex subset of $\mathbb{R}^n$.

In each case the geometry of the underlying space imposes a natural positive convex structure, and it invites itself to compare this structure with the one that can be constructed through the algebra. It turns out in each of these cases that the convex structure associated with the algebra is the natural one.

3.5.1 Monad Multiplication

We know that $\langle \mathfrak{S}(X), \mathfrak{m}_X \rangle$ is an algebra whenever X is a Polish space; this is actually a special case of the observation that $\langle \mathfrak{T}(x), \mathfrak{m}_x \rangle$ is a $\mathfrak{T}$-algebra in any monad $(\mathfrak{T}, \mathfrak{e}, \mathfrak{m})$; see Lemma 1.109. It shows for this specific case that each Polish space is associated in a natural fashion with a strongly convex structure. This association entails actually more than meets the eye: we will show in Section 3.6 that $X \mapsto \langle \mathfrak{S}(X), \mathfrak{m}_X \rangle$ is the object part of the left adjoint to the forgetful functor $\mathfrak{Alg} \to \mathfrak{Pol}$.

PROPOSITION 3.22
The pair $\langle \mathfrak{S}(X), \mathfrak{m}_X \rangle$ is an algebra for each Polish space X.

Since

$$\mathfrak{m}_X(\alpha_1 \cdot \mu_1 + \cdots + \alpha_n \cdot \mu_n) = \alpha_1 \cdot \mathfrak{m}_X(\mu_1) + \cdots + \alpha_n \cdot \mathfrak{m}_X(\mu_n),$$

the positive convex structure induced on $\mathfrak{S}(X)$ by this algebra is the natural one.

3.5.2 The Finite Case

The finite case can also easily be characterized: there are no algebras for $\{1, \ldots, n\}$ unless $n = 1$. This will be shown now. Since the base space needs to be connected for entertaining an algebra, we obtain a simple geometric description as a necessary condition for the existence of algebras as a byproduct.

We need a wee bit elementary topology for this.

DEFINITION 3.23   *A metric space A is called* connected *iff the decomposition $A = A_1 \cup A_2$ with disjoint open sets A_1, A_2 implies $A_1 = \emptyset$ or $A_2 = \emptyset$.*

Thus a connected space cannot be decomposed into two nontrivial open sets, so that the only clopen sets are the empty set and the space itself. The connected subspaces of the real line $\mathbb{R}$ are just the open, half-open or closed finite or infinite intervals. The rational numbers $\mathbb{Q}$ are not connected. A subset $\emptyset \neq A \subseteq \mathbb{N}$ of the natural numbers which carries the discrete topology (because we assume that it is a Polish space) is connected as a subspace iff $A = \{n\}$ for some $n \in \mathbb{N}$.

The following elementary facts about connected spaces are well known and readily established.

LEMMA 3.24

Let A be a metric space.

a. *If A is connected, and $f : A \to B$ is a continuous and surjective map to another metric space B, then B is connected.*

b. *If two arbitrary points in A can be joined through a connected subspace of A, then A is connected.*

This has as a consequence a geometric description of the space underlying a monad.

COROLLARY 3.25

If $\langle X, h \rangle$ is an algebra for the Giry monad, then X is connected.

PROOF If $\mu_1, \mu_2 \in \mathfrak{S}(X)$ are arbitrary probability measures on X, then the line segment $\{c \cdot \mu_1 + (1 - c) \cdot \mu_2 \mid 0 \leq c \leq 1\}$ is a connected subspace which joins μ_1 and μ_2. This is so because it is the image of the connected unit interval $[0, 1]$ under the continuous map $c \mapsto c \cdot \mu_1 + (1 - c) \cdot \mu_2$. Thus $\mathfrak{S}(X)$ is connected by Lemma 3.24. Since h is onto, its image X is connected. $\quad\square$

Consequently it is hopeless to search for algebras for, say, the natural numbers or a nontrivial subset of it:

PROPOSITION 3.26

A subspace $A \subseteq \mathbb{N}$ has an algebra for the Giry monad iff A is a singleton set.

PROOF It is clear that a singleton set has an algebra. Conversely, if A has an algebra, then A is connected by Lemma 3.25, and this can only be the case when A is a singleton. $\quad\square$

3.5.3   The Unit Interval

The next example deals with the unit interval.

PROPOSITION 3.27
The map

$$h : \mathfrak{S}\left([0,1]\right) \ni \mu \mapsto \int_0^1 t \; \mu(dt) \in [0,1]$$

defines an algebra $\langle [0,1] , h \rangle$.

PROOF   In fact, $h(\mu) \in [0,1]$ because μ is a subprobability measure. It is clear that $h(\delta_x) = x$ holds, and — by the very definition of the weak topology — that $\mu \mapsto h(\mu)$ is continuous. Thus it remains to show by Proposition 3.6 that the partition induced by h is positive convex. This is a fairly simple calculation. Consequently, the partition induced by h is a G-partition, showing that h is indeed the morphism part of an algebra.

It is not difficult to see that the positive convex structure induced on $[0,1]$ is the natural one.

$$\square$$

This is the only algebra that has an integral representation through Lebesgue measure: suppose that

$$h^*(\mu) = \int_0^1 f(t) \; \mu(dt)$$

for some continuous f. Then $h^*(\delta_x) = f(x)$, from which $f(x) = x$ is inferred for each $x \in [0,1]$.

3.5.4   Barycenter

The final example has a more geometric touch to it and deals only with the probabilistic case. We work with bounded and closed subsets of some Euclidean space and show that the construction of a barycenter yields an algebra. Fix $X \subseteq \mathbb{R}^n$ as a bounded, closed and convex subset of the Euclidean space $\mathbb{R}^n$ (for example, X could be a closed ball or a cube in $\mathbb{R}^n$).

Denote for two vectors $x, x' \in \mathbb{R}^n$ by

$$x \bullet x' := \sum_{i=1}^n x_i \cdot x_i'$$

their inner product. Then $\lambda x . x \bullet x'$ constitutes a continuous linear map on $\mathbb{R}^n$ for fixed x'. In fact, each linear functional on $\mathbb{R}^n$ can be represented in this way.

DEFINITION 3.28   *The vector $x^* \in \mathbb{R}^n$ is called a* barycenter *of the probability measure $\mu \in \mathfrak{P}(X)$ iff*

$$x \bullet x^* = \int_X x \bullet y \; \mu(dy)$$

holds for each $x \in X$.

Because X is compact, the integrand is bounded on X, thus the integral is always finite.

Since the proofs for the existence and membership properties of a barycenter would lead us too far from our path of investigating stochastic relations (we would have to study the geometry of compact convex sets), we refer the reader to the literature. Basic facts about barycenters can be found, e.g., in the massive overview of measure theory assembled by Fremlin (Fremlin, 2003).

PROPOSITION 3.29

The barycenter of $b(\mu)$ of $\mu \in \mathfrak{P}(X)$ exists, it is uniquely determined, and it is an element of X. $\langle X, b \rangle$ is an algebra for the Giry monad.

PROOF   0. Once we know that the barycenter exists, uniqueness follows from the well-known fact that the linear functionals on $\mathbb{R}^n$ separate points. The existence of the barycenter is established in (Fremlin, 2003, Theorem 461 E); its membership in X follows from (Fremlin, 2003, Theorem 461 H). Granted this, the proof that $\langle X, h \rangle$ is an algebra is sketched now along the following lines.

1. From the construction and the uniqueness of the barycenter it is clear that $b(\delta_x) = x$ holds for each $x \in X$.

2. Assume that $(\mu_n)_{n \in \mathbb{N}}$ is a sequence in $\mathfrak{P}(X)$ with $\mu_n \rightharpoonup_w \mu_0$. Put $x_n^* := b(\mu_n)$ as the barycenter of μ_n, then $(x_n^*)_{n \in \mathbb{N}}$ is a sequence in the compact set X, thus has a convergent subsequence (which we take w.l.g. as the sequence itself). Let x_0^* be its limit. Then we have for all $x \in X$:

$$x \bullet x_n^* = \int_X x \bullet y \; \mu_n(dy) \to \int_X x \bullet y \; \mu_0(dy) = x \bullet x_0^*.$$

Hence b is continuous. Approximating μ through a convex combination of discrete measures, the above argumentation together with the convexity of X shows also that $b(\mu) \in X$.

3. It remains to show that the partition induced by b is convex. This, however, follows immediately from the linearity of $y \mapsto \lambda x.x \bullet y$. ⬜

Calculating the convex structure for b, we infer from affinity of the integral as a function of the measure and from

$$x \bullet b(\mu) = \int_X x \bullet y \; \mu(dy)$$

that $(0 \leq c \leq 1, \mu_i \in \mathfrak{P}(X))$

$$b(c \cdot \mu_1 + (1-c) \cdot \mu_2) = c \cdot b(\mu_1) + (1-c) \cdot b(\mu_2)$$

that the convex structure induced by b is the natural one.

It should be mentioned that this example can be generalized considerably to metrizable topological vector spaces (Fuchssteiner and Lusky, 1981; Fremlin, 2003). The terminological effort is, however, somewhat heavy, and the example remains essentially the same. Thus we refrain from a more general discussion.

Although the characterization of algebras in terms of positive convex structures yields a somewhat uniform approach, it becomes clear from these examples that the specific instances of the algebras provide a rather colorful picture unified only through the common abstract treatment.

3.6 The Left Adjoint

The identification of the algebras for the Giry monad and the observation from Proposition 3.22 that $\langle \mathfrak{S}(X), \mathfrak{m}_X \rangle$ is always an algebra puts us in a position where we are able to identify the left adjoint for the forgetful functor $\mathfrak{U} : \mathfrak{Alg} \to \mathfrak{Pol}$. Define

$$\mathfrak{L}(X) := \langle \mathfrak{S}(X), \mathfrak{m}_X \rangle,$$

for a Polish space X, and put

$$\mathfrak{L}(f) := \mathfrak{S}(f),$$

for the continuous map $f : X \to Y$. Then we know from Proposition 3.22 that $\mathfrak{L}(X)$ is an algebra. From Lemma 3.1 we see that $\mathfrak{L}(f) : \mathfrak{L}(X) \to \mathfrak{L}(Y)$, is a morphism in $\mathfrak{Alg}$, and since $\mathfrak{m} : \mathfrak{S}^2 \overset{\bullet}{\to} \mathfrak{S}$ is a natural transformation, $\mathfrak{L}(f)$ is an algebra morphism. Thus $\mathfrak{L} : \mathfrak{Pol} \to \mathfrak{Alg}$ is a functor.

We will write as usual $\mathfrak{C}(a, b)$ for the morphisms $a \to b$ in category $\mathfrak{C}$.

LEMMA 3.30

Let $\theta : \mathfrak{L}(X) \to \langle Y, h \rangle$ be a morphism in $\mathfrak{Alg}$, and put $\Theta(\theta)(x) := \theta(\delta_x)$. This defines a bijection $\Theta : \mathfrak{Alg}(\mathfrak{L}(X), \langle Y, h \rangle) \to \mathfrak{Pol}(X, Y)$.

PROOF   1. Since $x \mapsto \delta_x$ defines a continuous map $X \to \mathfrak{S}(X)$, and since the morphisms in $\mathfrak{Alg}$ are continuous as well, $\Theta(\theta) \in \mathfrak{Pol}(X, Y)$ whenever $\theta \in \mathfrak{Alg}(\mathfrak{L}(X), \langle Y, h \rangle)$.

2. Now suppose that $\Theta(\theta_1)(x) = \Theta(\theta_2)(x)$ holds for all $x \in X$, thus $\theta_1(\delta_x) = \theta_2(\delta_x)$ for all $x \in X$. Let $\tau = \sum_{i=1}^{m} \alpha_i \cdot \delta_{x_i}$ be a discrete subprobability measure,

then

$$\theta_1(\tau) = \theta_1\left(\sum_{i=1}^{m}\alpha_i\cdot\delta_{x_i}\right) = \sum_{1\le i\le m}^{\mathcal{P}}\alpha_i\cdot\theta_1(\delta_{x_i}) = \sum_{1\le i\le m}^{\mathcal{P}}\alpha_i\cdot\theta_2(\delta_{x_i}) = \theta_2(\tau).$$

Here $\mathcal{P}$ is the positive convex structure associated with the algebra $\langle Y, h\rangle$ by Proposition 3.20. Thus θ_1 agrees with θ_2 on all discrete measures. Since these measures are dense in the weak topology, and since θ_1 as well as θ_2 are continuous, we may conclude that $\theta_1(\tau) = \theta_2(\tau)$ holds for all $\tau \in \mathfrak{S}(X)$. Thus Θ is injective.

3. Let $f : X \to Y$ be continuous, and put $\widetilde{\theta} := h \circ \mathfrak{S}(f)$, the composition being formed in $\mathfrak{Alg}$. We claim that $\widetilde{\theta} \in \mathfrak{Alg}(\mathfrak{L}(X), \langle Y, h\rangle)$.

In fact, consider the diagram

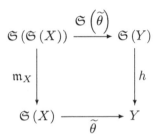

We have

$$\begin{aligned}
h \circ \mathfrak{S}\left(\widetilde{\theta}\right) &= h \circ \mathfrak{S}(h) \circ \mathfrak{S}(\mathfrak{S}(f)) \\
&= h \circ \mathfrak{m}_Y \circ \mathfrak{S}(\mathfrak{S}(f)) \quad \text{(because $\langle Y, h\rangle$ is an algebra)} \\
&= h \circ \mathfrak{S}(f) \circ \mathfrak{m}_X \quad \text{(since $\mathfrak{S}(f)$ is an $\mathfrak{Alg}$-morphism)} \\
&= \widetilde{\theta} \circ \mathfrak{m}_X
\end{aligned}$$

which implies that the diagram is commutative, establishing the claim. Since for each $x \in X$

$$\Theta(\widetilde{\theta})(x) = h(\mathfrak{S}(f)(\delta_x)) = h(\delta_{f(x)}) = f(x)$$

we conclude that $\Theta(\widetilde{\theta}) = f$, thus Θ is onto.   ⬜

In order to establish the properties of an adjunction, we need to establish the naturalness of $\Theta = \Theta_{X,\langle Y,h\rangle}$; see the remarks following Definition 1.103. This means that we have to establish the commutativity of the diagrams below, given the morphisms $f \in \mathfrak{Alg}(\langle Y, h\rangle, \langle Y', h'\rangle)$ and $g \in \mathfrak{Pol}(X', X)$.

The first diagram takes care of the covariant hom-set functor $\mathfrak{Alg}(\mathfrak{L}(X),-)$.

$$
\begin{array}{ccc}
\mathfrak{Alg}(\mathfrak{L}(X),\langle Y,h\rangle) & \xrightarrow{\ \Theta\ } & \mathfrak{Pol}(X,Y) \\
\Big\downarrow{\scriptstyle f_*} & & \Big\downarrow{\scriptstyle \mathfrak{U}(f)_*} \\
\mathfrak{Alg}(\mathfrak{L}(X),\langle Y',h'\rangle) & \xrightarrow[\ \Theta\ ]{} & \mathfrak{Pol}(X,Y)
\end{array}
$$

Here

$$f_* : \mathfrak{Alg}(\mathfrak{L}(X),\langle Y,h\rangle) \ni \theta \mapsto f\circ\theta \in \mathfrak{Alg}(\mathfrak{L}(X),\langle Y',h'\rangle)$$

is composition from the left, similarly $\mathfrak{U}(f)_*$; see Example 1.93. We see

$$\mathfrak{U}(f)_*(\Theta(\theta))(x) = (f\circ\Theta(\theta))(x) = f(\Theta(\theta)(x)) = f(\theta(\delta_x)),$$

and

$$\Theta(f_*(\theta))(x) = f_*(\theta)(x) = f(\theta(\delta_x)),$$

hence the diagram commutes. The second diagram takes care of the contravariant hom-set functor $\mathfrak{Alg}(-,\langle Y,h\rangle)$:

$$
\begin{array}{ccc}
\mathfrak{Alg}(\mathfrak{L}(X),\langle Y,h\rangle) & \xrightarrow{\ \Theta\ } & \mathfrak{Pol}(X,Y) \\
\Big\downarrow{\scriptstyle \mathfrak{L}(f)^*} & & \Big\downarrow{\scriptstyle f^*} \\
\mathfrak{Alg}(\mathfrak{L}(X'),\langle Y,h\rangle) & \xrightarrow[\ \Theta\ ]{} & \mathfrak{Pol}(X',Y)
\end{array}
$$

Here

$$f^* : \mathfrak{Pol}(X,Y) \ni g \mapsto g\circ f \in \mathfrak{Pol}(X',Y)$$

is composition from the right, similarly for $\mathfrak{L}(f)^*$. Because

$$f^*(\Theta(\theta))(x') = (\Theta(\theta)\circ f)(x') = \theta(\delta_{f(x')}),$$

and since

$$\Theta(\mathfrak{L}(f)^*(\theta))(x') = \mathfrak{L}(f)^*(\theta)(\delta_{x'}) = (\theta\circ\mathfrak{S}(f))(\delta_{x'}) = \theta(\delta_{f(x')})$$

we see that this diagram commutes as well.

Summarizing, we have established:

PROPOSITION 3.31

The functor $\mathfrak{L} : \mathfrak{Pol} \to \mathfrak{Alg}$ with $\mathfrak{L}(X) := \langle \mathfrak{S}(X), \mathfrak{m}_X\rangle$ and $\mathfrak{L}(f) := \mathfrak{S}(f)$ is left adjoint to the forgetful functor $\mathfrak{U} : \mathfrak{Alg} \to \mathfrak{Pol}$.

The probabilistic case is dealt with using the same arguments. The only essential place where the difference between subprobability measures and probability measures enters the discussion formally is in the proof of Lemma 3.30. Proving surjectivity of Θ, one has to take a convex combination of discrete measures, rather than a positive convex combination, as in the proof above. With this minor adjustment all proofs carry over verbatim.

We obtain for the category $\mathfrak{pAlg}$ of algebras for the probabilistic version of the Giry monad (the category has been introduced in Proposition 3.21).

PROPOSITION 3.32

The functor $\mathfrak{L}_{\text{prob}} : \mathfrak{Pol} \to \mathfrak{pAlg}$ *with* $\mathfrak{L}_{\text{prob}}(X) := \langle \mathfrak{P}(X), \mathfrak{m}_X \rangle$ *and* $\mathfrak{L}_{\text{prob}}(f) := \mathfrak{P}(f)$ *is left adjoint to the forgetful functor* $\mathfrak{U} : \mathfrak{pAlg} \to \mathfrak{Pol}$.

Hence the forgetful functor on the algebras for the Giry monad has the subprobability functor resp. the probability functor, both augmented by the monad's multiplication, as a left adjoint. This emphasizes the close ties between positive convex resp. convex structures and probabilities and sheds further light on these functors. It also adds a formal underpinning to the intuitive understanding prevailing in Computer Science, which often expresses the probability of an outcome as a convex combination of all the possible outcomes, see, e.g., (Morgan et al., 1996; van Breugel et al., 2002) for accounts in different fields. The interplay between convexity and probability is strikingly present for example in Heckmann's work (Heckmann, 1994) (in fact, he often interchanges both), but surprisingly not made explicit.

It seems that — roughly speaking — the rôle played by ordered structures in the context of the power set monad translates into one for positive convex structures for the subprobability based Giry monad.

3.7   Bibliographic Notes

The characterization of the algebras for the probability functor through convex structures has been known for the case that X is a compact Hausdorff space (Fedorchuk, 1991, 2.14) (the attribution to Swirszcz's work (Swirszcz, 1974) in (Fedorchuk, 1991) is slightly unclear). The methods for the proof are, however, rather different: the compact case makes essential use of the right adjoint of the probability functor, seen as a functor between the respective categories of compact Hausdorff spaces and compact convex sets. Thus Corollary 3.21 generalizes the known characterization to Polish spaces. In the subprobabilistic case that is of interest here convexity is evidently not strong enough and has to be replaced by positive convexity. The positive convex structures that have been of use in other investigations could be put to use here as well; we took the definition from Pumplün's work (Pumplün, 2003).

Chapter 4

The Existence of Semi-Pullbacks

4.1 Introduction

A category is said to have semi-pullbacks if, whenever $f : a \to b, g : c \to b$ is a pair of morphisms with the same target, there exists in the category an object d together with morphisms $r : d \to a, s : d \to c$ such that $f \circ r = g \circ s$. The existence of semi-pullbacks makes sure that bisimulations, which are defined as spans of morphisms, are transitive; bisimulations will be introduced in Chapter 5.

For stochastic relations, semi-pullbacks will be helpful when investigating the relationship between modal logics and bisimilar stochastic Kripke models. The argumentation goes like this: given two stochastic Kripke models $\mathcal{K}_1$ and $\mathcal{K}_2$ that accepts exactly the same formulas, we construct a third Kripke model $\mathcal{L}$ and morphisms Φ_1 and Φ_2 that form a co-span

$$\mathcal{K}_1 \xrightarrow{\;\Phi_1\;} \mathcal{L} \xleftarrow{\;\Phi_2\;} \mathcal{K}_2.$$

If the category under consideration has semi-pullbacks, then the upper left corner of the diagram below may be completed

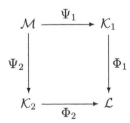

From this a bisimulation is constructed. The crucial point is not the construction of $\mathcal{L}$ (which nevertheless is quite involved in its own right) but rather the

existence of the semi-pullback, i.e., the existence of the object $\mathcal{M}$ together with the morphisms Ψ_1 and Ψ_2. This argumentation will be refined in Chapter 5.

The present chapter investigates the construction of semi-pullbacks in the category of stochastic relations over Polish resp. analytic spaces. The task of construction is rather formidable and requires some nontrivial constructions, using such marvels as the Hahn-Banach Theorem and the Riesz-Representation Theorem. It capitalizes on Alexandrov's famous embedding of a Polish space as a G_δ-set into the Hilbert cube (see Theorem 1.23) — the gist being that a Polish space can be embedded as a measurable set with compact closure into a compact set. As useful byproducts we obtain several extensions of stochastic relations from smaller to larger σ-algebras.

The problem of establishing the existence of semi-pullbacks has been solved for analytic spaces with universally measurable transition functions by A. Edalat (Edalat, 1999). We will refer occasionally to Edalat's approach and comment on the differences in the Bibliographic Notes at the end of this chapter. This positive result is complemented by a consideration of extending semi-pullbacks to pullbacks or at least to weak pullbacks. It would be nice if that could be brought to work, since then various techniques that have been useful for coalgebras could be made available also for the stochastic case. Unfortunately, this cannot be the case: we conclude with the negative result that not even weak pullbacks do exist for stochastic relations. Thus we do have to invent our own techniques for exploring, e.g., simple systems, rather than imitate and adapt the machinery from coalgebra.

Once the general availability of semi-pullbacks is ensured, a Hennessy-Milner Theorem on the equivalence of accepting the same formulas and bisimilarity can be established for a general class of modal logics; see Chapter 6, in particular Section 6.2. The investigation of bisimilarity of stochastic relations is facilitated through this result. For example it can be shown that two stochastic relations are bisimilar provided they have isomorphic factors; the converse holds under the assumption of compactness as well, so that one might speculate whether bisimilarity and having isomorphic factors are equivalent. This will be discussed in greater detail in Section 5.4.

4.2   A Road Map

In order to prepare for things to come, and to provide an antidote to getting the feeling that one gets lost in the measure-theoretic jungle, we will first discuss the problem and an outline of its solution in Section 4.2. At the heart of the solution lies a measure extension which is provided in Section 4.3, and Section 4.4 constructs a solution to the existence of semi-pullbacks for stochastic relations over Polish and then over analytic spaces. This looks somewhat unrelated to the problem at hand, but it turns out that the key to solving the existence problem will be just this. The reason will become apparent soon.

Let $\mathsf{K} = (X, Y, K)$ be a stochastic relation over the Polish spaces X and Y. Assume that K is the target of two morphisms

$$\mathsf{K}_1 \xrightarrow{\ \mathsf{f}_1\ } \mathsf{K} \xleftarrow{\ \mathsf{f}_2\ } \mathsf{K}_2$$

with, say, $\mathsf{f}_i = (\phi_i, \psi_i)$. We are looking for a stochastic relation L and two morphisms

$$\mathsf{K}_1 \xleftarrow{\ \mathsf{g}_1\ } \mathsf{L} \xrightarrow{\ \mathsf{g}_2\ } \mathsf{K}_2$$

such that $\mathsf{f}_1 \circ \mathsf{g}_1 = \mathsf{f}_2 \circ \mathsf{g}_2$ holds.

An expansion of the first flat diagram in terms of the defining properties yields the following commutative diagram:

$$
\begin{array}{ccccc}
X_1 & \xrightarrow{\ \phi_1\ } & X & \xleftarrow{\ \phi_2\ } & X_2 \\
{\scriptstyle K_1}\big\downarrow & & {\scriptstyle K}\big\downarrow & & \big\downarrow{\scriptstyle K_2} \\
\mathfrak{S}(Y_1) & \xrightarrow[\mathfrak{S}(\psi_1)]{} & \mathfrak{S}(Y) & \xleftarrow[\mathfrak{S}(\psi_2)]{} & \mathfrak{S}(Y_2)
\end{array}
$$

Written as a comprehensive diagram, the second flat diagram entails that in addition to $\mathsf{f}_1 \circ \mathsf{g}_1 = \mathsf{f}_2 \circ \mathsf{g}_2$ these diagrams should commute:

$$
\begin{array}{ccccc}
X_1 & \xleftarrow{\ \alpha_1\ } & A & \xrightarrow{\ \alpha_2\ } & X_2 \\
{\scriptstyle K_1}\big\downarrow & & {\scriptstyle L}\big\downarrow & & \big\downarrow{\scriptstyle K_2} \\
\mathfrak{S}(Y_1) & \xleftarrow[\mathfrak{S}(\beta_1)]{} & \mathfrak{S}(B) & \xrightarrow[\mathfrak{S}(\beta_2)]{} & \mathfrak{S}(Y_2)
\end{array}
$$

Here $\mathsf{L} = (A, B, L)$ is the relation involved, and $\mathsf{g}_i = (\alpha_i, \beta_i)$ is the morphism $\mathsf{g}_i : \mathsf{L} \to \mathsf{K}_i$ for $i = 1, 2$. We will define

$$A := \{\langle x_1, x_2\rangle \in X_1 \times X_2 \mid \phi_1(x_1) = \phi_2(x_2)\}$$
$$B := \{\langle y_1, y_2\rangle \in Y_1 \times Y_2 \mid \psi_1(y_1) = \psi_2(y_2)\},$$

then we will argue why A and B may be assumed to be Polish. Taking α_i, β_i as the projections

$$\alpha_i : A \ni \langle x_1, x_2\rangle \mapsto x_i \in X_i,$$
$$\beta_i : B \ni \langle y_1, y_2\rangle \mapsto y_i \in Y_i,$$

we need to find a stochastic relation $\mathsf{L} = (A, B, L)$ which makes $\mathbf{g}_i := (\alpha_i, \beta_i)$ into a morphism $\mathbf{g}_i : \mathsf{L} \to \mathsf{K}_i$ for $i = 1, 2$. Thus $L : A \rightsquigarrow B$ should satisfy the following constraints for all $\langle x_1, x_2 \rangle \in A$:

a. $L(x_1, x_2) \in \mathfrak{S}(B)$,

b. $\mathfrak{S}(\beta_1)(L(x_1, x_2)) = K_1(x_1)$,

c. $\mathfrak{S}(\beta_2)(L(x_1, x_2)) = K_2(x_2)$.

Reformulating again, we put

$$\Gamma(x_1, x_2) := \{\mu \in \mathfrak{S}(B) \mid \mathfrak{S}(\beta_1)(\mu) = K_1(x_1) \text{ and } \mathfrak{S}(\beta_2)(\mu) = K_2(x_2)\},$$

hence $L(x_1, x_2) \in \Gamma(x_1, x_2)$ for all $\langle x_1, x_2 \rangle \in A$. We want $L : A \to \mathfrak{S}(B)$ to be a measurable selector for Γ, thus $L : A \rightsquigarrow B$ is a stochastic relation (accounting for the *measurable* in measurable selector) such that $\forall a \in A :$ $L(a) \in \Gamma(a)$ (accounting for the *selector*).

Thus the problem is massaged into finding a measurable selector for the set-valued map Γ. The existence of such a selector can be asserted under the following conditions:

a. $\Gamma(x_1, x_2)$ is a closed subset of $\mathfrak{S}(B)$ for each $\langle x_1, x_2 \rangle \in A$,

b. the set $\{\langle x_1, x_2 \rangle \in A \mid \Gamma(x_1, x_2) \cap C \neq \emptyset\}$ is a measurable subset of A, whenever $C \subseteq \mathfrak{S}(B)$ is compact,

c. $\Gamma(x_1, x_2) \neq \emptyset$ for each and any $\langle x_1, x_2 \rangle \in A$.

These properties characterize Γ as a $\mathcal{C}$-measurable relation; see Section 1.4.

It will not be difficult to show that property a is satisfied, and property b will also easily be seen to be fulfilled. The real crucial property is the third one.

The Crucial Point. Fix $\langle x_1, x_2 \rangle \in A$. We will find without much ado a measure $\mu_0 \in \mathfrak{S}(B)$ such that these equations

$$\mathfrak{S}(\beta_1)(\mu_0)(E_1) = K_1(x_1)(E_1) \tag{4.1}$$
$$\mathfrak{S}(\beta_2)(\mu_0)(E_2) = K_2(x_2)(E_2) \tag{4.2}$$

are true whenever $E_i = \psi_i^{-1}[F_i]$ for some Borel set $F_i \subseteq Y$.

This is equivalent to saying that the measures $\mathfrak{S}(\beta_i)(\mu_0)$ and $K_i(x_i)$ coincide on the σ-algebra $\psi_i^{-1}[\mathcal{B}(Y)]$. The latter is usually a proper sub-σ-algebra of $\mathcal{B}(Y_i)$, hence we cannot guarantee offhand equality in the equations 4.1 and 4.2 on the full Borel sets $\mathcal{B}(Y_i)$ with $i = 1, 2$. If, however, we can find a measure, say $\mu_1 \in \mathfrak{S}(B)$, such that $\mathfrak{S}(\beta_i)(\mu_1)(E_i) = K_i(x_i)(E_i)$ for all E_i ranging OVER ALL OF $\mathcal{B}(Y_i)$, we may conclude $\mu_1 \in \Gamma(x_1, x_2)$, and then the latter set is in fact shown to be nonempty.

Thus the problem is reduced to finding a single measure on B that has suitable marginal distributions. We will demonstrate that we can extend the marginal distributions of μ_0 to a measure with the desired properties.

This measure extension can be done in two different ways, depending on the nature of the space Y which otherwise sits quietly in the background and serves patiently as a target space.

1. If Y is Polish, we can and do refer to a special case of Edalat's result, and we are done. This precludes, however, a more general solution that includes analytic spaces.

2. If Y is more generally a separable metric space, then still an extension can be found; this assumption will be crucial when the analytic case is targeted. The machinery from mathematical analysis for tackling this general case is well established, but somewhat heavy, including the Riesz Representation Theorem and the Hahn-Banach Theorem.

It is the latter way we will propose to go here because it opens up the avenue of establishing this result also for analytic spaces.

Discussion. The solution provided by Edalat can be considered constructive: it works with conditional distributions and their properties. These distributions do exist essentially due to the Radon-Nikodym Theorem which in turn leans heavily on the way the Lebesgue-Daniell integral is constructed. The basic idea for the more general solution lies in a measure extension process. It works in three steps: the first is to consider the integral of the given measure, which is a linear functional, then to formulate the extension problem in terms of extending this linear functional to a linear functional on a broader domain that corresponds to our needs, and finally to represent the extended functional as an integral again. Representing the measure as an integral is easy, since only the standard techniques of integration are involved; doing the extension is technically a bit more involved and requires the Hahn-Banach Theorem, and converting the linear functional back to a measure can be done through the Riesz Representation Theorem. It is this last step that gives us the measure we are looking for.

Through the use of the Hahn-Banach Theorem we make indirectly use of Zorn's Lemma, which in turn is known to be equivalent to the Axiom of Choice. Thus we propose a stronger solution but we pay more for it. This sounds probably a bit more dramatic and unusual than it really is: when constructing the product of an arbitrary family of sets or when looking into the ultrafilter extension of a Kripke model (Blackburn et al., 2001, Theorem 5.38) we silently take the Axiom of Choice for granted.

The Road Map. We will first delve into the problem of extending a measure with given marginal distributions; this is done in Section 4.3. Then we

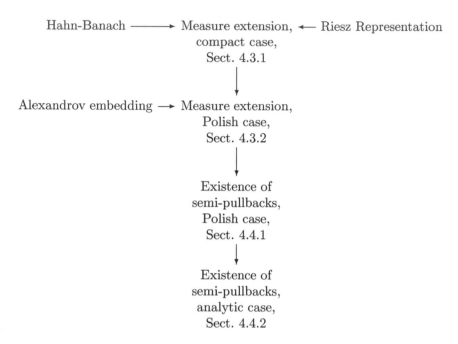

Figure 4.1: The Road Map

are in a good condition for tackling the problem along the lines sketched here in Section 4.4; see Figure 4.1.

4.3   Extending Semi-Pullbacks of Measures

The main argument in establishing the existence of a semi-pullback in the category of stochastic relations will be a selection argument: we will show that a certain set-valued map will have a (measurable) selector. This will require that this map always takes nonempty values. This section will be devoted to establishing a property of semi-pullbacks for measure spaces which in turn will be crucial for proving nonemptiness. Since it is rather technical in nature, it is convenient to encapsulate this development into a separate section.

We will consider the category $\mathfrak{Prob}$ of probability spaces which has as objects tuples $(X, \mathcal{A}, \mu)$ with $\mu \in \mathfrak{P}(X, \mathcal{A})$ for the measurable space $(X, \mathcal{A})$. Because for some arguments the σ-algebra is crucial, we abstain from the convention of dropping it from the notation for the space and write it down explicitly again. $\psi : (X, \mathcal{A}, \mu) \to (Y, \mathcal{B}, \nu)$ is a morphism in $\mathfrak{Prob}$ if $\psi : X \to Y$ is a surjective and $\mathcal{A} - \mathcal{B}$-measurable map which is measure preserving, i.e.,

$\nu = \mathfrak{P}(\psi)(\mu)$ holds. $\mathfrak{Prob}$ is closed under forming products (see Section 1.5), in particular it contains with two objects $(X, \mathcal{A}, \mu)$ and $(Y, \mathcal{B}, \nu)$ their product $(X \times Y, \mathcal{A} \otimes \mathcal{B}, \mu \otimes \nu)$, with $\mu \otimes \nu$ as the product measure which is uniquely determined through $(\mu \otimes \nu)(A \times B) = \mu(A) \cdot \nu(B)$.

We fix for the discussion the Polish spaces X_1 and X_2 with the respective Borel sets as σ-algebras. $(Z, \mathcal{C})$ is assumed to be a separable measurable space. Recall that $\mathcal{F}(X, \mathcal{A})$ is the linear space of all $\mathcal{A}$-$\mathcal{B}(\mathbb{R})$-measurable bounded functions $f : X \to \mathbb{R}$; $\mathcal{A} \mapsto \mathcal{F}(X, \mathcal{A})$ is monotone, hence $\mathcal{A} \subseteq \mathcal{B}$ implies that $\mathcal{F}(X, \mathcal{A})$ is a linear subspace of $\mathcal{F}(X, \mathcal{B})$.

Now let

$$(X_1, \mathcal{B}(X_1), \mu_1) \xrightarrow{\psi_1} (Z, \mathcal{C}, \nu) \xleftarrow{\psi_2} (X_2, \mathcal{B}(X_2), \mu_2)$$

be a pair of morphisms in $\mathfrak{Prob}$ with a common target, and assume that

$$(*) \quad \begin{array}{ccc} (S, \mathcal{A}, \theta) & \xrightarrow{\pi_2} & (X_2, \mathcal{B}(X_2), \mu_2) \\ \pi_1 \downarrow & & \downarrow \psi_2 \\ (X_1, \mathcal{B}(X_1), \mu_1) & \xrightarrow[\psi_1]{} & (Z, \mathcal{C}, \nu) \end{array}$$

is a semi-pullback diagram in $\mathfrak{Prob}$ with

$$\begin{aligned} S &:= \{\langle x_1, x_2 \rangle \mid \psi_1(x_1) = \psi_2(x_2)\} \in \psi_1^{-1}[\mathcal{C}] \otimes \psi_2^{-1}[\mathcal{C}] \\ \mathcal{A} &:= S \cap (\psi_1^{-1}[\mathcal{C}] \otimes \psi_2^{-1}[\mathcal{C}]) \\ &= S \cap (\psi_1 \times \psi_2)^{-1}[\mathcal{C} \otimes \mathcal{C}]. \end{aligned}$$

The π_i are again the projections. The last equality addressing $\mathcal{A}$ holds by Corollary 1.46; thus $\mathcal{A}$ is the smallest σ-algebra on S which makes

$$\psi_1 \times \psi_2 : \langle x_1, x_2 \rangle \mapsto \langle \psi_1(x_1), \psi_2(x_2) \rangle$$

measurable.

S is a Borel set, and the crucial step in the technical development will consist in "lifting" this pullback so that the object $(S, \mathcal{B}(S), \mu)$ for some suitable $\mu \in \mathfrak{P}(S, \mathcal{B}(S))$ stands in the upper left corner of the diagram. The essential difference is in the σ-algebras on S: starting with the initial σ-algebra with respect to $\psi_1 \times \psi_2$ we claim that we can find a measure μ on the Borel sets of S so that the properties of a semi-pullback will be preserved.

PROPOSITION 4.1

The semi-pullback $(*)$ *in* $\mathfrak{Prob}$ *may be extended to a semi-pullback*

$$
\begin{array}{ccc}
(S, \mathcal{B}(S), \mu) & \xrightarrow{\;\pi_2\;} & (X_2, \mathcal{B}(X_2), \mu_2) \\
\Big\downarrow{\scriptstyle \pi_1} & & \Big\downarrow{\scriptstyle \psi_2} \\
(X_1, \mathcal{B}(X_1), \mu_1) & \xrightarrow[\;\psi_1\;]{} & (Z, \mathcal{C}, \nu)
\end{array}
$$

in $\mathfrak{Prob}$.

This entails essentially an extension process, extending $\theta \in \mathfrak{P}(S, \mathcal{A})$ to a suitable $\mu \in \mathfrak{P}(S, \mathcal{B}(S))$. We establish the existence of this extension in two steps. The first step will assume that X_1 and X_2 are compact Polish spaces, and the second will show how to reduce the general case to the compact one.

We will need to make precise statements regarding the measurability of a Borel map in the course of the proof. For easier reference, the technical statement below is recorded:

PROPOSITION 4.2

Let X be a Polish space, $(Y, \mathcal{B})$ be a separable measurable space, and assume that $g : X \to Y$ is $\mathcal{B}(X)$-$\mathcal{B}$-measurable and onto. If $f : X \to Y$ is $\mathcal{B}(X)$-$\mathcal{B}$-measurable such that f is constant on the atoms of $g^{-1}[\mathcal{B}]$, then f is $g^{-1}[\mathcal{B}]$-$\mathcal{B}$ -measurable.

PROOF   Separability implies that $\{y\} \in \mathcal{B}$ for all $y \in Y$. The atoms of $g^{-1}[\mathcal{B}]$ are just the inverse images $g^{-1}[\{y\}]$ of the points $y \in Y$, because these sets are clearly atomic in that σ-algebra, and since they form a partition of X. Now let $B \in \mathcal{B}$ be a measurable set, then by assumption $f^{-1}[B]$ is a Borel set in X which is the union of atoms of $g^{-1}[\mathcal{B}]$. Thus the assertion follows from the Blackwell-Mackey-Theorem (Theorem 1.54). $\square$

4.3.1   The Compact Case

The line of attack for the case of a compact metric space will be as follows: we will construct a linear subspace of $\mathcal{F}(S, \mathcal{B}(S))$ which contains $\mathcal{F}(S, \mathcal{A})$ and some other functions of interest to us, and we will extend the positive linear functional $f \mapsto \int_S f \, d\theta$ linearly to this subspace.

The next step requires the Hahn-Banach Theorem for ordered linear spaces, which may be found in (Jacobs, 1978, Lemma IX.1.4), and which is quoted here for completeness.

THEOREM 4.3

Assume that $(H, \leq)$ is a partially ordered vector space with a linear subspace $H_0 \subseteq H$, and that $L_0 : H_0 \to \mathbb{R}$ is a linear map such that $L_0(f) \geq 0$ whenever $0 \leq f \in H_0$. Then there exists a linear map $L : H \to \mathbb{R}$ with the following properties:

a. *L extends L_0, thus if $f \in H_0$, then $L(f) = L_0(f)$,*

b. *L is positive, thus $f \geq 0$ implies $L(f) \geq 0$.*

A further extension using this Hahn-Banach Theorem brings us to a positive linear functional Λ on $\mathcal{F}(S, \mathcal{B}(S))$ which then can be represented through a measure $\mu \in \mathfrak{P}(S, \mathcal{B}(S))$, so that

$$\Lambda(f) = \int_S f \, d\mu$$

holds. Clearly, μ extends θ and is the measure we are looking for.

The commutativity of the diagram entails by the usual folklore arguments from measure theory that $(i - 1, 2)$

$$\forall f_i \in \mathcal{F}\left(X_i, \psi_i^{-1}[\mathcal{C}]\right) : \int_{X_i} f_i \, d\mu_i = \int_S f_i \circ \pi_i \, d\theta$$

holds, and by the same token it is sufficient to find for $\theta \in \mathfrak{P}(S, \mathcal{A})$ an extension $\mu \in \mathfrak{P}(S, \mathcal{B}(S))$ such that $(i = 1, 2)$

$$\forall f_i \in \mathcal{F}(X_i, \mathcal{B}(X_i)) : \int_{X_i} f_i \, d\mu_i = \int_S f_i \circ \pi_i \, d\mu$$

holds.

PROOF   (of Proposition 4.1)

1. Put for $i = 1, 2$

$$\mathcal{D}_i := \{f_i \circ \pi_i \mid f_i \in \mathcal{F}(X_i, \mathcal{B}(X_i))\},$$

then $\mathcal{D}_i \subseteq \mathcal{F}(S, \mathcal{B}(S))$, and

$$\Lambda_0(f_i \circ \pi_i) := \int_{X_i} f_i \, d\mu_i.$$

Then $\Lambda_0 : \mathcal{D}_1 \cup \mathcal{D}_2 \to \mathbb{R}$ is well defined. In fact, let $g \in \mathcal{D}_1 \cap \mathcal{D}_2$, thus there exist functions $f_i \in \mathcal{F}(X_i, \mathcal{B}(X_i))$ with

$$g = f_1 \circ \pi_1 = f_2 \circ \pi_2.$$

We claim that f_1 is constant on the atoms of $\psi_1^{-1}[\mathcal{C}]$. Take $x_1, x_1' \in X_1$ with $\psi_1(x_1) = \psi_1(x_1')$, then there exists $x_2 \in X_2$ such that $\langle x_1, x_2 \rangle \in S, \langle x_1', x_2 \rangle \in S$. Hence

$$f_1(x_1) = g(x_1, x_2) = f_2(x_2) = g(x_1', x_2) = f_1(x_1').$$

Thus f_1 is $\psi_1^{-1}[\mathcal{C}]$-measurable by Proposition 4.2, and consequently,

$$\int_S g \, d\theta = \int_S f_1 \circ \pi_1 \, d\theta = \int_{X_1} f_1 \, d\mu_1.$$

Similarly,

$$\int_S g \, d\theta = \int_{X_2} f_2 \, d\mu_2$$

is established. This implies that Λ_0 is well defined.

2. Let the linear functional $\Lambda_1 : \mathcal{F}(S, \mathcal{A}) \to \mathbb{R}$ be defined through

$$\Lambda_1(f) := \int_S f \, d\theta.$$

We will look for a joint extension of Λ_0 and Λ_1 to the linear space spanned by $\mathcal{F}(S, \mathcal{A}) \cup \mathcal{D}$, where $\mathcal{D} := \mathcal{D}_1 \cup \mathcal{D}_2$. This requires both functionals yielding the same value on the intersection $\mathcal{F}(S, \mathcal{A}) \cap (\mathcal{D}_1 \cup \mathcal{D}_2)$. Assume first that $g \in \mathcal{F}(S, \mathcal{A}) \cap \mathcal{D}_1$, thus $g = f_1 \circ \pi_1$ for some $f_1 \in \mathcal{F}(X_1, \mathcal{B}(X_1))$. Since g does not depend on the second component, we may infer from the definition of $\mathcal{A}$ that f_1 is even $\psi_1^{-1}[\mathcal{C}]$ − measurable, hence

$$\Lambda_1(g) = \int_S g \, d\theta = \int_S f_1 \circ \pi_1 \, d\theta = \int_{X_1} f_1 \, d\mu_1 = \Lambda_0(g).$$

The argumentation for $g \in \mathcal{F}(S, \mathcal{A}) \cap \mathcal{D}_2$ is similar.

Let Λ_2 be the joint linear extension of Λ_1 on $\mathcal{F}(S, \mathcal{A})$ and of Λ_0 on $\mathcal{D}$ to the linear space spanned by $\mathcal{F}(S, \mathcal{A})$ and $\mathcal{D}$.

From the construction it is clear that $\Lambda_2(1) = 1$ holds, and that Λ_2 is monotone.

3. The Hahn-Banach Theorem 4.3 for ordered linear spaces gives a positive linear operator $\Lambda : \mathcal{F}(S, \mathcal{B}(S)) \to \mathbb{R}$ that extends Λ_2. Since each continuous and bounded map $f : X_1 \times X_2 \to \mathbb{R}$ becomes a member of $\mathcal{F}(S, \mathcal{B}(S))$ when restricted to S, we obtain a positive linear operator $\Lambda'(f) := \Lambda(f \mid_S)$ on the linear space of all continuous maps $X_1 \times X_2 \to \mathbb{R}$. Because $X_1 \times X_2$ is compact, the Riesz Representation Theorem 1.76 yields a probability measure

$$\mu' \in \mathfrak{P}(X_1 \times X_2, \mathcal{B}(X_1 \times X_2))$$

with

$$\Lambda'(f) = \int_{X_1 \times X_2} f \, d\mu' = \int_S f \, d\mu'$$

for each $f \in \mathcal{F}(X_1 \times X_2, \mathcal{B}(X_1 \times X_2))$. Define for $B \in \mathcal{B}(S)$ the measure μ through restricting μ' to $\mathcal{B}(S)$, thus $\mu(B) := \mu'(B \cap S)$, then $\mu \in \mathfrak{P}(S, \mathcal{B}(S))$ will now be shown the measure we are looking for.

4. Let $f \in \mathcal{F}(S, \mathcal{A})$, then

$$\int_S f \, d\theta = \Lambda_1(f) = \Lambda_2(f) = \Lambda'(f) = \int_S f \, d\mu,$$

thus μ extends θ. Let $f_i \in \mathcal{F}(X_1, \mathcal{B}(X_i))$, then $f_i \circ \pi_i \in \mathcal{D}_i \subseteq \mathcal{D}$, hence

$$\int_{X_i} f_i \, d\mu_i = \Lambda_0(f_i \circ \pi_i) = \Lambda_2(f_i \circ \pi_i) = \Lambda'(f_i \circ \pi_i) = \int_S f_i \circ \pi_i \, d\mu,$$

rendering the diagram commutative.

□

The compactness assumption was used in the proof only to establish the existence of a measure, given a suitable linear functional on the space of continuous functions. This functional is then represented as the integral for this measure through the Riesz Theorem.

4.3.2 The General Polish Case

In the general case we do not have the Riesz Representation Theorem directly at our disposal, but compactness may nevertheless be capitalized upon since each Polish space may be embedded into a compact metric space as a measurable subspace. In particular, a Polish space is a measurable and dense subset of a compact metric space by Theorem 1.23. We will capitalize on this: X_1 and X_2 will be embedded into compact metric spaces, and this embedding will take ψ_1, ψ_2 and the measure θ with it. We then apply the extension procedure for the compact case. Restricting what we got from there to the original scenario, we conclude that the assertion holds also for the noncompact case.

PROOF (of Proposition 4.1)

1. X_i is a dense measurable subset of a compact metric space $\widetilde{X}_i$ by Alexandrov's Theorem 1.23, and $\psi_i : X_i \to Z$ may be extended to a Borel measurable map $\widetilde{\psi}_i : \widetilde{X}_i \to Z$ by (Srivastava, 1998, Proposition 3.3.4).

Define $\widetilde{\mu}_i(B_i) := \mu_i(B_i \cap X_i)$ for $B_i \in \mathcal{B}(\widetilde{X}_i)$, and put

$$S_0 := \{\langle x_1, x_2 \rangle \in \widetilde{X}_1 \times \widetilde{X}_2 \mid \widetilde{\psi}_1(x_1) = \widetilde{\psi}_2(x_2)\}.$$

Then $S_0 = \left(\widetilde{\psi}_1 \times \widetilde{\psi}_2\right)^{-1} \left[\Delta_{\widetilde{X}_1 \times \widetilde{X}_2}\right]$, thus

$$S_0 \in \left(\widetilde{\psi}_1 \times \widetilde{\psi}_2\right)^{-1} [\mathcal{B}(Z \times Z)]$$

$$= \widetilde{\psi}_1^{-1}[\mathcal{C}] \otimes \widetilde{\psi}_2^{-1}[\mathcal{C}].$$

Since $X_i \in \widetilde{\psi_i}^{-1}[\mathcal{C}]$, and since $S = S_0 \cap (X_1 \times X_2)$, we see that $S \in \widetilde{\psi_1}^{-1}[\mathcal{C}] \otimes \widetilde{\psi_2}^{-1}[\mathcal{C}]$. Now put $\widetilde{\theta}(E) := \theta(E \cap S)$ for $E \in \widetilde{\psi_1}^{-1}[\mathcal{C}] \otimes \widetilde{\psi_2}^{-1}[\mathcal{C}]$, then $\widetilde{\theta}(S_0 \setminus S) = 0$, because $\widetilde{\theta}$ is concentrated on S.

2. The construction shows that

$$
\begin{array}{ccc}
\left(S_0, \mathcal{A}_0, \widetilde{\theta}\right) & \xrightarrow{\;\widetilde{\pi_2}\;} & \left(\widetilde{X_2}, \mathcal{B}(\widetilde{X_2}), \widetilde{\mu_2}\right) \\
{\scriptstyle \widetilde{\pi_1}} \downarrow & & \downarrow {\scriptstyle \widetilde{\psi_2}} \\
\left(\widetilde{X_1}, \mathcal{B}(\widetilde{X_1}), \widetilde{\mu_1}\right) & \xrightarrow[\;\widetilde{\psi_1}\;]{} & (Z, \mathcal{C}, \nu)
\end{array}
$$

commutes, where

$$
\mathcal{A}_0 := \left(\widetilde{\psi_1}^{-1}[\mathcal{C}] \otimes \widetilde{\psi_2}^{-1}[\mathcal{C}]\right) \cap S_0.
$$

The compact case applies, hence we can find an extension $\widetilde{\mu} \in \mathfrak{P}\left(S_0, \mathcal{B}(S_0)\right)$ for $\widetilde{\theta} \in \mathfrak{P}\left(S_0, \mathcal{A}_0\right)$ which lets this diagram commute:

$$
\begin{array}{ccc}
\left(S_0, \mathcal{B}(S_0), \widetilde{\mu}\right) & \xrightarrow{\;\widetilde{\pi_2}\;} & \left(\widetilde{X_2}, \mathcal{B}(\widetilde{X_2}), \widetilde{\mu_2}\right) \\
{\scriptstyle \widetilde{\pi_1}} \downarrow & & \downarrow {\scriptstyle \widetilde{\psi_2}} \\
\left(\widetilde{X_1}, \mathcal{B}(\widetilde{X_1}), \widetilde{\mu_1}\right) & \xrightarrow[\;\widetilde{\psi_1}\;]{} & (Z, \mathcal{C}, \nu)
\end{array}
$$

3. We now roll back compactification. Put for the Borel set $B \subseteq S$

$$
\mu(B) := \widetilde{\mu}(B \cap S),
$$

then $\mu \in \mathfrak{P}\left(S, \mathcal{B}(S)\right)$, since

$$
\widetilde{\mu}(S_0 \setminus S) = \widetilde{\theta}(S_0 \setminus S) = 0.
$$

The other properties are obvious, so that we are done with the general case, too. $\Box$

The crucial point in this argumentation has been to prevent any mass from vanishing, i.e., to see that $\mu(S) = 1$ holds, which in turn could be established from the fact that $\widetilde{\mu}$ extends $\widetilde{\theta}$, and for which the incorporation of $\mathcal{F}(S, \mathcal{A})$ into the extension process was responsible.

We reformulate Proposition 4.1 in terms of subprobability distributions. It states that there exists sometimes a common distribution for two random variables with values in a Polish space with preassigned marginal distributions. This is a cornerstone for the construction leading to the proof of Theorem 4.9; it shows in particular where Edalat's work could enter the present discussion.

PROPOSITION 4.4

Let X_1 and X_2 be Polish spaces, $(Z, \mathcal{C})$ a separable measurable space, and assume that

$$\psi_i : X_i \to Z \ (i = 1, 2)$$

are measurable and surjective maps. Define

$$S := \{\langle x_1, x_2 \rangle \in X_1 \times X_2 \mid \psi_1(x_1) = \psi_2(x_2)\},$$

endow S with the trace $\mathcal{B}(S)$ of the product σ-algebra, and assume that sub-probability measures $\mu_1 \in \mathfrak{S}(X_1), \mu_2 \in \mathfrak{S}(X_2), \theta \in \mathfrak{S}(S)$ are given such that

$$\forall E_i \in \psi_i^{-1}[\mathcal{C}] : \mathfrak{S}(\pi_i)(\theta)(E_i) = \mu_i(E_i) \ (i = 1, 2)$$

holds, where $\pi_1 : S \to X_1, \pi_2 : S \to X_2$ are the projections. Then there exists $\mu \in \mathfrak{S}(S)$ such that

$$\forall E_i \in \mathcal{B}(X_i) : \mathfrak{S}(\pi_i)(\mu)(E_i) = \mu_i(E_i) \ (i = 1, 2)$$

holds.

PROOF   1. We want to apply Proposition 4.1, so we need to show how to construct diagram (*) from page 163. From the assumption we see that

$$\mu_1(X_1) = \theta(\pi_1^{-1}[X_1]) = \theta(S \cap (X_1 \times X_2)) = \theta(S),$$

similarly for μ_2, so that $\mu_1(X_1) = \mu_2(X_2) = \theta(S)$. If $\theta(S) = 0$ the assertion is pretty obvious, so we may assume that $\theta(S) > 0$, hence it is no loss of generality to assume that all measures are probability measures.

2. Let $C \in \mathcal{C}$, then

$$\mu_1(\psi_1^{-1}[C]) = \mathfrak{P}(\pi_1)(\theta)(\psi_1^{-1}[C]) = \mathfrak{P}(\pi_2)(\theta)(\psi_2^{-1}[C]) = \mu_2(\psi_2^{-1}[C]),$$

since $\pi_1 \circ \psi_1 = \pi_2 \circ \psi_2$ holds on S. So put $\nu(C) := \mu_1(\psi_1^{-1}[C])$, then $\nu \in \mathfrak{P}(Z, \mathcal{C})$ such that

$$\psi_i : (X_i, \mathcal{B}(X_i), \mu_i) \to (Z, \mathcal{C}, \nu)$$

is a morphism in $\mathfrak{Prob}$ for $i = 1, 2$. The assumption implies that

$$\pi_i : (S, \mathcal{A}, \theta) \to (X_i, \mathcal{B}(X_i), \mu_i)$$

is a morphism for $i = 1, 2$, where $\mathcal{A}$ is the trace of the σ-algebra $\psi_1^{-1}[\mathcal{C}] \otimes \psi_2^{-1}[\mathcal{C}]$ on S. Consequently the assertion follows from Proposition 4.1.   ⬜

In important special cases, there are other ways of establishing the Proposition, as will be discussed briefly.

REMARK 4.5   1. If Z is also a Polish space, and if $\psi_i : X_i \to Z$ are bijections, then the Blackwell-Mackey Theorem (Theorem 1.54) shows that $\psi_i^{-1}[\mathcal{C}] = \mathcal{B}(X_i)$. In this case the given measure $\theta \in \mathfrak{S}(S)$ is the desired one. This is so since the trace σ-algebra $\mathcal{A}$ equals $\mathcal{B}(S)$:

$$\begin{aligned}
\mathcal{A} &= S \cap \psi_1^{-1}[\mathcal{C}] \otimes \psi_2^{-1}[\mathcal{C}] \\
&= S \cap \mathcal{B}(X_1) \otimes \mathcal{B}(X_2) \\
&= S \cap \mathcal{B}(X_1 \times X_2) \\
&= \mathcal{B}(S),
\end{aligned}$$

hence θ has the desired properties on the proper σ-algebra.

2. The maps $\psi_i : X_i \to Z$ are morphisms in Edalat's category of probability measures on Polish spaces (Edalat, 1999), provided Z is a Polish space. The assertion can then be deduced from tracing the development in (Edalat, 1999, Cor. 5.4). The proof given above applies to Edalat's situation as well, but it should be clear that the present proof is independent of Edalat's. The development for the latter one depends substantially on the theory of regular conditional probabilities on analytic spaces, so that the impression might arise that the existence of the measure in question depends on these probabilities, too. The proof for Proposition 4.1 shows that this is not the case, that rather a straightforward proof can be given. Hence we are in the lucky position of having two independent proofs *for the Polish case.* Which one is preferred is largely a matter of taste: Edalat's proof working in analytic spaces, or the one proposed here depending on the Hahn-Banach Theorem as a classical tool in analysis (but making use of the sometimes dreaded axiom of choice).   ⫿

We will need an extension theorem for stochastic relations in order to secure the existence of semi-pullbacks for analytic spaces. We begin with a statement on the extension of a probability measure on a sub-σ-algebra. Note that we do not claim the uniqueness of the extension. This is different from the usual measure extensions in measure theory.

LEMMA 4.6

Let $\mathcal{A}$ be a sub-σ-algebra of the Borel sets of a Polish space X, and assume that θ is a probability measure on $\mathcal{A}$. Then θ can be extended to a probability measure on all of $\mathcal{B}(X)$.

PROOF   0. We need only to sketch the proof, since the main work has already been done in the proof of Proposition 4.1. Although the assertion is a bit different, the pattern of the argumentation is very similar to the one presented already.

1. First X is assumed to be compact, then a combination of the Hahn-Banach-Theorem and the Riesz Representation Theorem yields the existence of the desired measure.

2. If X is not compact, it is embedded as above as a measurable subset into a compact metric space. There the existence of an extension is established, and exactly the same technique as above moves that measure to the Borel sets of X. □

The application interesting us here is the possibility to establish an extension to probabilistic relations. Before we state and prove a corresponding property, we remark that each probability on the product of two Polish spaces can be decomposed into a stochastic relation and a measure on one of the factors; see Section 1.5.3.

But let us continue with the discussion of our extension problem. We will combine disintegration and the possibility of extending a measure from a sub-σ-algebra to a larger one in order to obtain

PROPOSITION 4.7

Let X and Y be Polish spaces, assume that $\mathcal{B} \subseteq \mathcal{B}(Y)$ is a countably generated σ-algebra, and let $K_0 : (X, \mathcal{B}(X)) \rightsquigarrow (Y, \mathcal{B})$ be a stochastic relation. Then K_0 can be extended to a stochastic relation $K : (X, \mathcal{B}(X)) \rightsquigarrow (Y, \mathcal{B}(Y))$.

PROOF   0. We will construct a probability measure on the product $\mathcal{B}(Y) \otimes \mathcal{B}$, extend this measure and then obtain the desired extension to the probabilistic relation through disintegration.

1. Let μ be a probability measure on $\mathcal{B}(X)$, and define for $D \in \mathcal{B}(Y) \otimes \mathcal{B}$ the measure

$$\mu_0(D) := \int_X K_0(x)(D_x) \, \mu(dx),$$

where, as usual, $D_x := \{y \in Y \mid \langle x, y \rangle \in D\}$, and by standard arguments $D_x \in \mathcal{B}$ for any $D \in \mathcal{B}(Y) \otimes \mathcal{B}$. Let μ_1 be an extension of μ_0 to all of $\mathcal{B}(X \times Y)$. This extension exists by Lemma 4.6. Since μ_1 is a measure on the product of two Polish spaces, there exists by Theorem 1.84 a stochastic relation $K_1 : (X, \mathcal{B}(X)) \rightsquigarrow (Y, \mathcal{B}(Y))$ such that

$$\mu_1(D) = \int_X K_1(x)(D_x) \, \mu(dx)$$

holds for all $D \in \mathcal{B}(X \times Y)$. K_1 is not uniquely determined, so we have to smooth this relation somewhat.

2. Let $\mathcal{B}_0 := \{B_n \mid n \in \mathbb{N}\}$ be a countable generator of the σ-algebra $\mathcal{B}$; we may and do assume that $\mathcal{B}_0$ is closed under finite intersections (otherwise form

all finite intersections of elements of $\mathcal{B}_0$, then this is still a countable generator with the desired property). Now let $E \in \mathcal{B}$, then $\mu_0(E \times Y) = \mu_1(E \times Y)$ holds by the construction of this extension, thus there exists for each $E \in \mathcal{B}$ a set $N(E) \in \mathcal{B}(X)$ with $\mu(N(E)) = 0$ such that

$$\forall x \in X \setminus N(E) : K_0(x)(E) = K_1(x)(E).$$

Now put $N := \bigcup_{n \in \mathbb{N}} N(B_n)$ as the set of all possibly violating x, then $N \in \mathcal{B}(X)$, and $\mu(N) = 0$ holds.

3. We claim that for any $x \in X \setminus N$ the equality $K_0(x)(E) = K_1(x)(E)$ holds for every Borel set $E \in \mathcal{B}$. In fact, put

$$\mathcal{E} := \{E \in \mathcal{B} \mid \forall x \in X \setminus N : K_0(x)(E) = K_1(x)(E)\},$$

then $\mathcal{B}_0 \subseteq \mathcal{E}$ by construction, $\mathcal{E}$ contains Y, and $\mathcal{E}$ is closed under complementation and disjoint countable unions. Thus $\mathcal{E} = \mathcal{B}$ is inferred by the $\pi - \lambda$-Theorem 1.1. Now let μ_2 be an arbitrary probability measure on $\mathcal{B}(Y)$, and define the stochastic relation K by cases as follows:

$$K(x)(D) := \begin{cases} K_1(x)(D), & x \notin N, D \in \mathcal{B}(Y) \\ K_0(x)(D), & x \in N, D \in \mathcal{B} \\ \mu_2(D), & x \in N, D \notin \mathcal{B}. \end{cases}$$

This relation has the desired properties.   □

The result applies directly to making a pair of surjective and measurable maps into morphisms under a rather weak condition of separability.

LEMMA 4.8

Let $\mathsf{M} := (A, B, M)$ be a stochastic relation between the measurable spaces A and B, and assume that B is separable. If X and Y are Polish spaces with measurable and surjective maps $\phi : X \to A, \psi : Y \to B$, then there exists a stochastic relation $\mathsf{K} := (X, Y, K)$ which makes $\mathsf{f} := (\phi, \psi)$ a morphism $\mathsf{f} : \mathsf{K} \to \mathsf{M}$.

PROOF   Let $\mathcal{B}$ be the σ-algebra on B, then $\psi^{-1}[\mathcal{B}]$ is a countably generated sub-σ-algebra of $\mathcal{B}(Y)$. Define for $x \in X$ and $D \in \mathcal{B}$ $K_0(x)(\psi^{-1}[D]) := M(\phi(x))(D)$, then $K_0 : (X, \mathcal{B}(X)) \rightsquigarrow (Y, \psi^{-1}[\mathcal{B}])$ is a stochastic relation which can be extended to a stochastic relation $K : (X, \mathcal{B}(X)) \rightsquigarrow (Y, \mathcal{B}(Y))$ by Proposition 4.7. It is plain from the construction that $\mathfrak{S}(\psi) \circ K = M \circ \phi$ holds.   □

4.4 The Existence of Semi-Pullbacks

We will show now that semi-pullbacks exist in a rather general setting, generalizing the constructions in (Edalat, 1999). This will ultimately lead to showing that semi-pullbacks exist for analytic objects, and it will turn out that the object underlying such a semi-pullback is Polish.

4.4.1 The Polish Case

The central Lemma reads as follows:

LEMMA 4.9

Let K_i be Polish objects, and assume that $\mathsf{K} = (X, Y, K)$ is a stochastic relation, where X, Y are separable measurable spaces. In $\mathfrak{Stoch}$ each diagram

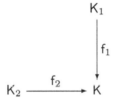

has a semi-pullback

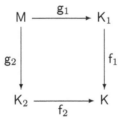

with a Polish object M.

PROOF 1. Assume $\mathsf{K}_i = (X_i, Y_i, K_i)$ with $\mathsf{f}_i = (\phi_i, \psi_i), i = 1, 2$. In view of Lemma 1.45 we may and do assume that the respective σ-algebras on X and Y are the Borel sets of second countable metric spaces. Because of Proposition 1.28 we may assume that the respective σ-algebras on X_1 and X_2 are obtained from Polish topologies which render ϕ_1 and K_1 as well as ϕ_2 and K_2 continuous. These topologies are fixed for the proof. Put

$$A := \{\langle x_1, x_2 \rangle \in X_1 \times X_2 \mid \phi_1(x_1) = \phi_2(x_2)\},$$
$$B := \{\langle y_1, y_2 \rangle \in Y_1 \times Y_2 \mid \psi_1(y_1) = \psi_2(y_2)\},$$

then both A and B are closed, hence Polish. $\alpha_i : A \to X_i$ and $\beta_i : B \to Y_i$ are the projections, $i = 1, 2$. The diagrams

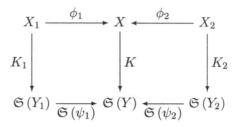

are commutative by assumption, thus we know that for $x_i \in X_i$

$$K(\phi_1(x_1)) = \mathfrak{S}(\psi_1)(K_1(x_1)),$$
$$K(\phi_2(x_2)) = \mathfrak{S}(\psi_2)(K_2(x_2))$$

both hold. The construction implies that $(\psi_1 \circ \beta_1)(y_1, y_2) = (\psi_2 \circ \beta_2)(y_1, y_2)$ is true for $\langle y_1, y_2 \rangle \in B$, and $\psi_1 \circ \beta_1 : B \to Y$ is surjective.

2. Fix $\langle x_1, x_2 \rangle \in A$. Separability of the target spaces now enters: Corollary 1.46 shows that the image of a surjective map under $\mathfrak{S}$ is onto again, so that there exists $\mu_0 \in \mathfrak{S}(B)$ with $\mathfrak{S}(\psi_1 \circ \beta_1)(\mu_0) = K(\phi_1(x_1))$, consequently, $\mathfrak{S}(\psi_i \circ \beta_i)(\mu_0) = \mathfrak{S}(\psi_i)(K_i(x_i))$ $(i = 1, 2)$. But this means

$$\forall E_i \in \psi_i^{-1}[\mathcal{B}(Y)] : \mathfrak{S}(\beta_i)(\mu_0)(E_i) = K_i(x_i)(E_i) \ (i = 1, 2).$$

Put

$$\Gamma(x_1, x_2) := \{\mu \in \mathfrak{S}(B) \mid \mathfrak{S}(\beta_1)(\mu) = K_1(x_1) \wedge \mathfrak{S}(\beta_2)(\mu) = K_2(x_2)\},$$

then Proposition 4.4 shows that $\Gamma(x_1, x_2) \neq \emptyset$.

3. Since K_1 and K_2 are continuous, $\Gamma : A \to \mathbb{F}(\mathfrak{S}(B))$ is easily established. The set

$$\exists \Gamma(C) = \{\langle x_1, x_2 \rangle \in A \mid \Gamma(x_1, x_2) \cap C \neq \emptyset\}$$

is closed in A for compact $C \subseteq \mathfrak{S}(B)$. In fact, let $(\langle x_1^{(n)}, x_2^{(n)} \rangle)_{n \in \mathbb{N}}$ be a sequence in this set with $x_i^{(n)} \to x_i$, as $n \to \infty$ for $i = 1, 2$, thus $\langle x_1, x_2 \rangle \in A$. There exists $\mu_n \in C$ such that $\mathfrak{S}(\beta_i)(\mu_n) = K_i(x_i^{(n)})$. Because C is compact, there exists a converging subsequence $\mu_{s(n)}$ and $\mu \in C$ with $\mu = \lim_{n \to \infty} \mu_{s(n)}$ in the topology of weak convergence. Continuity of K_i implies that $\mathfrak{S}(\beta_i)(\mu) = K_i(x_i)$, consequently $\langle x_1, x_2 \rangle \in \exists \Gamma(C)$; thus this set is closed, hence measurable.

4. From Proposition 1.57 it is now inferred that there exists a measurable map $M : A \to \mathfrak{S}(B)$ such that $M(x_1, x_2) \in \Gamma(x_1, x_2)$ holds for every $\langle x_1, x_2 \rangle \in A$. Thus $M : A \rightsquigarrow B$ is a stochastic relation with

$$K_1 \circ \alpha_1 = \mathfrak{S}(\beta_1) \circ M,$$
$$K_2 \circ \alpha_2 = \mathfrak{S}(\beta_2) \circ M.$$

Hence $\mathsf{M} := (A, B, M)$ is the desired semi-pullback. ▯

This statement includes several interesting special cases:

THEOREM 4.10
Semi-pullbacks exist in the category 𝔓𝔬𝔩𝔖𝔱𝔬𝔠𝔥 *of stochastic relations over Polish spaces.*

PROOF   This follows immediately from Lemma 4.9. ▯

COROLLARY 4.11
Suppose that in the diagram of Lemma 4.9 the target object K *is an analytic object. Then a Polish semi-pullback of that diagram exists in* 𝔖𝔱𝔬𝔠𝔥.

PROOF   This also follows immediately from Lemma 4.9. ▯

4.4.2   The Analytic Case

We obtain from Theorem 4.10 together with Corollary 4.11 the generalization to analytic spaces that is of interest to us through essentially an extension argument. Suppose that we have an analytic object M, then we can find a Polish object K and a morphism $\mathsf{g} : \mathsf{K} \to \mathsf{M}$ by Lemma 4.8. This is so since analytic spaces are surjective images of Polish spaces under measurable maps, and since analytic spaces are — as measurable spaces — separable. This observation has as a somewhat unexpected consequence that semi-pullbacks do exist for analytic spaces:

COROLLARY 4.12
Let M_i *be an analytic object for* $i = 1, 2$ *and assume that* $\mathsf{M} := (A, B, M)$ *is a stochastic relation for the separable measurable spaces* A, B. *For the pair*

$$\mathsf{M}_1 \xrightarrow{\ \ \mathsf{f}_1\ \ } \mathsf{M} \xleftarrow{\ \ \mathsf{f}_2\ \ } \mathsf{M}_2$$

there exist both a Polish object K *and morphisms*

$$\mathsf{M}_1 \xleftarrow{\ \ \mathsf{g}_1\ \ } \mathsf{K} \xrightarrow{\ \ \mathsf{g}_2\ \ } \mathsf{M}_2$$

forming a semi-pullback.

PROOF   We can find Polish objects K_i and morphisms in 𝔖𝔱𝔬𝔠𝔥 extending the diagram

$$
\begin{array}{ccccc}
K_1 & & & & K_2 \\
\downarrow h_1 & & & & \downarrow h_2 \\
M_1 & \xrightarrow{f_1} & M & \xleftarrow{f_2} & M_2
\end{array}
$$

by Lemma 4.8. Now, using Lemma 4.9, find a semi-pullback

$$
K_1 \xleftarrow{\ t_1\ } K \xrightarrow{\ t_2\ } K_2
$$

for the diagram

$$
K_1 \xrightarrow{h_1 \circ f_1} M \xleftarrow{h_2 \circ f_2} K_2
$$

Here K is a Polish object, and t_1, t_2 are morphisms in $\mathfrak{Stoch}$. Putting $g_i :=$ $h_i \circ t_i$ for $i = 1, 2$ now establishes the claim. $\quad\square$

Thus we have in particular established:

THEOREM 4.13
The category an$\mathfrak{Stoch}$ *of stochastic relations over analytic spaces has semi-pullbacks. They may be chosen as Polish objects.*

We close with a negative result, indicating that it is not possible to strengthen the results obtained here towards the existence of weak pullbacks or to pullbacks. Recall that a semi-pullback $r : d \to a, s : d \to c$ for a pair of morphisms $f : a \to b, g : c \to b$ is a weak pullback iff the following holds: whenever $r' : d' \to a, s' : d' \to c$ forms a commutative diagram with f and g (i.e., $f \circ r' = g \circ s'$ holds), then there exists a morphism $h : d' \to d$ with $r' = r \circ h, s' = s \circ h$. If morphism h is uniquely determined, then d together with r and s is called a pullback.

The category $\mathfrak{PolProb}$ is the full subcategory of $\mathfrak{Prob}$ (cf. Section 4.3) which has probability spaces based on Polish spaces as objects.

PROPOSITION 4.14
Let (X, μ) and (Y, ν) be objects in $\mathfrak{PolProb}$, and assume that $\phi : (X, \mu) \to (Y, \nu)$ is a morphism in $\mathfrak{PolProb}$ such that $\phi : X \to Y$ is not bijective. Then the kernel pair

$$
(X, \mu) \xrightarrow{\ \phi\ } (Y, \nu) \xleftarrow{\ \phi\ } (X, \mu)
$$

does not have a weak pullback in $\mathfrak{PolProb}$.

PROOF   Assume that (P, ρ) is a weak pullback for that kernel pair with morphisms based on the maps $\pi_1 : P \to X$ and $\pi_2 : P \to X$. Because the category of Polish spaces with surjective Borel maps has finite products and equalizers, we conclude that

$$P = \{\langle x_1, x_2 \rangle \in X \times X \mid \phi(x_1) = \phi(x_2)\}$$

and that π_1, π_2 are just the projections. Because the identity $id_X : (X, \mu) \rightarrow (X, \mu)$ is plainly a morphism with $f \circ id_X = f \circ id_X$, we find a morphism $\chi : (X, \mu) \rightarrow (P, \rho)$ such that $\pi_1 \circ \chi = \pi_2 \circ \chi$. Thus each element of P must be of the form $\langle x, x \rangle$, contradicting the fact that ϕ is not injective. ▯

Because $\mathfrak{PolProb}$ is a full subcategory of $\mathfrak{PolStoch}$, we may conclude

COROLLARY 4.15
Neither $\mathfrak{PolStoch}$ nor $\mathrm{an}\mathfrak{Stoch}$ have weak pullbacks.

When comparing the present situation for the probability functor with the scenario in general coalgebras, the reader may wish to consult the survey (Rutten, 2000) and to observe that many of the main theorems and constructions require the existence of a weak pullback for the functor governing the coalgebra. Corollary 4.15 is in part responsible for the fact that many constructions that are quite similar to the ones performed for coalgebras will have to develop their own, specific proof for stochastic relations, which does not agree or even resemble a coalgebraic argument. The discussion of simple systems in Section 5.7 will demonstrate this rather convincingly. This is also the deeper reason why modeling software architectures in categories like $\mathfrak{PolStoch}$ will be confined to linear models, because modeling decisions in them is difficult; see (Lajios, 2006), and the discussion in Section 2.4.7.1.

4.5 Bibliographic Notes

The problem to secure the existence of a semi-pullback has been addressed in a variety of ways, varying the base category suitably. Edalat (Edalat, 1999) considers a category of Markov processes with the state space an analytic space, and the transition probability function as universally measurable. The solution to this instance of the problem is essentially based on an explicit construction using regular conditional probabilities which are available due to the Polish descent of analytic spaces. Once this problem is solved, it becomes possible to tackle the equivalence of bisimilarity and accepting the same formulas for a whole family of modal logics with a countable number of diamonds, all interpreted over analytic spaces with a labeled Markov transition system based on universally measurable transition functions; see (Desharnais et al.,

2002). In (Doberkat, 2003) Markov processes over Polish spaces are considered, where the transition probability function is Borel measurable. The problem is transformed into finding a measurable selector for a set-valued map, as in the present discussion. Based on this solution, it is shown that bisimilarity and accepting the same set of formulas are equivalent for a simple negation free modal logic with a countable number of diamonds. The labeled Markov transition processes come from Borel functions based on Polish spaces, and the technical condition was that one of the processes is small, i.e., has a Borel section (Srivastava, 1998).

We did refine here the technique employed in (Doberkat, 2003) in order to show that semi-pullbacks exist in categories of Markov processes over analytic spaces when the transition probability functions are Borel measurable (rather than universally measurable). For practical purposes, the difference between universal and Borel measurability is probably negligible; for structural purposes it is not. Borel measurability is defined in terms of the inverse image of Borel sets (in exactly the same way as continuity is defined in terms of the inverse image of the open sets, or as uniform continuity is in terms of the inverse image of neighborhoods). Universal measurability requires additionally the concept of completing the Borel sets through all finite measures, and thus requires the additional concept of a finite measure. Hence Borel measurability is conceptually simpler and appears as more fundamental.

Chapter 5

Congruences and Bisimulations

5.1   Introduction

This chapter will study smooth equivalence relations in greater detail. They constitute important tools for the investigation of stochastic relations. We have seen already that smooth relations occur naturally when looking at a classification of relations through algebras for the associated monad in Section 3.1. There we had a look at an equivalence relation that occurred as the kernel of a Borel map.

We will now investigate these kernels more systematic and from different angles. As it turns out, these relations occur quite naturally in another disguise: take for example a Kripke model $\mathcal{K}$ with state space S for the simple modal logic introduced in the Introduction with its set Φ of formulas. If the set A of actions and the set AP of atomic propositions are countable, Φ is countable as well. Now define

$$s \equiv s' \text{ iff } \forall \varphi \in \Phi : [\mathcal{K}, s \models \varphi \Leftrightarrow \mathcal{K}, s' \models \varphi].$$

Then the relation $\equiv$ is smooth, as we will show, and the invariant sets for this relation become of interest: A set $B \subseteq S$ is $\equiv$-invariant iff $s \in B$ and if s' satisfies exactly the same formulas as s together imply $s' \in B$, or, using $\equiv$, if $s \in B$ and $s' \equiv s$ imply $s' \in B$, thus iff B is the union of the $\equiv$-equivalence classes. The invariant Borel sets form a σ-algebra which uniquely identifies the relation. A variant of this theme will be interesting as well: Say that $s \equiv_F s'$ iff s and s' satisfy exactly the same formulas from $F \subseteq \Phi$. When investigating a continuous time stochastic logic in 6.4, we will expand the reach of $\equiv_F$ in the sense that we are looking for a larger set G of formulas so that $\equiv_F$ equals $\equiv_G$ (which means that it is enough to check the formulas in F in order to collect sufficient information regarding the formulas in G). Although it is far from evident, the invariant sets help here, too.

We will investigate these relations in this chapter, introduce congruences for stochastic relations based on smooth relations, and show how to factor stochastic relations with congruences. Factor systems will be an important tool for the investigation of bisimulations, which will also be introduced here. First, bisimulations will be introduced as spans of morphisms with an additional property that ties the bisimilar objects together. We will develop a sufficient condition for two stochastic relations to be bisimilar. This condition is based on congruences that in some sense are generating each other, a condition that we describe as simulation equivalence. Later we will see that a natural condition like the one in the Hennessy-Milner Theorem for modal logics can be derived from it. In a special case the condition can be shown to be sufficient: this requires the mediating system to be a compact metric space (of which finite systems are a special case).

The criterion says that we need only to look at the subsystems of two stochastic relations. If we detect subsystems that are isomorphic, then we do not only know that the systems are behavioral equivalent (since we have identified a cospan of morphisms), but that they are bisimilar. This hints at the very close relationship of bisimilarity and behavioral equivalence which will be exploited; simulation equivalent congruences play a mediating rôle; they will arise in Chapter 6 through the theories for the logics **CSL** and its close cousin μ**CSL**. Thus these congruences will be used as a peg to hang equivalent models on.

One of the techniques for the investigation of bisimilar systems is factoring a system. This is investigated here as well, building the bridge to classical algebraic systems like groups, modules and the like. Because we have congruences at our disposal, we may look at the factor system and see which properties it has, and how the morphisms relate to factoring. One of the results will be that for a morphism $f : K \rightarrow K'$ an isomorphism between $K/\ker(f)$ and K' can be established (morphisms are epis). If d is a congruence on the factor relation K/c, then we show that $(K/c)/d$ is isomorphic to $K/c \bullet d$, where $c \bullet d$ is a congruence of K that is coarser than c and encodes the properties of d on the level of the base system. This basically entails that all the properties of a factor system may be derived from the base system itself, so that iterative factoring does not lead to unpleasant surprises. This isomorphism is similar to the ones captured in classical group theory through the Second Isomorphism Theorem.

We deal in this chapter also with systems that are simple in the sense that they do not have any interesting subsystems. So a stochastic relation K is simple if a factor system K/c is either trivial or equal to K itself. These systems are of considerable coalgebraic interest, because — via final systems — they form the basis for the proof principle of coinduction. If we could identify final systems in a sensible way, then we could establish probabilistic coinduction as well. But this hope cannot be supported: final systems are much too simple to be of any interest. We identify simple systems through their bisimulations (this is truly in the coalgebraic spirit), but then have a look at what this entails with respect to final systems, provided they exist, and here coalgebraic hope will leave us.

5.2 Smooth Equivalence Relations

An equivalence relation ρ on an analytic space is smooth (or countably generated) iff it can be decided whether or not two elements are equivalent by looking at a countable family of Borel sets; see Definition 1.50.

The relation's invariant Borel sets will be a powerful tool for investigating smooth relations. In order to appreciate this σ-algebra fully, we show that it is often enough to establish a property related to the equivalence relation for the generators, then it will hold on the generated σ-algebra. This will be interesting when we investigate various logics: here the generators are determined by the formulas of the logic, yielding a rather captivating interplay of formulas in the logic and a σ-algebra determined by their semantics.

LEMMA 5.1
Assume that an equivalence relation $\sim$ is defined for a set M through

$$m \sim m' \text{ iff } \forall G \in \mathcal{G} : [m \in G \Leftrightarrow m' \in G],$$

where the elements of $\mathcal{G}$ are subsets of M. Then

$$m \sim m' \text{ iff } \forall G \in \sigma(\mathcal{G}) : [m \in G \Leftrightarrow m' \in G].$$

PROOF   It is not difficult to see that the set

$$\mathcal{H} := \{G \in \sigma(\mathcal{G}) \mid m \in G \Leftrightarrow m' \in G\}$$

is a σ-algebra, where $m, m' \in M$ are fixed. For example, if $G_1, G_2 \in \mathcal{H}$ and $m \in G_1 \cup G_2$, then $m \in G_1$ or $m \in G_2$. Depending on which case applies, $m' \in G_1$ or $m' \in G_2$, thus $m' \in G_1 \cup G_2$, and vice versa. But by assumption $\mathcal{G} \subseteq \mathcal{H}$, thus $\sigma(\mathcal{G}) \subseteq \mathcal{H}$, and the conclusion follows. ☐

This Lemma tells us that invariance with respect to an equivalence relation, which will be defined below, is carried over from a generator to its σ-algebra. This implies that we have some degrees of freedom when selecting a generator.

DEFINITION 5.2   *Let ρ be a smooth equivalence relation on an analytic space X.*

a. *A subset $A \subseteq X$ is called ρ-invariant iff $x \in A$ and $x \rho x'$ together imply $x' \in A$.*

b. *Denote by $\mathcal{INV}(\mathcal{B}(X), \rho)$ the σ-algebra of ρ-invariant Borel subsets of X.*

Thus a ρ-invariant set A can be written as the union of the equivalence classes of its sets, $A = \bigcup\{[x]_\rho \mid x \in A\}$. We see further that the invariant Borel subsets constitute the σ-algebra of $(A_n)_{n \in \mathbb{N}}$ which determines ρ. This will be investigated further in a moment.

These are quite simple examples:

LEMMA 5.3
The identity relation Δ_X and the universal relation U_X are for each Polish space X smooth equivalence relations with

$$\mathcal{INV}(\mathcal{B}(X), \Delta_X) = \mathcal{B}(X),$$
$$\mathcal{INV}(\mathcal{B}(X), U_X) = \{\emptyset, X\}.$$

PROOF   The assertion is trivial for the universal relation. One argues for the identity relation as follows: the Borel sets of X are countably generated, and one can find such a countable generator $\mathcal{G}$ that separates points. This implies that Δ_X has $\mathcal{G}$ as the determining family, and since $\sigma(\mathcal{G}) = \mathcal{B}(X)$, the assertion follows. $\square$

We have seen in Proposition 1.53 that factoring an analytic space with a smooth equivalence yields an analytic space again. This closure property above is fairly fundamental for the development of the algebraic theory of stochastic relations, being one of the reasons for sometimes preferring analytic spaces over Polish ones, since the latter ones are not closed under factoring through a smooth relation. It will enable factoring through a congruence (see Section 5.3) without running the risk that the arising structure will lose essential properties. This will be discussed in due course.

5.2.1   Invariant Borel Sets

The invariant Borel sets may be characterized through the factor map by the inverse image of the Borel sets of a factor space. This will give a fairly practical handle on the invariant sets. The next Lemma is a bit more general by considering general surjective Borel maps, and we will see that this is helpful indeed.

LEMMA 5.4

Let X, Y be analytic spaces, and assume that $f : X \to Y$ is a surjective Borel map. Then $f^{-1}[\mathcal{B}(Y)] = \mathcal{INV}(\mathcal{B}(X), \ker(f))$.

PROOF   1. Given $A \in \mathcal{INV}(\mathcal{B}(X), \ker(f))$, we show first that

$$f^{-1}[f[A]] = A$$

holds. In fact, $A \subseteq f^{-1}[f[A]]$ is always true. Let $s \in f^{-1}[f[A]]$, thus $f(s) = f(s')$ for some $s' \in A$. Since A is $\ker(f)$ invariant, this implies $s \in A$, accounting for the other inclusion.

2. Let again $A \in \mathcal{INV}(\mathcal{B}(X), \ker(f))$, then $f[A] \subseteq Y$ is analytic. We claim that

$$f[X \setminus A] = Y \setminus f[A]$$

holds. For, if $y \in f[S \setminus A]$, we can find $x \notin B$ with $f(x) = y$. Assuming that $y = f(x')$ for some $x' \in A$, we would infer that $x \in A$ due to the $\ker(f)$-invariance of A, and since $\langle x, x' \rangle \in \ker(f)$. This is a contradiction. This settles the nontrivial inclusion. From the representation just established we see that $Y \setminus f[A]$ is analytic, and from Souslin's Theorem (Theorem 1.39) we infer now that $f[A]$ is Borel in Y.

3. It is clear that for each $B \in \mathcal{B}(Y)$ its inverse image $f^{-1}[B]$ under f is a Borel set which is $\ker(f)$-invariant. On the other hand, if $A \in \mathcal{B}(X)$ is $\ker(f)$-invariant, we write $A = f^{-1}[f[A]]$ by part 1, and $f[A] \in \mathcal{B}(Y)$ by part 2. This implies the desired equality. $\square$

As a by-product we obtain a characterization of ρ-invariant Borel sets in analytic spaces through the generating sequence $(A_n)_{n \in \mathbb{N}}$. This result is known for Polish spaces; it seems to be new for the analytic case. As a consequence, we can characterize the ρ-invariant Borel set through the canonic projection η_ρ.

PROPOSITION 5.5

Let X be an analytic space with a smooth equivalence relation ρ, then the ρ-invariant Borel sets of X are exactly the inverse images of the canonic projection η_ρ, viz.,

$$\mathcal{INV}(\mathcal{B}(X), \rho) = \eta_\rho^{-1}[\mathcal{B}(X/\rho)]$$

holds. Moreover, if ρ is determined by the sequence $(A_n)_{n \in \mathbb{N}}$ of Borel sets $A_n \subseteq X$, then

$$\mathcal{INV}(\mathcal{B}(X), \rho) = \sigma(\{A_n \mid n \in \mathbb{N}\}).$$

PROOF   1. X/ρ is an analytic space, and $\eta_\rho : X \to X/\rho$ is surjective and onto. Thus the first assertion follows from Lemma 5.4 upon observing that $\rho = \ker(\eta_\rho)$ holds.

2. Let $A \in \mathcal{B}(X/\rho)$ be a Borel set in X/ρ. Plainly,

$$A = \bigcup\{\{[x]_\rho\} \mid [x]_\rho \in A\},$$

so it is enough to show that each $\{[x]_\rho\}$ constitutes an atom in the σ-algebra $\sigma(\{\eta_\rho[A_n] \mid n \in \mathbb{N}\})$.

Granted that, we can argue as follows: The Blackwell-Mackey-Theorem (Theorem 1.54) implies that $\mathcal{B}(X/\rho) = \sigma(\{\eta_\rho[A_n] \mid n \in \mathbb{N}\})$ holds, thus $C \in \mathcal{INV}(\mathcal{B}(X), \rho)$ iff $C = \eta_\rho^{-1}[B]$ for some

$$B \in \eta_\rho^{-1}[\sigma(\{\eta_\rho[A_n] \mid n \in \mathbb{N}\})] = \sigma(\{A_n \mid n \in \mathbb{N}\}).$$

3. It is easy to see that

$$\bigcap\{\eta_\rho[A_n] \mid t \in A_n\} \cap \bigcap\{X/\rho \setminus \eta_\rho[A_n] \mid t \notin A_n\}$$

contains the class $[x]_\rho$ as its only element, and that

$$(X/\rho) \setminus \eta_\rho[A_n] = \eta_\rho[T \setminus A_n],$$

because A_n is ρ-invariant, cp. part 2 of the proof of Lemma 5.4. Thus the atom $\{[x]_\rho\}$ is a member of $\sigma(\{\eta_\rho[A_n] \mid n \in \mathbb{N}\})$.   ⬚

We obtain as a Corollary that a smooth equivalence relation is determined uniquely by its invariant sets:

COROLLARY 5.6

If $C \subseteq \mathcal{B}(X)$ is a countably generated sub-σ-algebra of the Borel sets of X, then there exists a unique smooth equivalence relation ρ_C on X with $C = \mathcal{INV}(\mathcal{B}(X), \rho_C)$.

5.2.2   Operations on Smooth Relations

We will study briefly operations with smooth equivalence relations. These operations will shape useful tools for constructions on stochastic relations. The interplay between smooth relations and measurable maps is further illustrated by the technique of transporting a smooth relation backwards along a measurable map.

LEMMA 5.7

Let α be a smooth equivalence relation on the analytic space A so that $\alpha = \ker(h)$ for some measurable map $h : A \to W$, W being an analytic space. Define for the Polish space X and the Borel map $f : X \to A$ on X the smooth relation $\alpha_f := \ker(h \circ f)$. If $E \subseteq X$ is an α_f-invariant Borel set, then *a. $f[E]$ is an α-invariant Borel set in A,*

b. $E = f^{-1}[f[E]]$.

Consequently, the invariant Borel sets of α_f are just the inverse images of the invariant Borel set of α under f, viz.,

$$\mathcal{INV}(\mathcal{B}(X), \alpha_f) = f^{-1}[\mathcal{INV}(\mathcal{B}(A), \alpha)].$$

PROOF   1. Let $E_0 := f^{-1}[F]$ be the inverse image of an α invariant set $F \subseteq A$, and assume that $x \in E_0$ with $x \,\alpha_f\, x'$. Since $f(x) \in F$, and since $h(f(x)) = h(f(x'))$, we have $f(x') \in F$, thus $x' \in E_0$. Consequently, E_0 is α_f-invariant, and we have shown that

$$\mathcal{INV}(\mathcal{B}(X), \alpha_f) \supseteq f^{-1}[\mathcal{INV}(\mathcal{B}(A), \alpha)]$$

holds.

2. Let $E \in \mathcal{INV}(\mathcal{B}(X), \alpha_f)$, then we assert that

$$E' := f[E] \in \mathcal{INV}(\mathcal{B}(A), \alpha).$$

Since E is α_f-invariant, $f[E]$ is α-invariant by construction. The hard part is showing that E' is a Borel set. First it is clear that E' is an analytic set, because it is the image of a Borel set under a Borel map. We claim that $f[X \setminus E] = A \setminus f[E]$. From this we may conclude that E' is also co-analytic, thus is a Borel set by Souslin's famous theorem (Theorem 1.39).

We first repeat the argumentation in the proof of Lemma 5.4 in showing that $f[X \setminus E] \subseteq A \setminus f[E]$ holds: Suppose $a \in f[X \setminus E]$, then we can find $x \in X \setminus E$ with $a = f(x)$. If a would be a member of $f[E]$, we could find $x' \in E$ with $a = f(x')$. Since $x \,\alpha_f\, x'$ and since E is α_f-invariant, we would find $x \in E$, contradicting the choice of x. This establishes the desired equality and shows that E' is in fact a Borel set. But we can say more: $E = f^{-1}[f[E]]$ will be shown to hold. Let $x \in f^{-1}[f[E]]$, thus $f(x) \in f[E]$, hence $f(x) = f(x')$ for some $x' \in E$. But this implies $x \in E$ since the latter set is α_f-invariant, and $x \,\alpha_f\, x'$. The other inclusion is trivial again.

3. The argument shows that each element of $\mathcal{INV}(\mathcal{B}(X), \alpha_f)$ can be represented as the inverse image of an element from $\mathcal{INV}(\mathcal{B}(A), \alpha)$ under f, thus

$$\mathcal{INV}(\mathcal{B}(X), \alpha_f) \subseteq f^{-1}[\mathcal{INV}(\mathcal{B}(A), \alpha)]$$

is established.   □

It should be noted that the image of a Borel set under a Borel map is usually not a Borel set. This is an immediate consequence of the observation in Proposition 1.35 that there are strictly more analytic sets than Borel sets. Hence the first property in Lemma 5.7 indicates that invariance paired with smoothness is fairly strong a property.

When investigating different morphisms for a stochastic relation, a relation derived from the images will be helpful. Abbreviate for two maps $g_1, g_2 : V \to W$ with common domain V and common range W (V and W are for the time being arbitrary sets) the common product image by

$$\lfloor g_1 \| g_2 \rfloor := \{ \langle g_1(v), g_2(v) \rangle \mid v \in V \}.$$

If g_1 and g_2 model processes, $\lfloor g_1 \| g_2 \rfloor$ may be visualized as having the processes run in parallel. It will turn out that the equivalence relation generated from this relation is quite closely connected to the events common to these processes. Denote the smallest equivalence relation that contains a given relation R by $\ell(R)$.

The events common to two morphisms can be identified through $\ell(\lfloor \cdot \| \cdot \rfloor)$. We will need this representation when dealing with properties of morphisms for simple relations in Section 5.7.1.

LEMMA 5.8

Assume B is a Polish space, and let $\psi_1, \psi_2 : B \to Y$ be surjective Borel maps. Assume further that $\ell(\lfloor \psi_1 \| \psi_2 \rfloor)$ is smooth. Let

$$\mathcal{C} := \{ C \in \mathcal{B}(Y) \mid \psi_1^{-1}[C] = \psi_2^{-1}[C] \}$$

be the σ-algebra of events common to both ψ_1 and ψ_2. Then these common events are exactly the $\ell(\lfloor \psi_1 \| \psi_2 \rfloor)$-invariant Borel sets, thus

$$\mathcal{C} = \mathcal{INV}\left(\mathcal{B}(Y), \ell(\lfloor \psi_1 \| \psi_2 \rfloor)\right).$$

PROOF   1. It is not difficult to see that if $C \in \mathcal{C}$ is a common event then $y \in C$ and $\langle y, y' \rangle \in \ell(\lfloor \psi_1 \| \psi_2 \rfloor)$ together imply $y' \in C$. This is so since $\langle y, y' \rangle \in \ell(\lfloor \psi_1 \| \psi_2 \rfloor)$ implies that there exist $y_0, \ldots, y_n \in Y$ with $y_0 = y, y_n = y'$ and $\langle y_i, y_{i+1} \rangle \in \lfloor \psi_1 \| \psi_2 \rfloor$ for $1 \leq i \leq n - 1$. This yields

$$\mathcal{C} \subseteq \mathcal{INV}\left(\mathcal{B}(Y), \ell(\lfloor \psi_1 \| \psi_2 \rfloor)\right)$$

in terms of the σ-algebras involved.

2. Now let

$$C \in \mathcal{INV}\left(\mathcal{B}(Y), \ell(\lfloor \psi_1 \| \psi_2 \rfloor)\right) = \eta_{\ell(\lfloor \psi_1 \| \psi_2 \rfloor)}^{-1}\left[\mathcal{B}(Y)/\ell(\lfloor \psi_1 \| \psi_2 \rfloor)\right],$$

the latter equality holding by Proposition 5.5. Thus we can find

$$D \in \mathcal{B}(Y)/\ell(\lfloor \psi_1 \| \psi_2 \rfloor)$$

with $C = \eta_{\ell(\lfloor \psi_1 \| \psi_2 \rfloor)}^{-1}[D]$. Consequently,

$$\psi_1^{-1}[C] = \psi_2^{-1}[C] \Longleftrightarrow \left(\eta_{\ell(\lfloor \psi_1 \| \psi_2 \rfloor)} \circ \psi_1\right)^{-1}[D] = \left(\eta_{\ell(\lfloor \psi_1 \| \psi_2 \rfloor)} \circ \psi_2\right)^{-1}[D].$$

Since by construction $\eta_{\ell(\lfloor\psi_1\|\psi_2\rfloor)} \circ \psi_1 = \eta_{\ell(\lfloor\psi_1\|\psi_2\rfloor)} \circ \psi_2$, the latter equality follows. Hence C is a common event, establishing the nontrivial inclusion. ⃞

The invariant sets leave a trace on the product: consider for an invariant set P all pairs $\langle x, x' \rangle \in \alpha$ with $x \in P$. Properties on this trace include the extendability of measurable sets as well as measures on the base space to it in a natural way.

LEMMA 5.9

Let α be a smooth equivalence relation on the analytic space X. Then

a. *If $P \in \mathcal{INV}(\mathcal{B}(X), \alpha)$, then $(P \times X) \cap \alpha = (X \times P) \cap \alpha = (P \times P) \cap \alpha$*

b. *$\otimes[X, \alpha] := \{(P \times X) \cap \alpha \mid P \in \mathcal{INV}(\mathcal{B}(X), \alpha)\}$ is a σ-algebra on α,*

c. *If $\mu \in \mathfrak{S}(X, \mathcal{INV}(\mathcal{B}(X), \alpha))$ is a subprobability measure on the α-invariant Borel sets of X, then $\mu^{\bullet}((P \times X) \cap \alpha) := \mu(P)$ defines a subprobability measure on $\otimes[X, \alpha]$.*

PROOF   1. Part a is proved by a direct calculation, and part b is established by checking the defining properties of a σ-algebra.

2. For proving part c it is observed that $\mu^{\bullet}$ is well defined, since $(P_1 \times X) \cap \alpha = (P_2 \times X) \cap \alpha$ implies $P_1 = P_2$ for the α-invariant sets P_1, P_2. The properties of a finite measure are then easily established. ⃞

When dealing with bisimulations, we will closely look at properties of a smooth relation as a subset of the Cartesian product. Here Lemma 5.9 will come in handy.

Factoring a factor space through a smooth relation will not really bring new structural information: we will show that the iterated factor space is isomorphic to a factor space that can be obtained from a relation on the base space. This will be an occasion to introduce a kind of multiplicative operation on relations for later use. Then we will show that other operations such as sums, intersections and countable products of smooth relations will also lead to smooth relations.

Assume that ρ is a smooth equivalence relation on the analytic space X, and that τ is a smooth equivalence on X/ρ. Define for $x, x' \in X$

$$x \; (\tau \bullet \rho) \; x' \Leftrightarrow [x]_\rho \; \tau \; [x']_\rho.$$

PROPOSITION 5.10

The equivalence relation $\tau \bullet \rho$ is smooth, and the analytic spaces $X/\tau \bullet \rho$ and $(X/\rho)/\tau$ are Borel isomorphic.

PROOF   0. Since τ is smooth, there exists a sequence $(A_n)_{n \in \mathbb{N}}$ of Borel sets $A_n \subseteq X/\rho$ which determines it. Then $\left(\eta_\rho^{-1}[A_n]_{n \in \mathbb{N}}\right)$ determines $\tau \bullet \rho$. Its members are by construction Borel sets in X.

1. Define

$$g_{\rho,\tau}\left([x]_{\tau \bullet \rho}\right) := \left[[x]_\rho\right]_\tau,$$

then $g_{\rho,\tau} : X/\tau \bullet \rho \to (X/\rho)/\tau$ is well defined and turns out to be a bijection. The construction shows that $g_{\rho,\tau} \circ \eta_{\tau \bullet \rho} = \eta_\tau \circ \eta_\rho$ holds; putting $h_{\rho,\tau} := g_{\rho,\tau}^{-1}$, we see that $\eta_{\tau \bullet \rho} = h_{\rho,\tau} \circ \eta_\tau \circ \eta_\rho$. This is also noted for later use.

2. Let $E \subseteq (X/\rho)/\tau$ be a Borel set, we need to show that $g_{\rho,\tau}^{-1}[E]$ is a Borel set in $X/\tau \bullet \rho$, equivalently by Proposition 5.5, that $E_0 := \eta_{\tau \bullet \rho}^{-1}\left[g_{\rho,\tau}^{-1}[E]\right]$ is an $\tau \bullet \rho$-invariant Borel set in X. But $E_0 = (\eta_\rho \circ \eta_\tau)^{-1}[E]$, so that E_0 is a Borel set by the measurability of the projections, and this set is clearly $\tau \bullet \rho$-invariant. Thus we get the measurability of E_0 again from Lemma 5.5.

3. Let $F \subseteq X/\tau \bullet \rho$ be a Borel set, hence $F_0 := \eta_{\tau \bullet \rho}^{-1}[F]$ is a Borel set in X, thus there exists a Borel set $F_1 \subseteq (X/\rho)/\tau$, such that $F_0 = \eta_\rho^{-1}\left[\eta_\tau^{-1}[F_1]\right]$ since F_0 is ρ-invariant. Hence $F_1 = h_{\rho,\tau}^{-1}[F]$, so $h_{\rho,\tau}$ is measurable, establishing the claim.   ☐

The definition of $\tau \bullet \rho$ translates a partition of the ρ-classes into a partition of the base set. The generated $\tau \bullet \rho$-partition is coarser than the ρ-partition, since $\rho \subseteq \tau \bullet \rho$. The converse holds as well: whenever we have two partitions coming from smooth equivalence relations, we may find a factor in terms of $\bullet$ relating them to each other. Surprisingly, this result about the containment of equivalence relations may be used in Section 5.2.3 to show that two completely unrelated equivalences may give rise to some sort of confluence.

COROLLARY 5.11

The following conditions are equivalent for smooth equivalence relations ρ and σ on X:

a. $\rho \subseteq \sigma$,

b. there exists a smooth equivalence relation θ on X/ρ such that $\sigma = \theta \bullet \rho$.

PROOF   1. The direction $b \Rightarrow a$ is trivial, so we are left with the proof for $a \Rightarrow b$.

2. Define $f\left([x]_\rho\right) := [x]_\sigma$ for $[x]_\rho \in X/\rho$ then $f : X/\rho \to X/\sigma$ is well defined and surjective. We claim that f is $\mathcal{B}(X/\rho)$-$\mathcal{B}(X/\rho)$-measurable. In fact, we need to show that $f^{-1}[D] \in \mathcal{INV}(\mathcal{B}(X), \rho)$ whenever $D \in \mathcal{B}(X/\rho)$. Since by Corollary 5.5

$$D \in \mathcal{B}(X/\rho) \Leftrightarrow \eta_\sigma^{-1}[D] \in \mathcal{INV}(\mathcal{B}(X), \sigma),$$

the claim follows from $\eta_\rho^{-1}\left[f^{-1}[D]\right] = \eta_\sigma^{-1}[D]$ and from $\mathcal{INV}(\mathcal{B}(X),\sigma) \subseteq$
$\mathcal{INV}(\mathcal{B}(X),\rho)$, which is inferred from $\rho \subseteq \sigma$ by a trivial calculation. Since
Lemma 5.4 tells us that $\ker(f)$ is smooth, we can set $\theta := \ker(f)$ and are
done. □

We turn now to sums and products. Knowing that smooth equivalence
relations are closed under the sum operation is helpful when discussing congruences on the sum of stochastic relations. It is clear that the sum and the
product of a countable number of analytic spaces is analytic again; this can
easily be established through Proposition 1.33.

LEMMA 5.12
*Let X and Y be analytic spaces with smooth equivalence relations α resp. β.
Then $\alpha + \beta := \alpha \cup \beta$ is a smooth equivalence relation on $X + Y$.*

PROOF If $(A_n)_{n\in\mathbb{N}}$ and $(B_n)_{n\in\mathbb{N}}$ determine α resp. β, then the countable
set of Borel sets $\{A_n + B_m \mid n,m \in \mathbb{N}\}$ determines $\alpha + \beta$. □

Smooth equivalence relations are closed under intersections and under countably infinite products.

LEMMA 5.13
*If ρ, ρ' are smooth equivalence relations on the analytic space X, then $\rho \cap \rho'$
is smooth, and $\mathcal{INV}(\mathcal{B}(X), \rho \cap \rho') = \sigma\left(\mathcal{INV}(\mathcal{B}(X),\rho) \cup \mathcal{INV}(\mathcal{B}(X),\rho')\right)$.*

PROOF Assume that $(A_n)_{n\in\mathbb{N}}$ and $(A'_n)_{n\in\mathbb{N}}$ determine ρ resp. ρ'. Then
the sequence $(A_n \cap A'_m)_{n,m\in\mathbb{N}}$ determines $\rho \cap \rho'$. The representation for the
invariant Borel sets for $\rho \cap \rho'$ follows then easily from Proposition 5.5. □

The closure under countably infinite products will resort to the construction
of the Borel sets on an infinite product through cylinder sets (cp. Section 1.2).

LEMMA 5.14
*Assume that $(X_n)_{n\in\mathbb{N}}$ is a sequence of analytic spaces, and let ρ_n be a smooth
equivalence relation on X_n for each $n \in \mathbb{N}$. Define*

$$(a_n)_{n\in\mathbb{N}} \left(\times_{n\in\mathbb{N}} \rho_n\right) (a'_n)_{n\in\mathbb{N}} \iff \forall n \in \mathbb{N}: a_n \ \rho_n \ a'_n.$$

Then

a. $\times_{n\in\mathbb{N}} \rho_n$ is a smooth equivalence relation on $\prod_{n\in\mathbb{N}} X_n$.

b. $\mathcal{INV}\left(\mathcal{B}(\prod_{n\in\mathbb{N}} X_n), \times_{n\in\mathbb{N}} \rho_n\right) = \bigotimes_{n\in\mathbb{N}} \mathcal{INV}(\mathcal{B}(X_n), \rho_n).$

PROOF   1. Abbreviate the equivalence relation $\times_{n\in\mathbb{N}}\, \rho_n$ by ρ^∞. Assume that ρ_n is determined by the sequence $(Z_{n,m})_{m\in\mathbb{N}}$ of Borel sets $Z_{n,m} \subseteq X_n$. Define for $n_1 \ldots n_k \in \mathbb{N}^*$ the cylinder set

$$W_{n_1 \ldots n_k} := Z_{1,n_1} \times \cdots \times Z_{k,n_k} \times \prod_{j>k} X_j.$$

Then it is easy to see that

$$
\begin{aligned}
(a_n)_{n\in\mathbb{N}}\ \rho^\infty\ &(a'_n)_{n\in\mathbb{N}} \\
&\Leftrightarrow \forall v \in \mathbb{N}^* : [\langle a_1, \ldots, a_k \rangle \in W_v \iff \langle a'_1, \ldots, a'_k \rangle \in W_v].
\end{aligned}
$$

This implies that ρ^∞ is generated through a countable family of Borel sets.

2. Since for each index $v \in \mathbb{N}^*$ the set W_v is a ρ^∞-invariant Borel set which is comprised from ρ_n-invariant factors, we have

$$\mathcal{INV}\left(\mathcal{B}(\prod_{n\in\mathbb{N}} X_n), \rho^\infty\right) \subseteq \bigotimes_{n\in\mathbb{N}} \mathcal{INV}\left(\mathcal{B}(X_n), \rho_n\right);$$

on the other hand,

$$\bigotimes_{n\in\mathbb{N}} \mathcal{INV}\left(\mathcal{B}(X_n), \rho_n\right)$$

is generated by cylinder sets of the form

$$B_1 \times \ldots \times B_n \times \prod_{j>n} X_j,$$

which are ρ^∞-invariant. This implies the other inclusion. $\square$

A finite version is here available as well: the product of two smooth equivalence relations is smooth again, and the invariant Borel sets for the product are just the product of the Borel sets for the factors.

5.2.3   A Confluence Property

We will establish a confluence property that will be helpful for understanding the relationship between congruences and bisimulations in a special situation. Then this confluence property will be used in a crucial way for establishing that bisimilar relations have isomorphic factors. Quite apart from this, it gives some insight into the manipulation of relations, and it indicates a rather surprising connection to selections of set-valued maps.

An equivalence relation ρ on a set X can be viewed as a set-valued map $x \mapsto [x]_\rho$ that assigns element $x \in X$ its equivalence class $[x]_\rho$, hence a particular nonempty set. The existence of a measurable selector f for this set-valued map $[\cdot]_\rho$ implies that ρ is smooth, provided we know that X/ρ is an analytic space. This is so since

$$x \, \rho \, x' \Leftrightarrow [x]_\rho = [x']_\rho \Leftrightarrow f(x) = f(x'),$$

so that $\rho = \ker(f)$, hence Lemma 1.52 applies.

This is essentially the outline for the proof of the confluence property that relates two smooth relations on a compact metric space.

PROPOSITION 5.15

Let T be a compact metric space, and assume that $\rho = \ker(\phi), \sigma = \ker(\psi)$ for some continuous maps $\phi : T \to N$ and $\psi : T \to N'$ with metric spaces N, N'. There exist smooth equivalence relations θ on T/σ and θ' on T/ρ such that

a. $\theta' \bullet \rho = \theta \bullet \sigma$,

b. *θ and θ' are minimal: if $\theta'_0 \bullet \rho = \theta_0 \bullet \sigma$, for smooth equivalence relations θ_0 on T/σ and θ'_0 on T/ρ, then $\theta \subseteq \theta_0$ and $\theta' \subseteq \theta'_0$.*

The diagram visualizes this claim and suggests the characterization as a confluence property.

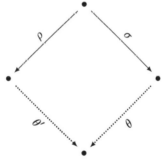

Note that the universal relations on the respective factor spaces would satisfy the first condition. The statement is then somewhat trivial, so minimality will make sure that we may apply it in a sensible way.

The proof will be broken into several parts. Because of Corollary 5.11 we will first find a smooth equivalence relation θ on T/σ such that $\rho \subseteq \theta \bullet \sigma$ holds. We will assume through the end of the proof of Proposition 5.15 that T is a compact metric space, and that $\rho = \ker(\phi), \sigma = \ker(\psi)$.

CLAIM 5.16

T/σ is a compact metric space when endowed with the final topology for $\eta_{|\sigma}$.

PROOF Let d' be the metric on N', and put for $t, t' \in T$

$$D([t]_\sigma, [t']_\sigma) := d'(\psi(t), \psi(t')),$$

then D is a metric on T/σ (since $\sigma = \ker(\psi)$ by assumption). Let $\mathcal{T}$ be the topology on T/σ induced by η_σ, then a set $G \subseteq T/\sigma$ which is D-open is also $\mathcal{T}$-open. This follows easily from the continuity of ψ. Conversely, let F be $\mathcal{T}$-closed, and assume that $([t_n]_\sigma)_{n\in\mathbb{N}}$ be a sequence in F such that $D([t_n]_\sigma, [t]_\sigma) \to 0$, as $n \to \infty$. Select an arbitrary $x_n \in [t_n]_\sigma$, thus $x_n \in \eta_\sigma^{-1}[F]$, the latter is a closed, hence compact set. Thus we can find a convergent subsequence (which we take w.l.g. the sequence itself), so that there exists $x^* \in \eta_\sigma^{-1}[F]$ with $x_n \to x^*$. By the continuity of ψ we may conclude $\psi(x_n) \to \psi(x^*)$. This implies $x^* \in [t]_\sigma$, thus $[t]_\sigma \in F$. Hence F is also metrically closed, and the topologies coincide.   ⬛

Claim 5.16 shows among others that the Borel sets in T/σ come from a compact metric space. This observation will make some arguments easier. When talking about the topology on T/σ, we refer interchangeably to the metric topology and the topology induced by the canonic projection.

CLAIM 5.17
Put
$$\zeta := \{\langle s, s'\rangle \mid s, s' \in T/\sigma, s \times s' \cap \rho \neq \emptyset\}.$$
Then $\zeta \subseteq (T/\sigma)^2$ is reflexive, symmetric, and a closed subset of $(T/\sigma)^2$.

PROOF   Since ρ is reflexive and symmetric, ζ is. Now let $\langle s_n, s'_n \rangle \in \zeta$ be a convergent sequence, say $s_n \to s, s'_n \to s'$. For s_n there exists by the construction of ζ a pair $\langle t_n, t'_n \rangle \in \rho$ with $t_n \in s_n, t'_n \in s'_n$. In particular, $\phi(t_n) = \phi(t'_n)$. Compactness implies the existence of a subsequence $(q(n))_{n\in\mathbb{N}}$ and of elements t, t' such that $t_{q(n)} \to t, t'_{q(n)} \to t'$, as $n \to \infty$. Continuity implies $\langle t, t' \rangle \in \ker(\phi) = \rho$, and $s = [t]_\sigma, s' = [t']_\sigma$. Thus ζ is closed.   ⬛

Now define inductively the n-fold composition of ζ:
$$\zeta^{(1)} := \zeta$$
$$\zeta^{(n+1)} := \zeta^{(n)} \circ \zeta,$$
where $\circ$ denotes the usual relational composition.

The following properties are easily established through a compactness argument using induction on n:

CLAIM 5.18
For each $n \in \mathbb{N}$

a. *$\zeta^{(n)} \subseteq T/\sigma$ is closed,*

b. *if $C \subseteq T/\sigma$ is compact, then the set*
$$\exists \zeta^{(n)}(C) = \{s \in T/\sigma \mid \exists s' \in C : \langle s, s' \rangle \in \zeta^{(n)}\}$$

is closed.

CLAIM 5.19
The transitive closure θ of ζ is a smooth equivalence relation on T/σ.

PROOF It is clear from the properties of ζ that θ is an equivalence relation. Smoothness needs to be shown, and we will exhibit a Borel measurable map into a Polish space with θ as its kernel. Write θ as

$$\theta = \bigcup_{n \in \mathbb{N}} \zeta^{(n)},$$

and let $C \subseteq T/\sigma$ be a compact set, then

$$\exists \theta(C) = \{s \in T/\sigma \mid \exists s' \in C : \langle s, s' \rangle \in \theta\}$$

is a measurable subset of T/σ by Claim 5.18. By Proposition 1.57 we can find a measurable selector w for the set valued map $s \mapsto \eta_\theta(s)$, hence a Borel measurable map $w : T/\sigma \to T/\sigma$ such that $w(s) \in [s]_\theta$ for each $s \in T/\sigma$. Now we know that T/σ is a Polish space, and from $\theta = \ker(w)$ we infer that θ is smooth. $\quad\Box$

We are in a position now to establish Proposition 5.15.

PROOF (of Proposition 5.15). 1. It is sufficient for the first part to establish $\rho \subseteq \theta \bullet \sigma$. In fact, let $t \, \rho \, t'$, then $[t]_\sigma \times [t']_\sigma \cap \rho \neq \emptyset$. This implies $\langle [t]_\sigma, [t']_\sigma \rangle \in \zeta \subseteq \theta$, which in turn establishes the inclusion and hence the Proposition.

2. Now assume that $\theta'_0 \bullet \rho = \theta_0 \bullet \sigma$, for smooth equivalence relations θ_0 on T/σ and θ'_0 on T/ρ holds, thus from Corollary 5.11 we infer $\sigma \subseteq \theta'_0 \bullet \rho$ and $\rho \subseteq \theta_0 \bullet \sigma$. In order to establish $\theta \subseteq \theta_0$ it is enough to show that $\zeta \subseteq \theta_0$. But if $\langle s, s' \rangle \in \zeta$, we know that $s \times s' \cap \theta_0 \bullet \sigma \neq \emptyset$, thus $\langle t, t' \rangle \in \theta_0 \bullet \sigma$ for some $t \in s = [t]_\sigma, t' \in s' = [t']_\sigma$. Consequently, $s \, \theta_0 \, s'$ holds. Interchanging the rôles of ρ and σ establishes that $\theta' \subseteq \theta'_0$ also holds. $\quad\Box$

For later use we record a property of θ-invariant Borel sets that characterizes these sets in terms of the equivalence relations from which θ is constructed. It gives an easy criterion on invariance and indicates that the relation θ will have some use in the discussions to follow.

LEMMA 5.20
Under the assumptions of Proposition 5.15, let $D \subseteq T/\sigma$ be a Borel set, where θ is defined as in Claim 5.19 as the equivalence relation generated through $\{\langle s, s' \rangle \mid s, s' \in T/\sigma, s \times s' \cap \rho \neq \emptyset\}$. Then these conditions are equivalent:

a. D is θ-invariant,

b. $\eta_\sigma^{-1}[D] \in \mathcal{INV}\left(\mathcal{B}(T), \sigma\right) \cap \mathcal{INV}\left(\mathcal{B}(T), \rho\right).$

PROOF   1. $a \Rightarrow b$: We know from Lemma 5.4 that $\eta_\sigma^{-1}[D] \in \mathcal{INV}\left(\mathcal{B}(T), \sigma\right)$ because $D \subseteq T/\sigma$ is a Borel set. Now let $t \in \eta_\sigma^{-1}[D]$ with $t \, \rho \, t'$. Then $[t]_\sigma \in D$ and $[t]_\sigma \times [t']_\sigma \cap \rho \neq \emptyset$, thus $\langle [t]_\sigma, [t']_\sigma \rangle \in \theta$, and, since D is θ-invariant, $t' \in \eta_\sigma^{-1}[D]$. Hence $\eta_\sigma^{-1}[D]$ is also ρ-invariant.

2. The implication $b \Rightarrow a$ is established through a routine argument using induction accounting for the construction of θ.   ⬜

5.2.4   Simulation Equivalence

We will construct a way of transporting a smooth equivalence relation along a map between equivalence classes. Such a map will be employed as a bridge between the relations; in particular we will transport vital properties along it. These properties will become apparent later, when we develop criteria for the bisimilarity of stochastic relations. Quite independent of this particular application the concept of spawning yields some insight into the nature of smooth equivalence relation in terms of the Borel structure on their factor spaces.

As a preparation for the definition of how two smooth relations relate to each other we will have a quick look at how the atoms of a countably generated σ-algebra are characterized through the generators.

LEMMA 5.21

Let $\mathcal{E} = \sigma(\{E_n \mid n \in \mathbb{N}\})$ be a countably generated σ-algebra over a set E. Define $A^1 := A, A^0 := E \setminus A$ for $A \subseteq E$, and put

$$E(s) := \bigcap_{n \in \mathbb{N}} E_n^{s(n)}$$

for $s \in \{0,1\}^{\mathbb{N}}$. Then there exists $\mathbf{F} \subseteq \{0,1\}^{\mathbb{N}}$ such that $\{E(s) \mid s \in \mathbf{F}\}$ are exactly the atoms of $\mathcal{E}$.

PROOF   0. We first note that as in the proof of Lemma 5.1

$$\mathcal{G}_{a,b} := \{A \subseteq E \mid a \in A \Leftrightarrow b \in A\}$$

is a σ-algebra. It is clear that the sets $E(\alpha)$ are elements of $\mathcal{E}$. Define $\mathbf{F}$ as the set of all indices s for which $E(s) \neq \emptyset$.

1. Now let $s \in \mathbf{F}$, then $E(s)$ is an atom of $\mathcal{E}$. Suppose it is not, then there exists $B \in \mathcal{E}$ with $\emptyset \neq B \subset E(s)$, so we can pick $a, b \in E(s)$ with $a \in B$ and

$b \notin B$. From part 0. we infer that there exists an index $m \in \mathbb{N}$ such that E_m contains exactly one of a, b. On the other hand, we see that the construction implies $[a \in E_n \Leftrightarrow b \in E_n]$ for each $n \in \mathbb{N}$. This is a contradiction, so $E(s)$ is an atom of $\mathcal{E}$.

2. Let A be an atom of $\mathcal{E}$, and put

$$h(A) := \bigcap \{E_n \mid A \subseteq E_n\} \cap \bigcap \{E \setminus E_n \mid A \cap E_n = \emptyset\},$$

then there exists $s \in \{0,1\}^{\mathbb{N}}$ such that $h(A) = E(s)$. Now $A \subseteq h(A)$ is immediate. Because $h(A)$ is an atom as well, we see $A = h(A) = E(s)$. $\Box$

We are poised to give a technical definition that permits stating how a smooth equivalence relation is transported through a map between classes in such a way that important properties are maintained. We call this *spawning*. This definition of spawning is at present on the level of equivalence relations. It will be later extended to incorporate congruences.

DEFINITION 5.22   *Let α and β be smooth equivalence relations on the analytic spaces X resp. Y, and assume that $\Upsilon : X/\alpha \rightarrow Y/\beta$ is a map between the equivalence classes. We say that α spawns β via $(\Upsilon, \mathcal{A}_0)$ iff $\mathcal{A}_0$ is a countable generator of $\mathcal{INV}(\mathcal{B}(X), \alpha)$ such that*

a. $\mathcal{A}_0$ *is closed under finite intersections,*

b. $\{\Upsilon_A \mid A \in \mathcal{A}_0\}$ *is a generator of $\mathcal{INV}(\mathcal{B}(Y), \beta)$, where $\Upsilon_A := \bigcup \{\Upsilon([x]_\alpha) \mid x \in A\}$.*

Thus if α spawns β, then the measurable structure induced by α on X is all we need for constructing the measurable structure induced by β on Y: the map Υ can be made to carry over the generator $\mathcal{A}_0$ from $\mathcal{INV}(\mathcal{B}(X), \alpha)$ to $\mathcal{INV}(\mathcal{B}(Y), \beta)$ and — in the light of Lemma 5.21 — to transport the atoms from one σ-algebra to the other. This is of particular interest since the atoms are just the equivalence classes. Hence α together with Υ and the generator $\mathcal{A}_0$ is all we may care to know or to learn about β.

The first condition reflects a measure-theoretic precaution: we will need to make sure, e.g., in the construction of the direct sum of stochastic relations that measures are uniquely determined by their values on a set of generators. This, however, can best be guaranteed if the generator is stable against taking finite intersections. Note that $\Upsilon_{A_1 \cap A_2} = \Upsilon_{A_1} \cap \Upsilon_{A_2}$ also holds, so that closedness under intersections is inherited through Υ.

Whenever we have two smooth equivalence relations such that one spawns the other we obtain on the sum of the underlying spaces a unique smooth relation the traces of which on the summands are just the given relations. Since we will introduce later on the sum of two relations, this effect will be studied now carefully.

LEMMA 5.23

Let X, Y, Z be analytic spaces, and assume that $f : X \to Z$ and $g : Y \to Z$ are surjective Borel maps. Assume $f(x) = g(y)$, then put $\Upsilon : [x]_{\ker(f)} \mapsto [y]_{\ker(g)}$ and $\Theta : [y]_{\ker(g)} \mapsto [x]_{\ker(f)}$. Let $\mathcal{C}$ be a countable generator of $\mathcal{B}(Z)$ that separates points and is closed under intersections. Then

a. $\ker(f)$ spawns $\ker(g)$ via $(\Upsilon, f^{-1}[\mathcal{C}])$,

b. $\ker(g)$ spawns $\ker(f)$ via $(\Theta, g^{-1}[\mathcal{C}])$.

PROOF   1. It is clear that both Υ and Θ are well defined, and that $f^{-1}[\mathcal{C}]$ as well as $g^{-1}[\mathcal{C}]$ are closed under intersections. Because $\mathcal{C}$ separates points, $f^{-1}[\mathcal{C}]$ determines $\ker(f)$, since $f(x) = f(x')$ is equivalent to $f(x) \in C \Leftrightarrow f(x') \in C$ for all $C \in \mathcal{S}$. Similarly for $\ker(g)$. Since the identity relation is a smooth equivalence relation on Z as well, Lemma 5.7 implies that

$$\mathcal{INV}\,(\mathcal{B}(X), \ker(f)) = \sigma(f^{-1}[\mathcal{C}]),$$
$$\mathcal{INV}\,(\mathcal{B}(Y), \ker(g)) = \sigma(g^{-1}[\mathcal{C}]).$$

2. Let $C \in \mathcal{C}$, then $\Upsilon_{f^{-1}[C]} = g^{-1}[C]$, since $y \in \Upsilon_{f^{-1}[C]}$ iff we can find $x \in f^{-1}[C]$ with $y \in \Upsilon([x]_{\ker(f)})$. Thus $f(x) = g(y) \in C$, hence $y \in g^{-1}[C]$. This implies the first part. The second part follows with exactly the same arguments.   □

We will use this Lemma later, e.g., in Section 5.4, for an investigation of the behavioral equivalence of stochastic relations.

LEMMA 5.24

Let α and β be smooth equivalence relations on the analytic spaces X resp. Y which spawn each other through the spawning maps $\Upsilon : X/\alpha \to Y/\beta$ resp. $\Xi : Y/\beta \to X/\alpha$. Then

a. There exists a unique smooth equivalence relation $\alpha \diamond \beta$ on $X + Y$ with these properties

i. $[x]_{\alpha \diamond \beta} \cap X = [x]_\alpha$ and $[x]_{\alpha \diamond \beta} \cap Y = \Upsilon([x]_\alpha)$ for all $x \in X$.

ii. $[y]_{\alpha \diamond \beta} \cap Y = [y]_\beta$ and $[y]_{\alpha \diamond \beta} \cap X = \Xi([y]_\beta)$ for all $y \in Y$.

b. Both X/α and Y/β are Borel isomorphic to $(X + Y)/\alpha \diamond \beta$.

PROOF   1. We consider the equivalence relation $\alpha \diamond \beta$ generated from $\{A_n + \Upsilon_{A_n} \mid n \in \mathbb{N}\}$ with $\mathcal{A}_0 = \{A_n \mid n \in \mathbb{N}\}$. Relation $\alpha \diamond \beta$ is evidently smooth, and it is uniquely determined through α and β. Let $x \in X$, then we can find $y \in Y$ with $\Upsilon([x]_\alpha) = [y]_\beta$, since Υ maps X/α to Y/β. It is easy to see that

$$\Upsilon([x]_\alpha) = \bigcap \{\Upsilon_{A_n} \mid A_n \in \mathcal{A}_0, x \in A_n\} \cap \bigcap \{Y \setminus \Upsilon_{A_n} \mid A_n \in \mathcal{A}_0, x \notin A_n\}.$$

2. The claim in the second part of part a. follows from the observation that β is the relation on Y which is generated from $\{\Upsilon_{A_n} \mid n \in \mathbb{N}\}$, and that α is the relation on X which is generated from $\{\Theta_{B_n} \mid n \in \mathbb{N}\}$ where $\mathcal{B}_0 = \{B_n \mid n \in \mathbb{N}\}$.

3. Define the map $\xi : X/\alpha \to (X+Y)/\alpha \diamond \beta$ through $\xi([x]_\alpha) := [x]_{\alpha \diamond \beta}$. Then ξ is well defined. It is onto: let $[y]_{\alpha \diamond \beta} \in (X+Y)/\alpha \diamond \beta$ for some $y \in Y$, thus

$$[y]_{\alpha \diamond \beta} \cap X = \Xi([y]_\beta) = [x]_\alpha$$

for some $x \in X$, so that

$$\xi([x]_\alpha) = [x]_{\alpha \diamond \beta} = [y]_{\alpha \diamond \beta}.$$

If $[x]_{\alpha \diamond \beta} = [x']_{\alpha \diamond \beta}$ for some $x, x' \in X$, then $x \, \alpha \, x'$; hence ξ is injective.

4. Now let $G \in \mathcal{B}((X+Y)/\alpha \diamond \beta)$ be a Borel subset of $(X+Y)/\alpha \diamond \beta$, so that $\eta_{\alpha \diamond \beta}^{-1}[G] \in \mathcal{INV}(\mathcal{B}(X+Y), \alpha \diamond \beta)$. Then

$$\eta_\alpha^{-1}\left[f^{-1}[G]\right] = \eta_{\alpha \diamond \beta}^{-1}[G] \cap X \in \mathcal{INV}(\mathcal{B}(X), \alpha),$$

which in turn implies that $f^{-1}[G] \in \mathcal{B}(X/\alpha)$. Thus ξ is Borel measurable. Take Borel set $H \in \mathcal{B}(X/\alpha)$, equivalently $\eta_\alpha^{-1}[H] \in \mathcal{INV}(\mathcal{B}(X), \alpha)$, then the argumentation above shows that $\eta_{\alpha \diamond \beta}^{-1}[H] = H \cup \Upsilon_H$. This is an $\alpha \diamond \beta$-invariant Borel set in $X+Y$, so that the image $f[H]$ of H is a Borel set. $\quad\square$

The isomorphism of two factor systems is an illustration of the concept of spawning.

PROPOSITION 5.25
Let T, T' be analytic spaces with smooth equivalence relations ρ resp. ρ'. Assume that $\Upsilon : T/\rho \to T'/\rho'$ is a Borel isomorphism, and let $\mathcal{A}$ be a countable generator of $\mathcal{INV}(\mathcal{B}(X), \rho)$ which is closed under finite intersections. Then ρ spawns ρ' via $(\Upsilon, \mathcal{A})$.

PROOF 0. The assumption that the generator $\mathcal{A}$ is closed under finite intersections is easily met: take an arbitrary countable generator $\mathcal{A}_0$, then

$$\left\{\bigcap \mathcal{F} \mid \mathcal{F} \subseteq \mathcal{A}_0 \text{ is finite}\right\}$$

is a countable generator which is closed under finite intersections.

1. If $A \in \mathcal{INV}\left(\mathcal{B}(T), \rho\right)$ is a ρ-invariant Borel set, then Υ_A is ρ'-invariant, and it is easily established that

$$\Upsilon_A = \eta_{\rho'}^{-1}\left[\Upsilon\left[\eta_\rho\left[A\right]\right]\right]$$

holds. From Proposition 5.5 we infer that $\mathcal{C}_1$ is a generator for $\mathcal{B}(T/\rho)$, where

$$\mathcal{C}_1 := \eta_\rho\left[\mathcal{INV}\left(\mathcal{B}(T), \rho\right)\right].$$

Consequently, $\{\Upsilon_A \mid A \in \mathcal{A}\}$ is a generator for the ρ'-invariant Borel sets $\mathcal{INV}\left(\mathcal{B}(T'), \rho'\right)$, because we may conclude

$$
\begin{aligned}
\mathcal{INV}\left(\mathcal{B}(T'), \rho'\right) &= \eta_{\rho'}^{-1}\left[\mathcal{B}(T'/\rho')\right] \text{ (by Lemma 5.4)} \\
&= \eta_{\rho'}^{-1}\left[\Upsilon\left[\mathcal{B}(T'/\rho')\right]\right] \text{ (since } \Upsilon \text{ is a Borel isomorphism)} \\
&= \sigma\left(\left\{\eta_{\rho'}^{-1}\left[\Upsilon\left[C\right]\right] \mid C \in \mathcal{C}_1\right\}\right) \text{ (by construction of } \mathcal{C}_1) \\
&= \sigma\left(\{\Upsilon_A \mid A \in \mathcal{INV}\left(\mathcal{B}(T), \rho\right)\}\right) \\
&= \sigma\left(\{\Upsilon_A \mid A \in \mathcal{A}\}\right) \text{ (since } \mathcal{A} \text{ generates } \mathcal{INV}\left(\mathcal{B}(T), \rho\right)).
\end{aligned}
$$

$\Box$

The proof is technically a bit laborious. The statement, however, will be most useful in permitting us to show that stochastic relations are bisimilar, provided they have isomorphic factors. Working with isomorphisms alone for characterizing bisimilarity may be too strong a condition. In the application to modal logic in Section 6.2 we will see that the equivalence relation which is induced on states through having the same logic satisfies the condition on spawning, but it is far from clear in this case whether or not the corresponding factor spaces are Borel isomorphic. Consequently, it seems to be worthwhile to work with the weaker condition.

5.3   Factoring

Observing a stochastic relation $\mathsf{K} = (X, Y, K)$, elements with equivalent behavior are identified. This leads to a pair (α, β) of equivalence relations on the inputs X resp. the outputs Y with the idea that equivalent inputs lead to equivalent outputs. While equivalent inputs can be described directly through α, the equivalence of outputs requires a description on the level of measurable sets. This leads then naturally to the notion of a congruence, which will be defined in this section. We will investigate congruent systems and present some technical properties that shed some light on the underlying invariant sets. This in turn will help us to investigate properties of specific congruences.

5.3.1 Congruences

Think of a stochastic relation $\mathsf{K} = (X, Y, K)$ as a model that relates inputs and outputs, and assume that there are equivalences α and β on inputs resp. outputs. Two inputs $x, x' \in X$ cannot be distinguished through α iff they are α-equivalent. It is less intuitive to describe distinguishing outputs through β, in particular when we do not have a handle on specific outputs but rather on sets of them. We argue that a set $B \subseteq Y$ cannot be distinguished through β iff whenever $y \in B$ and $y \mathrel{\beta} y'$ we have $y' \in B$ as well (or: it must not happen that y' with $y' \mathrel{\beta} y$ fails to be a member of B, but $y \in B$ holds). This entails that B is invariant with respect to β.

This consideration leads to the fairly fundamental definition of a congruence for K.

DEFINITION 5.26 *A congruence* $\mathsf{c} = (\alpha, \beta)$ *for the stochastic relation* $\mathsf{K} = (X, Y, K)$ *over the analytic spaces X and Y is a pair of smooth equivalence relations α on X and β on Y such that $K(x)(D) = K(x')(D)$ holds whenever $x \mathrel{\alpha} x'$ and D is an β-invariant measurable subset of Y.*

In algebraic theories, kernels of morphisms and congruences are basically the same thing. This is also true in the present case. Denote for the morphism $\mathsf{f} : \mathsf{K}_1 \to \mathsf{K}_2$ with $\mathsf{f} = (\phi, \psi)$ its kernel $\ker(\mathsf{f})$ by the pair $(\ker(\phi), \ker(\psi))$.

PROPOSITION 5.27
If $\mathsf{f} : \mathsf{K} \to \mathsf{K}'$ is a morphism for the stochastic relations K and K', then $\ker(\mathsf{f})$ is a congruence for K.

PROOF Let $\mathsf{K} = (X, Y, K)$ and $\mathsf{K}' = (X', Y', K')$ with $\mathsf{f} = (\phi, \psi)$. Let $x_1 \ker(\phi) x_2$ and $D \subseteq Y$ be a $\ker(\psi)$-invariant Borel subset of Y. Lemma 5.4 shows that $D = \psi^{-1}[D']$ for some Borel set $D' \subseteq Y'$. Thus

$$
\begin{aligned}
K(x_1)(D) &= K(x_1)(\psi^{-1}[D']) \\
&= (\mathfrak{S}(\psi) \circ K)(x_1)(D') \\
&= (K' \circ \phi)(x_1)(D') \\
&= K(\phi(x_1))(D') \\
&= K(\phi(x_2))(D') \\
&= K(x_2)(D),
\end{aligned}
$$

since $\mathsf{f} = (\phi, \psi)$ is a morphism. ☐

This construction permits introducing factor objects. They will be heavily used throughout.

PROPOSITION 5.28
Let $c = (\alpha, \beta)$ *be a congruence on the stochastic relation* $\mathsf{K} = (X, Y, K)$ *with analytic spaces* X *and* Y, *and define*

$$K_{\alpha,\beta}([x]_\alpha)(D) := K(x)(\eta_\beta^{-1}[D])$$

for $x \in X, D \in \mathcal{B}(Y/\beta)$, *then*

a. $K_{\alpha,\beta} : X/\alpha \rightsquigarrow Y/\beta$ *defines a stochastic relation* K/c *over the analytic spaces* X/α *and* Y/β,

b. $\eta_c := (\eta_\alpha, \eta_\beta) : \mathsf{K} \to \mathsf{K}/c$ *is a morphism.*

We call
$$\mathsf{K}/c := (X/\alpha, Y/\beta, K_{\alpha,\beta})$$

the factor object *(of* K *with respect to* c*).*

PROOF   0. We know from Proposition 1.53 that the factor spaces X/α and Y/β are analytic spaces.
  1. Given $D \in \mathcal{B}(Y/\beta)$, $\eta_\beta^{-1}[D]$ is an invariant Borel set, thus

$$x \mapsto K(x)(\eta_\beta^{-1}[D])$$

does depend only on the α-class of $x \in X$. Consequently, $K_{\alpha,\beta}$ is well defined.
  2. $K_{\alpha,\beta} : X/\alpha \rightsquigarrow Y/\beta$ is a stochastic relation. In fact, it is plain that $K_{\alpha,\beta}([x]_\alpha)$ is a subprobability measure on $\mathcal{B}(Y/\beta)$, so it remains to show that $t \mapsto K_{\alpha,\beta}(t)(D)$ is a $\mathcal{B}(X/\alpha)$-measurable map for each $D \in \mathcal{B}(Y/\beta)$. Fix such a D and a Borel set $F \subseteq \mathbb{R}$, then

$$F_D := \{x \in X \mid K(x)(\eta_\beta^{-1}[D]) \in F\}$$

is a Borel set in X, and since $\eta_\beta^{-1}[D]$ is β-invariant, F_D is α-invariant with

$$\{t \in X/\alpha \mid K_{\alpha,\beta}(t)(D) \in F\} = \eta_\alpha[F_D] \in \mathcal{B}(X/\alpha)$$

by Corollary 5.5. This establishes measurability.
  3. The construction of $K_{\alpha,\beta}$ yields $K_{\alpha,\beta} \circ \eta_\alpha = \mathfrak{S}(\eta_\beta) \circ K$, hence η_c is a morphism.   ∎

Let us see what happens if the second component β is the universal relation. If fact, let $\mathsf{K} = (X, Y, K)$ be a Polish object such that for simplicity $K(x)(Y) = 1$ holds for each $x \in X$. Since we know that for the universal relation U_Y on Y the invariant Borel sets are just $\{\emptyset, Y\}$ (see page 182), it is clear that (α, U_Y) is a congruence for an arbitrary smooth relation α on X. But it says only that $K(x)(Y) = K(x')(Y)$ and $K(x)(\emptyset) = K(x')(\emptyset)$ hold, whenever $x \, \alpha \, x'$, so it is quite trivial.

DEFINITION 5.29   *Call a congruence* $c = (\alpha, \beta)$ *for a stochastic relation* $K = (X, Y, K)$ nontrivial *iff* $\beta \neq U_Y$.

We will have to take care of the nontriviality of congruences when investigating the problem of the bisimilarity of relations.

Restricting a stochastic relation to the invariant sets of a congruence yields a stochastic relation again; this will be of use when discussing sets of states that accept the same formula of a logic (cp. Section 6.4). Interestingly, the converse holds as well, as we will see now.

LEMMA 5.30

Let α and β be smooth equivalence relations on the analytic spaces X resp. Y, and assume $K : X \rightsquigarrow Y$ is a stochastic relation. Then the following conditions are equivalent:

a. (α, β) is a congruence for K.

b. $K : (X, \mathcal{INV}\,(\mathcal{B}(X), \alpha)) \rightsquigarrow (Y, \mathcal{INV}\,(\mathcal{B}(Y), \beta))$ is a stochastic relation.

PROOF   1.   We need to establish for $a \Rightarrow b$ the following: If $B \in \mathcal{INV}\,(\mathcal{B}(Y), \beta)$ is a β-invariant Borel set in Y, and $E \subseteq \mathbb{R}_+$ is a Borel set in the real line, then

$$(K(\cdot)(B))^{-1}\,[E] = \{x \in X \mid K(x)(B) \in E\}$$

is an α-invariant Borel set in X. In fact, let $K(x)(B) \in E$, and assume that $x \, \alpha \, x'$. Since $B \in \mathcal{INV}\,(\mathcal{B}(Y), \beta)$, we know that $K(x)(B) = K(x')(B)$. This establishes the assertion, since $\{x \in X \mid K(x)(B) \in E\}$ is a Borel set on account of $K : (X, \mathcal{B}(X)) \rightsquigarrow (Y, \mathcal{B}(Y))$ being a stochastic relation.

2.   Assume for $b \Rightarrow a$ that $x \, \alpha \, x'$, and that $B \in \mathcal{INV}\,(\mathcal{B}(Y), \beta)$ is a β-invariant Borel set in Y. We show $K(x)(B) = K(x')(B)$. Since $K(\cdot)(B)$ is an $\mathcal{INV}\,(\mathcal{B}(X), \alpha)$-measurable function by assumption, we know that the set $\{\widehat{x} \in X \mid K(\widehat{x})(B) \leq q\}$ is α-invariant for each real number q. Thus we see $K(x)(B) \leq q$ iff $K(x')(B) \leq q$. Since q is arbitrary, we may conclude that in fact $K(x)(B) = K(x')(B)$ holds.   ⬜

We obtain as a consequence that the integral for functions that are measurable with the respect to the invariant sets have some invariance properties.

COROLLARY 5.31

Let (α, β) be a congruence for the stochastic relation $K : X \rightsquigarrow Y$. Assume furthermore that $f : Y \to \mathbb{R}$ is a bounded real-valued function which is $\mathcal{INV}\,(\mathcal{B}(Y), \beta)$-$\mathcal{B}(\mathbb{R})$-measurable. Then

$$\int_W f \, dK(x) = \int_W f \, dK(x'),$$

whenever $x \; \alpha \; x'$.

PROOF   Because f can be decomposed into a positive and a negative part, we may and do assume that $f : Y \to \mathbb{R}_+$ holds. We know from Proposition 1.5 that f can be approximated from below by step functions, i.e., by functions of the form $f_n := \sum_{i=0}^{k_n} q_{n,i} \cdot \chi_{A_{n,i}}$ with coefficients $q_{n,i} \geq 0$ and $A_{n,i} \in \mathcal{INV}(\mathcal{B}(Y), \beta)$. Thus $f_1(y) \leq f_2(y) \leq f_3(y) \dots$, and $f(y) = \sup_{n \in \mathbb{N}} f_n(y)$ holds for all $y \in Y$. But then we obtain from the Bounded Convergence Theorem for $x \; \alpha \; x'$:

$$\int_Y f \; dK(x) = \lim_{n \to \infty} \int_Y f_n \; dK(x)$$

$$= \lim_{n \to \infty} \sum_{i=0}^{k_n} q_{n,i} \cdot K(x)(A_{n,i})$$

$$= \lim_{n \to \infty} \sum_{i=0}^{k_n} q_{n,i} \cdot K(x')(A_{n,i})$$

$$= \lim_{n \to \infty} \int_Y f_n \; dK(x')$$

$$= \int_Y f \; dK(x').$$

This settles the assertion.   ▯

REMARK 5.32   An alternative to the proof above uses the Choquet representation from Proposition 1.61

$$\int_Y f \; dK(x) = \int_0^\infty K(x)(\{f > t\}) \; dt,$$

where $\{f > t\} := \{y \in Y \mid f(y) > t\}$. and $f \geq 0$. The latter is a β-invariant Borel subset of Y. Because of this set's invariance, we see that

$$K(x)(\{f > t\}) = K(x')(\{f > t\})$$

holds for each t, establishing the claim.   ▯

This section has provided us with a small set of quite effective tools. Questions pertaining to smooth equivalence relations will occur over and over again, so that we provide here a concise, central locus of information.

5.3.2   Isomorphism Theorems

Now fix an analytic object $\mathsf{K} = (X, Y, K)$, and let $\mathsf{c} = (\rho, \tau)$ be a congruence on K. Assume that $\mathsf{d} = (\kappa, \lambda)$ is a congruence of K/c. Define $\mathsf{d} \bullet \mathsf{c} := (\kappa \bullet \rho, \lambda \bullet \tau)$.

PROPOSITION 5.33
d $\bullet$ c *is a congruence on* K, *and* K/d $\bullet$ c *is isomorphic to* (K/c)/d.

PROOF 1. The first assertion follows from Corollary 5.27 together with the observation that $(\kappa \bullet \rho, \lambda \bullet \tau) = (\ker (\eta_\kappa \circ \eta_\rho), \ker (\eta_\lambda \circ \eta_\tau))$ holds.

2. Construct the Borel isomorphisms $g_{\rho,\kappa} : X/\kappa \bullet \rho \to (X/\rho)/\kappa$ and $g_{\tau,\lambda} : Y/\lambda \bullet \tau \to (Y/\tau)/\lambda$ with their respective inverses $h_{\rho,\kappa}$ and $h_{\tau,\lambda}$ as in the proof of Proposition 5.10. We show that the inner and the outer diagram

$$
\begin{array}{ccc}
X/\kappa \bullet \rho & \underset{h_{\rho,\kappa}}{\overset{g_{\rho,\kappa}}{\rightleftarrows}} & (X/\rho)/\kappa \\[2pt]
{\scriptstyle K_{\kappa\bullet\rho,\lambda\bullet\tau}}\Big\downarrow & & \Big\downarrow {\scriptstyle (K_{\rho,\tau})_{\kappa,\lambda}} \\[2pt]
\mathfrak{S}\,(Y/\lambda \bullet \tau) & \underset{\mathfrak{S}\,(g_{\tau,\lambda})}{\overset{\mathfrak{S}\,(h_{\tau,\lambda})}{\rightleftarrows}} & \mathfrak{S}\,((Y/\tau)/\lambda)
\end{array}
$$

both commute.

3. Let $B \in \mathcal{B}((Y/\tau)/\lambda)$, a Borel set in $(Y/\tau)/\lambda$, then

$$
\begin{aligned}
K_{\kappa\bullet\rho,\lambda\bullet\tau}([x]_{\kappa\bullet\rho}) \left(g_{\tau,\lambda}^{-1}[B] \right) &= K(x)(\eta_{\lambda\bullet\tau}^{-1} \left[g_{\tau,\lambda}^{-1}[B] \right]) \\
&= K(x)(\eta_\tau^{-1} \left[\eta_\lambda^{-1}[B] \right]) \\
&= K_{\rho,\tau}([x]_\rho)(\eta_\lambda^{-1}[B]) \\
&= (K_{\rho,\tau})_{\kappa,\lambda} \, (g_{\rho,\kappa}([x]_\rho))(B),
\end{aligned}
$$

because $g_{\tau,\lambda} \circ \eta_{\lambda\bullet\tau} = \eta_\lambda \circ \eta_\tau$. Thus the outer diagram commutes. This implies that

$$
\mathsf{g} := (g_{\rho,\kappa}, g_{\beta,\tau}) : \mathsf{K}/\mathsf{d} \bullet \mathsf{c} \to (\mathsf{K}/\mathsf{c})/\mathsf{d}
$$

is a morphism.

4. Suppose that $G \in \mathcal{B}(Y/\lambda \bullet \tau)$ is a Borel set, then

$$
\begin{aligned}
K_{\kappa\bullet\rho,\lambda\bullet\tau}(h_{\rho,\kappa}(\left[[x]_\rho\right]_\kappa)(G) &= K_{\kappa\bullet\rho,\lambda\bullet\tau}([x]_{\kappa\bullet\rho})(G) \\
&= K(x)(\eta_{\lambda\bullet\tau}^{-1}[G]) \\
&= K_{\rho,\tau}([x]_\rho)(\eta_\lambda^{-1} \left[h_{\tau,\lambda}^{-1}[G] \right]) \\
&= (K_{\rho,\tau})_{\kappa,\lambda} \, (\left[[x]_\rho\right]_\lambda)(h_{\beta,\tau}^{-1}[G]).
\end{aligned}
$$

This is so since $\eta_{\lambda\bullet\tau} = h_{\tau,\lambda} \circ \eta_\lambda \circ \eta_\tau$ holds (see the proof of Proposition 5.10). Thus the inner diagram commutes. This implies that

$$
\mathsf{h} := (h_{\rho,\kappa}, h_{\beta,\tau}) : (\mathsf{K}/\mathsf{c})/\mathsf{d} \to \mathsf{K}/\mathsf{d} \bullet \mathsf{c}
$$

is a morphism. It is plain that h is left- and right inverse to g. $\quad\square$

Factoring a stochastic relation with a congruence entails identifying inputs resp. outputs that have been observed as representing identical behavior. Proposition 5.33 says then that identifying identical behavior in observing the factor system amounts to a system that can also be obtained through a single observational step from the original system. This means that there are no arbitrary long chains of factor systems which could not have been obtained directly from the original system, or, that factoring does not change the fundamental behavior of a system (after all, a system is bisimilar to its factor systems, bisimilarity requesting the existence of a span of morphisms, as we will see in Section 5.4).

Algebraically, this proposition is quite similar to the well-known Second Isomorphism Theorem of Group Theory, cp. (Lang, 1965, § I.4): Factoring the quotient of a normal subgroup gives a group isomorphic to a factor. A similar but slightly stronger construction for coalgebras is carried out by Rutten (Rutten, 2000, Theorem 7.4) in the context of bisimulation relations for coalgebras. Proposition 5.33 and Rutten's Theorem are not directly comparable, however, since the functor underlying the coalgebra is assumed to preserve weak pullbacks (which is no realistic assumption for stochastic relations by Corollary 4.15), and since the relationship between bisimulations and congruences is slightly less involved in the coalgebraic case.

Let (α, β) and (α', β') be pairs of equivalence relations, and define

$$(\alpha, \beta) \preceq (\alpha', \beta') \Leftrightarrow \alpha \subseteq \alpha' \text{ and } \beta \subseteq \beta'.$$

Thus $(\alpha, \beta) \preceq (\alpha', \beta')$ iff α refines α' and β refines β' simultaneously. It is clear that $c \preceq d \bullet c$ for each congruence d.

PROPOSITION 5.34

Assume that $f : K \to K'$ *is a morphism, and let* c *be a congruence on* K *such that* $c \preceq \ker(f)$. *Then there exists a unique morphism* $f_c : K/c \to K'$ *with* $f = f_c \circ \eta_c$.

PROOF   1. Let $K = (X, Y, K), K' = (X', Y', K')$ with $\phi : X \to X', \psi : Y \to Y'$ constituting morphism f, and $c = (\alpha, \beta)$. Because $\alpha \subseteq \ker(\phi), \beta \subseteq \ker(\psi)$, the maps

$$\phi_\alpha([x]_\alpha) := \phi(x),$$
$$\psi_\beta([y]_\beta) := \psi(y)$$

are well defined. Since ϕ is $\mathcal{B}(X)$-$\mathcal{B}(X')$-measurable, and since $\mathcal{B}(X)/\alpha$ is the final σ-algebra on X/α with respect to η_α, $\mathcal{B}(X)/\alpha$-$\mathcal{B}(X')$-measurability of ϕ_α is inferred. A similar argument is used for ψ_β. Clearly, these maps are onto.

2. It remains to show that $f_c := (\phi_\alpha, \psi_\beta)$ is a morphism. In fact, let $D' \subseteq Y'$ be a Borel set, then

$$K'(\phi_\alpha([x]_\alpha))(D') = K'(\phi(x))(D') = K(x)(\psi^{-1}[D']) =$$
$$K_{\alpha,\beta}([x]_\alpha)(\psi_\beta^{-1}[D']) = (\mathfrak{S}(\psi_\beta) \circ K_{\alpha,\beta})([x]_\alpha)(D'),$$

because $\psi^{-1}[D'] = \eta_\beta^{-1}\left[\psi_\beta^{-1}[D']\right]$, and because $(\eta_\alpha, \eta_\beta)$ is a morphism. Consequently, the equality $K' \circ \phi_\alpha = \mathfrak{S}(\psi_\beta) \circ K_{\alpha,\beta}$ has been established. Uniqueness follows, since η_c is an epi. ∎

COROLLARY 5.35

Assume that $f : K \to K'$ *is a morphism. Then there exists a unique isomorphism* $f^\sharp : K/\ker(f) \to K'$ *with* $f = f^\sharp \circ \eta_{\ker(f)}$.

PROOF Define $f^\sharp := f_{\ker(f)}$, then the maps constituting this morphism are bijective Borel maps, so by (Srivastava, 1998, Proposition 4.5.1) they are Borel isomorphisms. The equations establishing the morphism property for $f_{\ker(f)}$ show that the inverses also constitute a morphism. ∎

COROLLARY 5.36

Let c *and* d *be congruences on* K, *then the following statements are equivalent:*

a. $c \prec d$

b. $d = e \bullet c$ *for some congruence* e *on* K.

PROOF The implication $b \Rightarrow a$ is obvious. Assume that $c \preceq d = \ker(\eta_d)$ holds. Then the assertion follows from Proposition 5.34 together with Corollary 5.27. ∎

This property is somewhat surprising in that it relates the refinement of congruences to factor spaces. If congruence c is finer than congruence d, then d can be obtained through observing and factoring the behavior in the factor system for c (so that the original system does not have to be observed but rather a simplified one).

5.4   Bisimulations

Bisimulations are introduced as spans of morphisms such that common events exist. They relate two systems in terms of their elements, hence in terms of nondeterministic relations of their state spaces. In fact, assume that $(S, (\to_a)_{a \in A})$ and $(S', (\to'_a)_{a \in A})$ are two labeled transition systems, then a relation $R \subseteq S \times S'$ is called a bisimulation iff

- Whenever $\langle s, s' \rangle \in R$ and $s \to_a s_1$, then there exists s'_1 with $s' \to'_a s'_1$ and $\langle s_1, s'_1 \rangle \in R$.

- Whenever $\langle s, s' \rangle \in R$ and $s' \to'_a s'_1$, then there exists s_1 with $s \to_a s_1$ and $\langle s_1, s'_1 \rangle \in R$.

Interpreting a labeled transition system as a coalgebra (S, α_S) for the functor $\mathfrak{F} := \mathfrak{Pow}(A \times \cdot)$, it is an easy exercise to show that R is a congruence iff there exists a coalgebraic structure α_R on R such that this diagram commutes:

$$
\begin{array}{ccccc}
S & \xleftarrow{\pi_S} & R & \xrightarrow{\pi_{S'}} & S' \\
\Big\downarrow{\alpha_S} & & \Big\downarrow{\alpha_R} & & \Big\downarrow{\alpha_{S'}} \\
\mathfrak{F}(S) & \xleftarrow{\mathfrak{F}(\pi_S)} & \mathfrak{F}(R) & \xrightarrow{\mathfrak{F}(\pi_{S'})} & \mathfrak{F}(S')
\end{array}
$$

In Section 5.6 we will specialize the discussion to the case that the morphisms are projections, and relate the different notions of bisimulations to each other. The present section is devoted to the general case, which turns out to be rich enough.

DEFINITION 5.37   *The stochastic relations* $\mathsf{K} = (X, Y, K)$ *and* $\mathsf{L} = (V, W, L)$ *are called* bisimilar *iff there exists a stochastic relation* $\mathsf{M} = (A, B, M)$ *and morphisms* $\mathsf{f} = (\phi, \psi) : \mathsf{M} \to \mathsf{L}, \mathsf{g} = (\gamma, \delta) : \mathsf{M} \to \mathsf{L}$ *such that*

a. the diagram

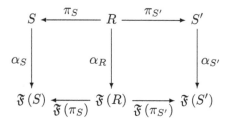

is commutative,

b. the σ-algebra $\psi^{-1}[\mathcal{B}(Y)] \cap \delta^{-1}[\mathcal{B}(W)]$ is nontrivial, i.e., contains not only $\emptyset$ and B.

The relation M *is called* mediating.

The first condition on bisimilarity states that f and g form a span of $\mathfrak{Stoch}$-morphisms

$$\mathsf{K} \xleftarrow{\quad f \quad} \mathsf{M} \xrightarrow{\quad g \quad} \mathsf{L},$$

thus we have for each $a \in A, D \in \mathcal{B}(Y), E \in \mathcal{B}(W)$ the equalities

$$K(\phi(a))(D) = (\mathfrak{S}\,(\phi) \circ M)(a)(D) = M(a)(\phi^{-1}\,[D])$$

and

$$L(\psi(a))(E) = (\mathfrak{S}\,(\psi) \circ M)(a)(E) = M(a)(\psi^{-1}\,[E]).$$

The second condition states that we can find an event $C^* \in \mathcal{B}(B)$ which is common to both K and L in the sense that

$$\psi^{-1}\,[D] = C^* = \delta^{-1}\,[E]$$

for some $D \in \mathcal{B}(Y)$ and $E \in \mathcal{B}(W)$ such that both $C^* \neq \emptyset$ and $C^* \neq B$ hold (note that for $C^* = \emptyset$ or $C^* = W$ we can always take the empty and the full set, resp.). Given such a C^* with D and E from above we get for each $a \in A$

$$
\begin{aligned}
K(\phi(a))(D) &= M(a)(\psi^{-1}\,[D]) \\
&= M(a)(C^*) \\
&= M(a)(\delta^{-1}\,[E]) \\
&= L(\gamma(a))(E),
\end{aligned}
$$

thus the event C^* ties K and L together. Loosely speaking, $\psi^{-1}\,[\mathcal{B}(Y)] \cap \delta^{-1}\,[\mathcal{B}(W)]$ can be described as the σ-algebra of common events, which is required to be nontrivial. If $Y = W$, another interpretation of common events is discussed in Lemma 5.8 in terms of invariant Borel sets. The discussion there, however, addresses different issues than in the present context.

Note that without the second condition two relations K and L which are strictly probabilistic (i.e., for which the entire space is always be assigned probability one) would always be bisimilar: Put $A := X \times V, B := Y \times W$ and set for $\langle x, v \rangle \in A$ as the mediating relation $M(x, v) := K(x) \otimes L(v)$, then the projections will make the diagram commutative. It is also clear that this argument does not work for the subprobabilistic case. This curious behavior of probabilistic relations is a bit surprising, but these relations step out of line in other situations as well: e.g., it will be shown that the full subcategory of probabilistic relations in $\mathfrak{anStoch}$ has a final object, while $\mathfrak{anStoch}$ itself does not have one; see Section 5.7, in particular Corollary 5.59 and the discussion leading to it. The second condition in Definition 5.37 serves to prevent this somewhat anomalous behavior; it is technically not too restrictive, as we will see below.

An important instance of congruences and factor spaces is furnished through simulation equivalent congruences.

DEFINITION 5.38   *Let* $\mathsf{K} = (X, Y, K)$ *and* $\mathsf{K}' = (X', Y', K')$ *be Polish objects with congruences* $\mathsf{c} = (\alpha, \beta)$ *and* $\mathsf{c}' = (\alpha', \beta')$, *respectively.*

a. *Congruence* c *simulates* c' *(symbolically* $\mathsf{c} \propto \mathsf{c}'$*) iff* α *spawns* α' *via* $(\Upsilon, \mathcal{A}_0)$, β *spawns* β' *via* $(\Theta, \mathcal{B}_0)$ *such that*

$$\forall x \in X \forall x' \in \Upsilon([x]_\alpha) \forall B \in \mathcal{B}_0 : K(x)(B) = K'(x')(\Theta_B).$$

b. *Call these congruences* simulation equivalent *iff both* $\mathsf{c} \propto \mathsf{c}'$ *and* $\mathsf{c}' \propto \mathsf{c}$ *hold.*

Thus simulation equivalent congruences behave in exactly the same way. The same behavior is exhibited on each atom, i.e., equivalence class, as far as the input is concerned, and on the respective invariant output sets. It becomes visible now that a characterization of equivalent behavior through congruences exhibits the double face of congruences: it is certainly necessary to use the equivalence relation on the input spaces; but since the behavior on the output spaces is modeled through probabilities, we need also the invariant Borel sets for a characterization.

We will show now how simulation equivalent congruences on stochastic relations give rise to a factor object built on their sum. This construction will be of use in Proposition 5.39 for investigating the bisimilarity of stochastic relations.

Assume that c and c' are simulation equivalent congruences on the Polish objects $\mathsf{K} = (X, Y, K)$, and $\mathsf{K}' = (X', Y', K')$, respectively. Construct for K and K' the direct sum

$$\mathsf{K} \oplus \mathsf{K}' := (X + X', Y + Y', K \oplus K'),$$

where the only nonobvious construction is $K \oplus K'$: put for the Borel set $E \subseteq Y + Y'$

$$(K \oplus K')(z)(E) := \begin{cases} K(z)(E \cap Y), & \text{if } z \in X \\ K'(z)(E \cap Y'), & \text{if } z \in X', \end{cases}$$

then clearly $K \oplus K' : X + X' \rightsquigarrow Y + Y'$. Define on $X + X'$ resp. $Y + Y'$ the σ-algebras

$$\mathcal{G} := \{C + C' \mid C \in \mathcal{INV}\,(\mathcal{B}(X), \alpha), C' \in \mathcal{INV}\,(\mathcal{B}(X'), \alpha')\}$$
$$\mathcal{H} := \{D + D' \mid D \in \mathcal{INV}\,(\mathcal{B}(Y), \beta), D' \in \mathcal{INV}\,(\mathcal{B}(Y'), \beta')\},$$

then $\mathcal{G}$ and $\mathcal{H}$ are countably generated sub-σ-algebras of the respective Borel sets. Because the σ-algebras in question are countably generated, so is their sum, and because the congruences are simulation equivalent, we claim that $z \, (\alpha \diamond \alpha') \, z'$ implies $(K \oplus K')(z)(F) = (K \oplus K')(z')(F)$ for all $F \in \mathcal{H}$. To establish this, fix $z \in X, z' \in X'$, and consider

$$\mathcal{S} := \{F \in \mathcal{H} \mid (K \oplus K')(z)(F) = (K \oplus K')(z')(F)\}.$$

Since the congruences are simulation equivalent, this is a σ-algebra containing the generator $\{D_n + \Theta_{D_n} \mid n \in \mathbb{N}\}$, where β spawns β' via $(\Theta, \{D_n \mid n \in \mathbb{N}\})$. Since the generator is closed under finite intersections, measures are uniquely determined by the π-λ-Theorem 1.1. This implies $\mathcal{H} \subseteq \sigma(\mathcal{S})$, thus $\mathcal{H} = \mathcal{S}$. Consequently,

$$\mathcal{G} = \mathcal{INV}\left(\mathcal{B}(X + X'), \alpha \diamond \alpha'\right)$$
$$\mathcal{H} = \mathcal{INV}\left(\mathcal{B}(Y + Y'), \beta \diamond \beta'\right),$$

and $c \diamond c' := (\alpha \diamond \alpha', \beta \diamond \beta')$ is a congruence on $K \oplus K'$.

The factor object $(K \oplus K')/(c \diamond c')$ constructed in this way will be investigated more closely in Proposition 5.39 below. There we will establish that K and K' are bisimilar, provided they have simulation equivalent nontrivial congruences.

Simulation equivalent congruences give rise to bisimilar stochastic relations; thus if we are presented with two stochastic relations for which we can establish the existence of nontrivial simulation equivalent congruences, then the relations are bisimilar. This is a rather far-reaching generalization of the by now well-known characterization of bisimilarity of labeled Markov transition systems through mutually equivalent states, which will be discussed at length in Section 6.2. Note that we give here is an *intrinsic* characterization of bisimilarity: we investigate the relations and their congruences on their own, but we do not need an external instance (like a logic) to determine bisimilarity. The technical tool for establishing this property is the existence of semi-pullbacks, which we have established in Theorem 4.13.

PROPOSITION 5.39

If there exists nontrivial congruences c_i on the Polish objects K_i for $i = 1, 2$ that are simulation equivalent, then

a. there are morphisms

$$K_1 \xrightarrow{\quad f_1 \quad} (K_1 \oplus K_2)/(c_1 \diamond c_2) \xleftarrow{\quad f_2 \quad} K_2$$

b. K_1 and K_2 are bisimilar.

PROOF　1. Assume $K_i = (X_i, Y_i, K_i)$ and $c_i = (\alpha_i, \beta_i)$ for $i = 1, 2$. Construct the sum $K_1 \oplus K_2$ as above, and let (κ_i, λ_i) be the corresponding injections, which are, however, no morphisms. Let $(\eta_{\alpha_1 \diamond \alpha_2}, \eta_{\beta_1 \diamond \beta_2}) : K_1 \oplus K_2 \to (K_1 \oplus K_2)/(c_1 \diamond c_2)$ be the factor map, then $(\eta_{\alpha_1 \diamond \alpha_2} \circ \kappa_i, \eta_{\beta_1 \diamond \beta_2} \circ \lambda_i)$ constitutes a morphism $K_i \to (K_1 \oplus K_2)/(c_1 \diamond c_2)$, as will be shown now. Surjectivity has to be established, and we have to show that the σ-algebra of common events is nontrivial.

2. Each equivalence class $a \in (X_1 + X_2)/(\alpha_1 \diamond \alpha_2)$ can be represented as $a = [x_1]_{\alpha_1} + [x_2]_{\alpha_2}$ for some suitably chosen $x_1 \in X_1, x_2 \in X_2$. Similarly, each

equivalence class $b \in (Y_1 + Y_2)/(\beta_1 \diamond \beta_2)$ can be written as $b = [y_1]_{\beta_1} + [y_2]_{\beta_2}$ for some $y_1 \in Y_1, y_2 \in Y_2$. Conversely, the sum of classes is a class again. This follows from Lemma 5.24.

3. Now we have the following diagram:

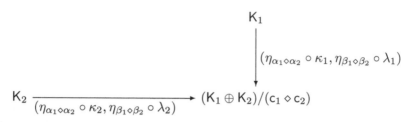

This yields part a.

4. The semi-pullback of the pair of morphisms with a joint target constructed in the first step exists by Corollary 4.12. It is a Polish object (A, B, M), where

$$A := \{\langle x_1, x_2 \rangle \in X_1 \times X_2 \mid [x_1]_{\alpha_1 \diamond \alpha_2} = [x_2]_{\alpha_1 \diamond \alpha_2}\},$$
$$B := \{\langle y_1, y_2 \rangle \in Y_1 \times Y_2 \mid [y_1]_{\beta_1 \diamond \beta_2} = [y_2]_{\beta_1 \diamond \beta_2}\}.$$

We need to establish that there are indeed nontrivial common events. Since c is nontrivial, we can find an invariant Borel set $D \in \mathcal{INV}(\mathcal{B}(Y_1), \beta_1)$ with $\emptyset \neq D \neq Y_1$. Assume that β_1 spawns β_2 via $(\Theta, \{D_n \mid n \in \mathbb{N}\})$, then $\emptyset \neq \Theta_D \neq Y_2$ also holds. Because D is β_1-invariant,

$$\pi_{1,Y_1}^{-1}[D] = \{\langle y_1, y_2 \rangle \mid y_1 \in D\} = \{\langle y_1, y_2 \rangle \mid y_2 \in \Theta_D\} = \pi_{2,Y_2}^{-1}[\Theta_D]$$

thus

$$\pi_{1,Y_1}^{-1}[D] \in \pi_{1,Y_1}^{-1}[\mathcal{B}(Y_1)] \cap \pi_{2,Y_2}^{-1}[\mathcal{B}(Y_2)],$$

and we are done once it is shown that $\pi_{1,Y_1}^{-1}[D] \neq B$. Since $D \neq Y_1$ is invariant, there exists y_1 with

$$[y_1]_{\beta_1 \diamond \beta_2} \cap D = [y_1]_{\beta_1} \cap D = \emptyset.$$

Let $[y_2]_{\beta_2} := \Theta([\psi_1]_{\beta_1})$, then $[y_2]_{\beta_1 \diamond \beta_2} \cap \Theta_D = [y_2]_{\beta_2} \cap \Theta_D = \emptyset$. Consequently, $\langle y_1, y_2 \rangle \in B \setminus \pi_{1,Y_1}^{-1}[D]$. This shows that $\pi_{1,Y_1}^{-1}[\mathcal{B}(Y_1)] \cap \pi_{2,Y_2}^{-1}[\mathcal{B}(Y_2)]$ is nontrivial. $\Box$

We note for later use with a view back at Lemma 5.24:

COROLLARY 5.40
   Under the conditions of Proposition 5.39, $(K_1 \oplus K_2)/(c_1 \diamond c_2)$ is isomorphic to K_1/c_1 and to K_2/c_2.

The strategy of the proof to Proposition 5.39 has been to make sure that the classes associated with the congruences are distributed evenly among the summands in the sense that each class in the sum is the sum of appropriate classes. This then implies that we can construct surjective maps, and from them morphisms through some general mechanisms. The idea works in particular with isomorphic factor spaces.

PROPOSITION 5.41

Let K and K' be analytic objects such that K/c is isomorphic to K'/c' for some nontrivial congruences c and c'. Then

a. c and c' are simulation equivalent,

b. K and K' are bisimilar.

PROOF 0. Let $K = (X, Y, K)$ with $c = (\alpha, \beta)$, similar for K' and c'. Assume that $f = (\Phi, \Psi)$ is the isomorphism $K/c \to K'/c'$ which is composed of the Borel isomorphisms $\Phi : X/\alpha \to X'/\alpha'$ and $\Psi : Y/\beta \to Y'/\beta'$. Let moreover $\mathcal{A}$ and $\mathcal{B}$ be countable generators of $\mathcal{INV}(\mathcal{B}(X), \alpha)$ and $\mathcal{INV}(\mathcal{B}(Y), \beta)$ which are closed under finite intersections. We know from Proposition 5.25 that α spawns α' via $(\Phi, \mathcal{A})$, and that β spawns β' via $(\Psi, \mathcal{B})$. Hence we have to establish for each $x \in X, x' \in \Phi([x]_\alpha)$ and for each β-invariant Borel subset $B \subseteq Y$ that $K(x)(B) = K'(x')(\Psi_B)$ holds. This will imply that c simulates c'; interchanging the rôles of c and c' then will yield simulation equivalence.

1. Given $B \in \mathcal{INV}(\mathcal{B}(Y), \beta)$ we know from Lemma 5.4 that we can find a Borel set $B_1 \in \mathcal{B}(Y/\beta)$ such that $B = \eta_\beta^{-1}[B_1]$. Since Ψ is a Borel isomorphism, we find $B_2 \in \mathcal{B}(Y'/\beta')$ with $B_1 = \Psi^{-1}[B_2]$. A routine calculation shows that $\Psi_B = \eta_{\beta'}^{-1}[B_2]$. Now assume that $x \in X, x' \in \Phi([x]_\alpha)$, then the following chain of equations is obtained from the argumentation above, and from the assumption that f is an isomorphism.

$$K(x)(B) = K(x)(\eta_\beta^{-1}\left[\Psi^{-1}[B_2]\right])$$
$$= K_{\alpha,\beta}([x]_\alpha)(\Psi^{-1}[B_2])$$
$$= K'_{\alpha',\beta'}(\Phi([x]_\alpha)(B_2)$$
$$= K'(x')(\eta_{\beta'}^{-1}[B_2])$$
$$= K'(x')(\Psi_B).$$

This establishes the desired relation $c \propto c'$ and completes the proof for the first part.

2. Bisimilarity now follows by Proposition 5.39. ☐

Thus isomorphic factor spaces make sure that the relations are bisimilar. These factor spaces arise, e.g., when considering blocks that partition the target spaces of a relation into pairs of pieces that have the same size each. This idea is expressed in the following definition.

DEFINITION 5.42   *Let* $\mathsf{K} = (X, Y, K)$ *and* $\mathsf{L} = (X, Z, L)$ *be analytic objects. Call* $\mathcal{J} = \{\langle B_i, C_i \rangle \mid i \in I\}$ *a* block *for* K, L *iff*

a. *the index set* $I \neq \emptyset$ *is at most countable,*

b. $\{B_i \mid i \in I\}$ *and* $\{C_i \mid i \in I\}$ *are partitions of* Y *resp.* Z *into Borel sets,*

c. $K(x)(B_i) = L(x)(C_i)$ *holds for all* $x \in X$ *and all* $i \in I$.

Thus a block $\mathcal{J}$ cuts Y and Z into the same number of nonempty pieces, and corresponding pieces have the same probability for all $x \in X$.

The existence of a block makes sure that K and L are bisimilar.

COROLLARY 5.43

The analytic objects $\mathsf{K} = (X, Y, K)$ *and* $\mathsf{L} = (X, Z, L)$ *are bisimilar, provided there exists a block of size at least two for them.*

PROOF   0. We will show that the existence of a block enables us to construct isomorphic factor spaces for suitable nontrivial congruences. This will then imply the assertion through Proposition 5.41.

1. Let $\mathcal{J} = \{\langle B_i, C_i \rangle \mid i \in I\}$ be the block which contains at least two elements. The partition $\{B_i \mid i \in I\}$ induces a smooth equivalence relation β on Y such that the equivalence classes are exactly the partition elements. Thus $[y]_\beta = B_i$ iff $y \in B_i$. The invariant Borel sets for β are isomorphic to the power set of I,

$$\mathcal{INV}\,(\mathcal{B}(Y), \beta) = \{\bigcup_{i \in I_0} B_i \mid I_0 \subseteq I\}.$$

This is so because $\mathcal{INV}\,(\mathcal{B}(Y), \beta) = \sigma(\{B_i \mid i \in I\})$, and because the B_i form a partition of Y.

Put $\mathsf{c} := (\Delta_X, \beta)$, then c is a nontrivial congruence for K. Computing the factor relation $K_{\Delta_X, \beta}$, we see that

$$K_{\Delta_X, \beta}(x)(E) = \sum_{i \in I_0} K(x)(B_i),$$

provided $\eta_\beta^{-1}[E] = \bigcup_{i \in I_0} B_i$ holds for the Borel set $E \in \mathcal{B}(Y/\beta)$; see Proposition 5.5. Similarly, define the smooth equivalence γ on Z through the partition $\{C_i \mid i \in I\}$, then $\mathsf{d} := (\Delta_X, \gamma)$ is a nontrivial congruence for L. We have

$$L_{\Delta_X, \gamma}(x)(F) = \sum_{i \in I_0} K(x)(C_i),$$

whenever $\eta_\gamma^{-1}[F] = \bigcup_{i \in I_0} C_i$ holds for the Borel set $F \subseteq Z/\gamma$.

2. Now team up each element in the partition for Y with its partner in Z: put

$$\psi : Y/\beta \ni B_i \mapsto C_i \in Z/\gamma,$$

then $\psi : Y/\beta \to Z/\gamma$ is a measurable bijection, and it is immediate from the definition of a block that $(id_X, \psi) : \mathsf{K}/\mathsf{c} \to \mathsf{L}/\mathsf{d}$ is an isomorphism. $\square$

This result is intuitively quite satisfactory: two relations defined over the same source which divide up their respective targets in exactly the same way are certainly prime candidates for being bisimilar.

EXAMPLE 5.44

Let $K : S \rightsquigarrow S$ and $L : S \rightsquigarrow S$ be stochastic relations over the analytic space S such that

a. $K(s)(S) = L(s)(S) \neq 0$ for all $s \in S$,

b. there exists points $s_K \neq s_L$ in S with $K(s)(\{s_K\}) = L(s)(\{s_L\})$ for all $s \in S$.

Then (S, S, K) and (S, S, L) are bisimilar. This follows at once from Corollary 5.43, because

$$\{\langle\{s_K\}, \{s_L\}\rangle, \langle S \setminus \{s_K\}, S \setminus \{s_L\}\rangle\}$$

is a block for these relations.

We have seen in Proposition 5.41 that isomorphic factor spaces make sure that the relations are bisimilar. The natural question is whether or not the converse also holds: given bisimilar relations, do they have isomorphic factor spaces? A first step towards an answer is done in

PROPOSITION 5.45

If the Polish objects K *and* K' *are bisimilar such that the mediating object is compact with continuous morphisms, then* K *and* K' *have isomorphic non-degenerate factor spaces.*

PROOF   1. Let

$$\mathsf{K} \xleftarrow{\quad f \quad} \mathsf{M} \xrightarrow{\quad f' \quad} \mathsf{K}'$$

be the span of morphisms constituting bisimilarity. Because K is isomorphic to $\mathsf{M}/\ker(\mathsf{f})$ by Corollary 5.35, we may restrict our attention to factors of M. Thus we assume that $\mathsf{K} = \mathsf{M}/\mathsf{c}, \mathsf{K}' = \mathsf{M}/\mathsf{c}'$, where both c and c' are the kernels of continuous morphisms. Suppose that we can find congruences d and d' such that $d \bullet c = d' \bullet c'$. Then

$$K/d = (M/c)/d$$
$$\cong M/d \bullet c \text{ (by Proposition 5.33)}$$
$$= M/d' \bullet c'$$
$$= (M/c')/d'$$
$$\cong K'/d'$$

($\cong$ indicating isomorphism), and we are done, provided K/d is shown to be nondegenerate, or, equivalently, d not to have the universal relation as its second component. When looking for suitable congruences d and d', in view of Corollary 5.36 it is sufficient to find a congruence $d' = (\gamma, \delta)$ with $c \preceq d' \bullet c'$ for the given congruences c and c', and δ is not universal.

2. Assume $M = (X, Y, M)$, and suppose $c = (\alpha, \beta), c' = (\alpha', \beta')$. We know that there exist smooth equivalence relations γ and δ with $\alpha \subseteq \gamma \bullet \alpha'$ and $\beta \subseteq \delta \bullet \beta'$; moreover we know for a δ-invariant Borel subset $D \in \mathcal{INV}(Y/\beta', \delta)$ that $\eta_{\beta'}^{-1}[D] \in \mathcal{INV}(\mathcal{B}(Y), \beta) \cap \mathcal{INV}(\mathcal{B}(Y), \beta')$. This was shown in Proposition 5.15 and Lemma 5.20.

We show that $d = (\gamma, \delta)$ is a congruence, thus we have to show that $K_{\alpha', \beta'}(s)(D) = K_{\alpha', \beta'}(s')(D)$, whenever D is a δ-invariant Borel subset of Y/β', and $s \; \gamma \; s'$.

Assume first that

$$\langle s, s' \rangle \in \gamma_0 := \{ \langle t, t' \rangle \mid t, t' \in X/\alpha', t \times t' \cap \alpha \neq \emptyset \}.$$

Then we can find $\langle x, x' \rangle \in \alpha$ such that $s = [x]_{\alpha'}$, $s' = [x']_{\alpha'}$, and $[x]_\alpha = [x']_\alpha$. Thus we obtain from D's invariance properties

$$K_{\alpha', \beta'}(s)(D) = K(x)(\eta_{\beta'}^{-1}[D]) = K(x')(\eta_{\beta'}^{-1}[D]) = K_{\alpha', \beta'}(s')(D).$$

This means that the assertion is true for all $\langle s, s' \rangle \in \gamma_0$.

Now consider

$$\hat{\gamma} := \{ \langle t, t' \rangle \mid t, t' \in X/\alpha', K_{\alpha', \beta'}(t)(D) = K_{\alpha', \beta'}(t')(D) \},$$

then $\hat{\gamma}$ is an equivalence relation which contains γ_0, and consequently it contains γ, as the construction of γ as the transitive closure of γ_0 shows (see Section 5.2.3, Claim 5.19 on page 193).

3. Since M/c and M'/c' are bisimilar, we can find $F \in \mathcal{INV}(\mathcal{B}(Y), \beta) \cap \mathcal{INV}(\mathcal{B}(Y), \beta')$ with $\emptyset \neq F \neq Y$ (this is so since e.g. $\mathcal{INV}(\mathcal{B}(Y), \beta) = \eta_\beta^{-1}[\mathcal{B}(Y/\beta)]$ by Lemma 5.4). Now minimality of the construction leading to Proposition 5.15 enters the argumentation: from Lemma 5.20 we infer that

$$\eta_{\beta'}^{-1}[\mathcal{INV}(\mathcal{B}(Y/\beta'), \delta)] = \mathcal{INV}(\mathcal{B}(Y), \beta) \cap \mathcal{INV}(\mathcal{B}(Y), \beta')$$

holds, thus we can find $F_0 \in \mathcal{INV}\left(\mathcal{B}(Y/\beta'), \delta\right)$ with $\emptyset \neq F_0 \neq Y/\beta'$. Consequently, δ is not universal, and we are done. □

Summarizing, we have established the following characterization of bisimilarity through congruences:

THEOREM 5.46
Consider for analytic objects K *and* K′ *the statements*

a. *There exist simulation equivalent congruences* c *and* c′ *on* K *resp.* K′.

b. *There exist nontrivial congruences* c *and* c′ *on* K *resp.* K′ *such that* K/c *and* K′/c′ *are isomorphic.*

c. K *and* K′ *are bisimilar.*

Then $a \Leftrightarrow b \Rightarrow c$ *holds always, and* $c \Rightarrow b$ *holds in case the mediating object is compact and the associated morphisms are continuous.*

This is an intrinsic characterization of bisimilarity through congruences, because it suffices to look only at the stochastic relations and decide whether they are bisimilar. It would be most valuable to lift the rather strong condition on compactness. The proofs given above, in particular in Section 5.2.3, Claims 5.16 through 5.19, rely on compact spaces via the possibility to extract a converging subsequence from each sequence (hence on sequential compactness, to be specific). Otherwise smoothness cannot be guaranteed, but smoothness is crucial since it makes sure that the factor space is analytic.

Conjecture. The characterization of bisimilarity through isomorphic factor spaces is valid for all stochastic relations over analytic spaces.

5.5 Behavioral Equivalence and a Portmanteau

While bisimilar stochastic relations are related through a span of morphisms, we think of behavioral equivalent relations as relations for which a cospan exists.

DEFINITION 5.47 *Let* K_1 *and* K_2 *be stochastic relations, then* K_1 *and* K_2 *are called* behavioral equivalent *iff there exists a stochastic relation* L *and morphisms*

$$K_1 \xrightarrow{\;f_1\;} L \xleftarrow{\;f_2\;} K_2$$

Assume $f_i = (f_i, g_i), K_i = (X_i, Y_i, K_i)$ and $L = (X, Y, L)$. The condition says that we can find for each given $x_1 \in X_1$ an element $x_2 \in X_2$ such that $f_1(x_1) = f_2(x_2)$ with this property: for each $B \in \mathcal{B}(Y)$ with $B_i = g_i^{-1}[B], i = 1, 2$, we have due to f_1 and f_2 being morphisms,

$$
\begin{aligned}
K_1(x_1)(B_1) &= K_1(x_1)(g_1^{-1}[B]) \\
&= L(f_1(x_1))(B) \\
&= L(f_2(x_2))(B) \\
&= K_2(x_2)(B_2).
\end{aligned}
$$

Hence the probability of hitting through K_1 the set $B_1 = g_1^{-1}[B]$ starting from x_1 equals the probability of hitting the set $B_2 = g_2^{-1}[B]$ from x_2 through K_2, if the observations for x_1 and for x_2 coincide.

We will see that behavioral equivalent relations have simulation equivalent congruences, and, using factoring just as in the proof of Proposition 5.39, that relations having simulation equivalent congruences are behavioral equivalent. Thus dealing with a cospan of morphisms seems to be easier than dealing with a span. Looking behind the curtain, this is not really a surprise, since the construction of a cospan requires factoring, whereas the construction of a span requires handling a semi-pullback. Nevertheless, the difference in complexity for solving apparently symmetric problems is striking.

LEMMA 5.48

If the analytic objects K_1 and K_2 are behavioral equivalent with morphisms $f_i : K_i \to L$ for some analytic object L, then $\ker(f_1)$ and $\ker(f_2)$ are simulation equivalent congruences.

PROOF   0. Assume $K_i = (X_i, Y_i, K_i)$ with $f_i = (f_i, g_i), i = 1, 2$. Assume furthermore that $L = (A, B, L)$.

1. We know that $\ker(f_1)$ and $\ker(f_2)$ are congruences (Proposition 5.27). It is inferred from Lemma 5.23 that they spawn each other.

2. Let $\mathcal{C}$ be a countable generator of $\mathcal{B}(B)$ that also separates points, so that $\ker(g_1)$ spawns $\ker(g_2)$ via $(\Xi, g_1^{-1}[\mathcal{C}])$ with $\Xi : [y_1]_{\ker(g_1)} \mapsto [y_2]_{\ker(g_2)}$ iff $g_1(y_1) = g_2(y_2)$ as in the proof for Lemma 5.23. If $x_1 \in X_1, x_2 \in X_2$ with $f_1(x_1) = f_2(x_2)$, then we have for all $C \in \mathcal{C}$

$$
K_1(x_1)(g_1^{-1}[C]) = L(f_1(x_1))(C) = L(f_2(x_2))(C) = K_2(x_2)(g_2^{-1}[C]).
$$

Since $\Xi_{g_1^{-1}[C]} = g_2^{-1}[C]$, this implies that $\ker(f_1)$ and $\ker(f_2)$ are simulation equivalent. ∎

As a companion to Theorem 5.46 we obtain:

PROPOSITION 5.49

For analytic objects K *and* K$'$ *these statements are equivalent:*

a. there exist simulation equivalent congruences c *and* c$'$ *on* K *resp.* K$'$,

b. K *and* K$'$ *are behavioral equivalent*

PROOF We infer $b \Rightarrow a$ from Lemma 5.48; the implication $a \Rightarrow b$ is obtained from part a. of Proposition 5.39. $\Box$

The question arises under which condition all these notions of characterizing behavior coincide. A glimpse at the topological conditions in Theorem 5.46 suggests that additional properties will be necessary for such a characterization. We will deal in Chapter 6 with yet another notion of equivalence, viz., logical equivalence. It will be formulated for Kripke models, and we will relate it to bisimilarity and behavioral equivalence, since it will turn out this additional notion of equivalence can be subsumed under the existence of simulation equivalent congruences.

A Portmanteau. The diagram in Figure 5.1 gives the implications that we found between the existence of simulation equivalent congruences, behavioral equivalence and bisimilarity of stochastic relations. Thus, in order to investigate bisimilarity of stochastic relations or for finding out about their behavioral equivalence, it is helpful to find congruences and to show that they are simulation equivalent. This will then permit constructing a span or a cospan of morphisms. The span will be constructed through a semi-pullback as in Section 4.4; the cospan will be constructed through factoring as in Section 5.4.

We will see that these results for stochastic relations need to be adapted for the situation at hand, viz., for constructing models for modal and continuous time logics, that we will investigate in Chapter 6. Hence we will look into these specific constructions in order to transform the corresponding stochastic relation into the model we are looking for.

5.6 2-Bisimulations

A bisimulation between the stochastic relations K_1 and K_2 has been defined in Section 5.4 through a stochastic relation M (the *mediating object*) together with two morphisms

$$K_1 \xleftarrow{\quad f_1 \quad} M \xrightarrow{\quad f_2 \quad} K_2$$

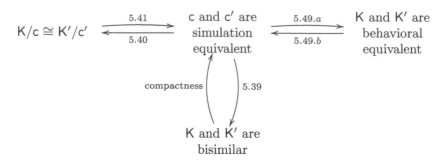

Figure 5.1:   Bisimilarity and behavioral equivalence vs. simulation
equivalence

that share some common event, cp. Definition 5.37. If K_1 and K_2 coincide,
this is called a *bisimulation on* K_1. We will in this section specialize this
notion: rather than considering general morphisms, we consider projections.
Consequently, we have domains and ranges of the mediating object as relations. These relations may have interesting properties, for example when we
discuss bisimulations on a single relation. There the question arises whether
or not the underlying sets are equivalence relations, are smooth, etc.

Formally, let

$$M = ((A, \mathcal{X}), (B, \mathcal{Y}), M)$$

be the mediating object with suitable σ-algebras $\mathcal{X}$ and $\mathcal{Y}$ on A resp. B.
If A and B are measurable subsets of $X_1 \times X_2$ resp. $Y_1 \times Y_2$, and if $f_1 = (\pi_{1,X_1}, \pi_{1,Y_1})$, $f_2 = (\pi_{2,X_2}, \pi_{2,Y_2}) - \pi$ indicating the projections — then the
bisimulation is called a *2-bisimulation*. Thus a 2-bisimulation renders this
diagram commutative:

$$
\begin{array}{ccccc}
X_1 & \xleftarrow{\;\pi_{1,X_1}\;} & A & \xrightarrow{\;\pi_{2,X_2}\;} & X_2 \\
{\scriptstyle K_1}\downarrow & & {\scriptstyle M}\downarrow & & \downarrow{\scriptstyle K_2} \\
\mathfrak{S}(Y_1, \mathcal{A}_1) & \xleftarrow[\mathfrak{S}(\pi_{1,Y_1})]{} & \mathfrak{S}(B, \mathcal{Y}) & \xrightarrow[\mathfrak{S}(\pi_{2,Y_2})]{} & \mathfrak{S}(Y_2, \mathcal{B}_2)
\end{array}
$$

We require for 2-bisimulations A and B only to be measurable subsets of
$X_1 \times X_2$ resp. $Y_1 \times Y_2$, and the σ-algebras $\mathcal{X}$ and $\mathcal{Y}$ chosen so that the
projections are morphisms, i.e., surjective and measurable maps. Note also
that the condition on a nontrivial σ-algebra of common events now reads that
there exists Borel sets $C_1 \subseteq Y_1, C_2 \subseteq Y_2$ with

$$\emptyset \neq B \cap (C_1 \times Y_2) = B \cap (Y_1 \times C_2) \neq B.$$

Having congruences available permits specializing the notion of a bisimulation further (and these specializations will be used later on when characterizing simple systems).

DEFINITION 5.50 *Let α and β be smooth equivalence relations on X resp. Y.*

a. *A 2-bisimulation* $\mathsf{M} = (\alpha, \beta, M)$ *on* K *is called a* smooth 2-bisimulation *on* K.

b. *If the stochastic relation* $\mathsf{N} = ((\alpha, \mathcal{B}(\alpha)), (\beta, \otimes [Y, \beta]), N)$ *has the property that*

$$(\mathfrak{S}(\pi_{1,Y}) \circ N(a_1, a_2))(E) = K(a_1)(E) \text{ and}$$
$$(\mathfrak{S}(\pi_{2,Y}) \circ N(a_1, a_2))(E) = K(a_2)(E)$$

hold whenever $\langle a_1, a_2 \rangle \in \alpha$ and E is a β-invariant Borel set of Y, then N *is called a* weak 2-bisimulation *on* K.

Let $\beta \neq U_Y$, then there exist a β-invariant Borel set $\emptyset \neq P \neq Y$ (since there exists $y_1, y_2 \in Y$ with $\langle y_1, y_2 \rangle \notin \beta$, one may take $P := [y_1]_\beta$). Because by Lemma 5.9, part a,

$$\emptyset \neq \beta \cap (P \times Y) = \beta \cap (P \times P) = \beta \cap (Y \times P) \neq \beta,$$

we see that the σ-algebra of common events is in this case not empty.

Smooth 2-bisimulations correspond to the bisimulation equivalences studied in coalgebras, as we will see soon. *Weak* 2-bisimulations restrict their attention to the β-invariant Borel sets of Y (rather than on all Borel sets), $N(a)((B \times Y) \cap \beta)$ is defined for $a \in \alpha$ and for the β-invariant Borel set $B \in \mathcal{B}(Y)$; see Lemma 5.9, part b. This looks of course much more restrictive than for a smooth 2-bisimulation: Clearly a smooth 2-bisimulation is a weak one, and we will show in Proposition 5.51 that we can even produce a smooth 2-bisimulation from a weak one, provided the relation K is a Polish object.

We will begin with an observation relating congruences, smooth and weak 2-bisimulations. Fix for the discussion that follows the stochastic relation $\mathsf{K} = (X, Y, K)$ and a pair $\mathsf{c} = (\alpha, \beta)$ of smooth equivalence relations on the analytic spaces X resp. Y.

PROPOSITION 5.51

Consider the following conditions:

a. $\mathsf{c} = (\alpha, \beta)$ *is a nontrivial congruence on* K.

b. *There exists* $N : (\alpha, \mathcal{B}(\alpha)) \rightsquigarrow (\beta, \otimes [Y, \beta])$ *such that the stochastic relation* $((\alpha, \mathcal{B}(\alpha)), (\beta, \otimes [Y, \beta]), N)$ *is a weak 2-bisimulation on* K.

c. There exists $M : \alpha \rightsquigarrow \beta$ such that (α, β, M) is a smooth 2-bisimulation on K.

Then the following holds:

a. $c \Rightarrow b \Rightarrow a$ is true for the analytic spaces X and Y,

b. If both X and Y are Polish, then all conditions are equivalent.

PROOF   0. $c \Rightarrow b$ is quite obvious, since each smooth 2-bisimulation is a weak one, so for the general case the implication $b \Rightarrow a$, and for the Polish case the implication $a \Rightarrow c$ needs to be established.

1. $b \Rightarrow a$: Let $C \in \mathcal{INV}\,(\mathcal{B}(Y), \beta)$ be a β-invariant Borel subset of Y, then

$$(C \times Y) \cap \beta = (Y \times C) \cap \beta = (C \times C) \cap \beta$$

has been established in Lemma 5.9, part a. Thus we obtain for $\langle x, x' \rangle \in \alpha$ the following chain of equations from $((\alpha, \mathcal{B}(\alpha)), (\beta, \otimes [Y, \beta]), N)$ being a 2-bisimulation

$$
\begin{aligned}
K(x)(C) &= K(\pi_{1,X}(x, x'))(C) \\
&= \mathfrak{S}\,(\pi_{1,Y})\,(N(x, x'))(C) \\
&= N(x, x')((C \times Y) \cap \beta) \\
&= N(x, x')((Y \times C) \cap \beta) \\
&= \mathfrak{S}\,(\pi_{2,Y})\,(N(x, x'))(C) \\
&= K(\pi_{2,X}(x, x'))(C) \\
&= K(x')(C).
\end{aligned}
$$

2. $a \Rightarrow c$: This part is harder. We need to construct a stochastic relation $M : \alpha \rightsquigarrow \beta$ so that (α, β, M) forms a 2-bisimulation. The plan is very similar to the plan pursued for the existence of semi-pullbacks in Section 4.4.1, in particular for the proof of the central Lemma 4.9. There are subtle differences in the respective scenarios, so we adapt the proof *mutatis mutandis*; the central arguments, however, remain in each case the same. The plan goes as follows: we show that this problem can again be considered a selection problem. For this, we define on α a suitable set-valued map Γ that takes on closed sets of measures on β and that satisfies the conditions of Proposition 1.57 for the existence of a selector. The main difficulty will again lie in showing that Γ takes in fact nonempty values, and here invariant sets come in. Before doing all that, it is shown that the stage we are working on can be set up through closed sets and continuous maps.

Since β is smooth, there exists a Polish space W and a Borel measurable map $g : Y \to W$ such that $\beta = \ker(g)$ by Lemma 1.52. We can find by Proposition 1.28 a finer Polish topology on Y with the same Borel sets $\mathcal{B}(Y)$ that makes g continuous. Thus β may be assumed a closed subset of $Y \times Y$.

Since Y is a Polish space, the space $\mathfrak{S}(Y)$ is Polish as well. Because α is smooth, we find a Polish space V and a Borel measurable map $h : X \to V$ such that $\alpha = \ker(h)$. Applying the same argument as above, we can find a Polish topology on X which makes $h : X \to Y$ as well as $K : X \to \mathfrak{S}(Y)$ continuous maps, rendering in particular α a closed, hence Polish, subset of $X \times X$.

Given $\langle x_1, x_2 \rangle \in \alpha$, the set

$$\Gamma(x_1, x_2) := \{\mu \in \mathfrak{S}(\beta) \mid \mathfrak{S}(\pi_{1,Y})(\mu) = K(x_1), \mathfrak{S}(\pi_{2,Y})(\mu) = K(x_2)\}$$

will be scrutinized with the goal of finding a measurable selector for Γ. It is immediate that it is a closed subset of $\mathfrak{S}(\beta)$, because the projections induce continuous maps on the respective spaces of subprobabilities. Whenever $C \subseteq \mathfrak{S}(\beta)$ is compact, the weak inverse

$$\exists \Gamma(C) = \{\langle x_1, x_2 \rangle \in \alpha \mid \Gamma(x_1, x_2) \cap C \neq \emptyset\}$$

of C is a closed subset of α: let $(\langle x_{1,n}, x_{2,n} \rangle_{n \in \mathbb{N}}) \subseteq \exists \Gamma(C)$ be a sequence with

$$\lim_{n \to \infty} \langle x_{1,n}, x_{2,n} \rangle = \langle x_1, x_2 \rangle.$$

For each $n \in \mathbb{N}$ there exists $\gamma_n \in C$ with $\gamma_n \in \Gamma(x_{1,n}, x_{2,n})$. Since C is compact, there exists a subsequence s and $\gamma \in C$ with $\gamma_{s(n)} \to \gamma$. Continuity of K and closedness of α together imply that $\gamma \in \Gamma(x_1, x_2)$, thus $\langle x_1, x_2 \rangle \in \exists \Gamma(C)$.

We show first that $\Gamma(x_1, x_2) \neq \emptyset$, whenever $\langle x_1, x_2 \rangle \in \alpha$. For this the techniques developed in Section 4.2 are used. Put $Z := Y/\beta$ with $\mathcal{C} := \mathcal{B}(Y/\beta)$, then $(Z, \mathcal{C})$ is an analytic space; hence it is separable; see 1.3.3. The map $\psi : y \mapsto [y]_\beta$ is measurable from Y onto Z, and we have

$$S := \{\langle y_1, y_2 \rangle \mid \psi(y_1) = \psi(y_2)\} = \beta.$$

We know moreover from Proposition 5.5 that $\eta_\beta^{-1}[\mathcal{C}] = \mathcal{INV}(\mathcal{B}(Y), \beta)$ holds. Now fix $\langle x_1, x_2 \rangle \in \alpha$ and put $\nu_1 := K(x_1), \nu_2 := K(x_2)$, then a measure θ_1 on the σ-algebra $\otimes[Y, \beta]$ is defined through

$$(*) \quad \theta_1((B \times B) \cap \beta) = \nu_1(B)(= \nu_2(B));$$

see Lemma 5.9, part c. An appeal to Proposition 4.4 yields an extension of θ_1 to a measure θ which is defined on all of $\mathcal{B}(S)$. Thus we have now $\theta \in \mathfrak{S}(S)$ such that

$$\forall E_i \in \psi^{-1}[\mathcal{C}] : \mathfrak{S}(\pi_{i,Y})(\theta)(E_i) = \nu_i(E_i), i = 1, 2.$$

From Proposition 4.4 we obtain a measure $\mu \in \mathfrak{S}(S)$ such that

$$\forall E_i \in \mathcal{B}(S) : \mathfrak{S}(\pi_{i,Y})(\mu)(E_i) = \nu_i(E_i), i = 1, 2.$$

But this means that $\Gamma(x_1, x_2) \neq \emptyset$, thus we can apply the selection theorem and obtain through Proposition 1.57 a measurable selector M for Γ, consequently, $M : \alpha \rightsquigarrow \beta$. Thus $\mathsf{M} := (\alpha, \beta, M)$ is a stochastic relation. From M being a selector to Γ one sees that M is a 2-bisimulation for K, since

$$(\mathfrak{S}(\pi_{1,Y}) \circ M)(x_1, x_2) = K(x_1)$$
$$(\mathfrak{S}(\pi_{2,Y}) \circ M)(x_1, x_2) = K(x_2)$$

is true for all $\langle x_1, x_2 \rangle \in \alpha$.   ⬛

Thus we have established a very close relationship between congruences and 2-bisimulations for stochastic relations. The basic idea has been again to extend a stochastic relation that is defined on a small and fairly easy to handle σ-algebra to a larger one. But this is complicated, because we do not have direct access to the Borel sets, when we need it: the Borel sets are defined in terms of a closure operation and not through some explicit procedure, so we cannot put a handle on them directly (in fact, this is a white lie: the Borel sets can be defined stepwise through transfinite induction, see, e.g., (Srivastava, 1998) or (Aumann, 1952); but this process is rather complicated and will not help us here at all). Hence we have to walk a by-path again: we show through a selection argument that such a measure must exist.

Albeit there are subtle variations here and in Section 4.4.1, both arguments work essentially as follows:

a. We know that the situation is easily managed on a small σ-algebra which we start from (this is like the begin of a proof by induction: the picture is nice and clear in the beginning).

b. We know also that our request for an extension is not unreasonable, since our map Γ has some reasonable properties (this is like the induction hypothesis).

c. From this we conclude that we can find an extension through a selector (this is much like the inductive step itself).

Quite apart from the involved technical development, this close relationship between bisimilarity and congruences is somewhat akin to the scenario for general coalgebras. The situation cannot be mirrored, however, since for coalgebras one usually requires a functor which preserves weak pullbacks; see, e.g., (Rutten, 2000). The structure for the subprobability functor $\mathfrak{S}$ is slightly more involved because the hope for establishing weak pullbacks is vain. Consequently it seems to be difficult to fit general coalgebras and stochastic relations too tightly under one common roof.

Anyway, Proposition 5.51 provides us with a considerable degree of freedom. It will be of use when investigating simple relations: we can select the proper scenario in investigating simple relations without having to be afraid that we lose important properties, as will be seen in Section 5.7. This holds at least in the Polish case. In the case of an analytic object we have to be a bit careful, but Proposition 5.51 tells us as well where to install watch dogs.

A partial converse to Proposition 5.51 is furnished through

LEMMA 5.52
Let α and β be smooth equivalence relations on X resp. Y. Assume that

$$\mathsf{M} := ((\alpha, \mathcal{B}(\alpha)), (\beta, \otimes [Y, \beta]), M)$$

is a weak 2-bisimulation on K. *Then (α, β) is a congruence of* K.

PROOF   Let $B \in \mathcal{INV}\left(\mathcal{B}(Y), \ell(\beta)\right)$, then we know from Lemma 5.9, part a that $(B \times Y) \cap \beta = (Y \times B) \cap \beta$ holds. Thus we get from the assumption that M is a bisimulation on K the equality $K(x_1)(B) = K(x_2)(B)$ for all $\langle x_1, x_2 \rangle \in \alpha$. ▯

Proposition 5.51 builds the much needed bridge between congruences and bisimulations. Quite apart from being of considerable interest unto its own, we will cross this bridge when investigating simple systems.

5.7   Simple Relations

An algebraic structure which is isomorphic to each of its nontrivial factor spaces is called simple. Take, e.g., a simple and nontrivial group G and an epimorphism $\phi : G \to H$, then ϕ is an isomorphism (Lang, 1965, p. 104). Since simple systems do not have nontrivial subsystems, a system S is simple if each epimorphism $S \to T$ is an isomorphism. The very close connection between simple systems and trivial bisimulations is well known in the theory of coalgebras: a system is simple iff it has only trivial bisimulations.

Simple systems will be characterized both for Polish and analytic spaces. We deal first with the Polish case which is a bit easier to handle, and turn then to the analytic case. A technique for reducing the analytic to the Polish case is developed, so that we may capitalize on previous results. A complete characterization of simple relations can be given for the analytic case.

Call a congruence $\mathsf{c} = (\alpha, \beta)$ on X and Y *plain* iff both equivalence relations are the identity, viz., iff both $\alpha = \Delta_X$ and $\beta = \Delta_Y$ hold. Similarly, call a smooth or weak 2-bisimulation *plain* iff the underlying congruence is plain.

DEFINITION 5.53   *A stochastic relation* K *is called* simple *iff each morphism with domain* K *is an isomorphism.*

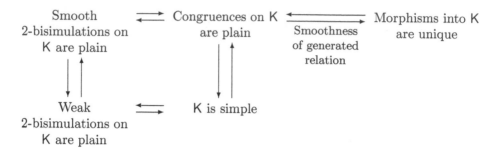

Figure 5.2:   Simple systems: the Polish case

This definition looks a bit stronger than usual, since usually *epi*morphisms emanating from a simple structure are assumed to be isomorphisms. But since all our morphisms are epis, we deal only with surjective maps, thus the common definition applies in this context as well.

Looking first at relations based on Polish spaces, things are rather satisfyingly characterized through equivalences of smooth and weak 2-bisimulations and plain congruences. There is even a characterization through morphisms going into the relation in question. The case of analytic relations is a bit more involved, since the equivalence of smooth and weak 2-bisimulations is not guaranteed; so it is relegated to a separate discussion.

5.7.1   The Polish Case

We characterize simple systems if both spaces on which the relation is defined are Polish. The following characterization is summarized in Figure 5.2.

THEOREM 5.54
Consider these statements for the Polish object K

(a). K *is simple.*

(b). Each smooth 2-bisimulation on K *is plain.*

(c). Each weak 2-bisimulation on K *is plain.*

(d). Let $\mathsf{f}_1, \mathsf{f}_2 : \mathsf{M} \to \mathsf{K}$ *be morphisms, where* M *is a Polish object, then* $\mathsf{f}_1 = \mathsf{f}_2$.

(e). Each congruence on K *is plain.*

Then

a. These implications hold always: $(a) \Leftrightarrow (b) \Leftrightarrow (c) \Leftrightarrow (e) \Leftarrow (d)$.

b. Let in (d) $\mathsf{f}_i = (\phi_i, \psi_i)$. *If both* $\ell(\lfloor \phi_1 \| \phi_2 \rfloor)$ *and* $\ell(\lfloor \psi_1 \| \psi_2 \rfloor)$ *are smooth, then* $(e) \Rightarrow (d)$ *holds as well.*

The **proof** for Theorem 5.54 is broken into several pieces:

$(e) \Rightarrow (a)$: Let $f : K \to L$ be a morphism, then f can be factored through $K/\ker(f)$ as $f = f' \circ \eta_{\ker(f)}$ with an isomorphism f' by Corollary 5.35. $\ker(f)$ is a congruence which is plain by assumption. Thus f is an isomorphism.

$(a) \Rightarrow (e)$: If c is a congruence on K, then $\eta_c : K \to K/c$ is a morphism.

$(b) \Rightarrow (e)$: This is a special case of Proposition 5.51.

$(d) \Rightarrow (b)$: Let $M := (A, B, M)$ be a smooth bisimulation on K, then

$$(\pi_{1,X}, \pi_{1,Y}), (\pi_{2,X}, \pi_{2,Y}) : M \to K$$

are morphisms which are equal by assumption.

This settles the proof of part a. Turning to the proof of part b, assume that (e) holds in addition to $(\ell(\lfloor \phi_1 \| \phi_2 \rfloor), \ell(\lfloor \psi_1 \| \psi_2 \rfloor))$ being smooth. We note from the proof of Lemma 5.8 that a $\ell(\lfloor \psi_1 \| \psi_2 \rfloor)$-invariant Borel set $D \subseteq Y$ has the property that $\psi_1^{-1}[D] = \psi_2^{-1}[D]$ holds, hence that D is an event common to ψ_1 and ψ_2. Now define the equivalence relation

$$R_D := \{\langle x_1, x_2 \rangle \mid K(x_1)(D) = K(x_2)(D)\},$$

then $\lfloor \psi_1 \| \psi_2 \rfloor \subseteq R_D$ follows from $f_1, f_2 : M \to K$ being morphisms: suppose $\langle x_1, x_2 \rangle = \langle \phi_1(a), \phi_2(a) \rangle$, and $E = \psi_1^{-1}[D] = \psi_2^{-1}[D]$, we obtain

$$\begin{aligned} K(x_1)(D) &= (K \circ \phi_1)(a)(D) \\ &= (\mathfrak{S}(\psi_1) \circ M)(a)(D) \\ &= M(a)(E) \\ &= (\mathfrak{S}(\psi_2) \circ M)(a)(D) \\ &= K(x_2)(D). \end{aligned}$$

Since R_D is an equivalence relation for each D, and $\ell(\lfloor \phi_1 \| \phi_2 \rfloor)$ is the smallest equivalence relation containing $\lfloor \phi_1 \| \phi_2 \rfloor$, this implies

$$\ell(\lfloor \phi_1 \| \phi_2 \rfloor) \subseteq \bigcap \{R_D \mid D \in \mathcal{INV}(\mathcal{B}(Y), \ell(T))\}$$

which in turn yields that $(\ell(\lfloor \phi_1 \| \phi_2 \rfloor), \ell(\lfloor \psi_1 \| \psi_2 \rfloor))$ is a congruence on K. This congruence is plain by assumption, yielding $f_1 = f_2$, as desired.

5.7.2 The Analytic Case

We will reduce the case of relations on analytic spaces to the one where we have Polish spaces at our disposal, and we have seen that we can move smooth equivalence relations along arrows (albeit reversing the direction) in Lemma 5.7. This will be used now to move congruences.

PROPOSITION 5.55

Let $K = (X, Y, K)$ be a Polish object, $L = (A, B, L)$ be an analytic object, assume that $f = (\phi, \psi) : K \to L$ is a morphism, and that $c = (\alpha, \beta)$ is a congruence on L. Then $c_f := (\alpha_\phi, \beta_\psi)$ is a congruence on K.

PROOF   We know from the constructions that both α_ϕ and β_ψ are smooth equivalence relations. Now let $a\ \alpha_\phi\ a'$, thus $\phi(a)\ \alpha\ \phi(a')$. Assume that $E \subseteq Y$ is a β_ψ-invariant Borel set; from Lemma 5.7 we infer that $E = \psi^{-1}[E_0]$ for some β-invariant Borel set $E_0 \subseteq B$. Then

$$
\begin{aligned}
K(a)(E) &= K(a)\left(\psi^{-1}[E_0]\right)\\
&= \left(\mathfrak{S}\left(\psi\right) \circ K\right)(a)(E_0)\\
&= L(\phi(a))(E_0)\\
&= L(\phi(a'))(E_0)\\
&= K(a)(E),
\end{aligned}
$$

because (ϕ, ψ) is a morphism, and because (α, β) is a congruence on L. This shows that c_f is in fact a congruence on K. $\quad\Box$

For a characterization of simple stochastic relations analogous to Theorem 5.54, we fix an analytic object $K = (X, Y, K)$ together with Polish spaces and surjective Borel maps $f : X_0 \to X$ and $g : Y_0 \to Y$ which define the analytic structure on X resp. Y. We establish for K the following property:

PROPOSITION 5.56
These conditions are equivalent for K:

a. Each weak 2-bisimulation on K is plain.

b. Each congruence on K is plain.

PROOF   0. Since each weak 2-bisimulation is defined on a congruence, the implication $b \Rightarrow a$ is obvious from Lemma 5.52. In order to establish the other implication, we will construct from a given congruence $c = (\alpha, \beta)$ on K together with the derived pair $c_{f,g} := (\alpha_f, \beta_g)$ a stochastic relation $K_0 := (X_0, Y_0, K_0)$ on which $c_{f,g}$ is a congruence. Then construct a smooth 2-bisimulation $M_0 = (\alpha_f, \beta_g, M_0)$ on K_0, and use this for constructing a weak 2-bisimulation $M = (\alpha, \beta, M)$ on K.

1. The relations α_f and β_g are smooth equivalence relations on X_0 resp. Y_0. Define for $E \in \mathcal{INV}(\mathcal{B}(Y), \beta)$ and $x_0 \in X_0$

$$
K_0'(x_0)(g^{-1}[E]) := K(f(x_0))(E),
$$

then we see from Lemma 5.7 that $K_0' : (X_0, \mathcal{B}(X_0)) \rightsquigarrow (Y_0, \mathcal{INV}(\mathcal{B}(Y_0), \beta_g))$ is a stochastic relation, so by Proposition 4.7 we can find a stochastic relation

$$
K_0 : (X_0, \mathcal{B}(X_0)) \rightsquigarrow (Y_0, \mathcal{B}(Y_0))
$$

extending K_0'. Then $c_{f,g}$ is a congruence on K_0: let $\langle x_0, x_1 \rangle \in \alpha_f$, and $E_0 \in \mathcal{INV}(\mathcal{B}(Y_0), \beta_g)$ be an invariant Borel set in Y_0. We know then that

$\langle f(x_0), f(x_1) \rangle \in \alpha$, and that $E_0 = g^{-1}[g[E_0]]$ with $g[E_0] \in \mathcal{INV}(\mathcal{B}(Y), \beta)$.
Hence

$$
\begin{aligned}
K_0(x_0)(E_0) &= K_0(x_0)(g^{-1}[g[E_0]]) \\
&= K(f(x_0))(g[E_0]) \\
&= K(f(x_1))(g[E_0]) \\
&= K_0(x_1)(E_0).
\end{aligned}
$$

From Proposition 5.51 we get a smooth 2-bisimulation $M_0 = (\alpha_f, \beta_g, M_0)$ on K_0. We show that this implies

$$
M_0(x_0, x_1)((P \times Y_0) \cap \beta_g) = M_0(x_0', x_1')((P \times Y_0) \cap \beta_g),
$$

provided $P \in \mathcal{INV}(\mathcal{B}(Y_0), \beta_g)$ is a β_g-invariant Borel set in Y_0, and we have $\langle x_0, x_1 \rangle, \langle x_0', x_1' \rangle \in \alpha_f$ with $f(x_0) = f(x_0')$ or $f(x_1) = f(x_1')$. This is done through the bisimulation property for M_0: assume that $f(x_0) = f(x_0')$, then

$$
\begin{aligned}
M_0(x_0, x_1)((P \times Y_0) \cap \beta_g) &= M_0(x_0, x_1)(\pi_{1,Y_0}^{-1}[P]) \\
&= (\mathfrak{S}(\pi_{1,Y_0}) \circ M_0)(x_0, x_1)(P) \\
&= (K_0 \circ \pi_{1,X_0})(x_0, x_1)(P) \\
&= K_0(x_0)(P) \\
&\overset{(*)}{=} K_0(x_0')(P) \\
&= M_0(x_0', x_1')((P \times Y_0) \cap \beta_g).
\end{aligned}
$$

Eq. $(*)$ follows from the observation that $f(x_0) = f(x_0')$ implies $\langle x_0, x_0' \rangle \in \alpha_f$.

Now introduce the stochastic relation $M = ((\alpha, \mathcal{B}(\alpha)), (\beta, \otimes[Y, \beta]), M)$ by defining the subprobability

$$
M(a, a')((B \times Y) \cap \beta) := M_0(x_0, x_0')((g^{-1}[B] \times Y_0) \cap \beta_g)
$$

for $\langle a, a' \rangle = \langle f(x_0), f(x_0') \rangle \in \alpha$ and for $B \in \mathcal{INV}(\mathcal{B}(Y), \beta)$.

The discussion above shows that M is well defined, provided we can establish that $(g^{-1}[B] \times Y_0) \cap \beta_g \in \otimes[Y_0, \beta_g]$ is true. But we know that $g^{-1}[B] \in \mathcal{INV}(\mathcal{B}(Y_0), \beta_g)$ holds.

2. It remains to show that M is indeed a weak 2-bisimulation. Let $\langle a, a' \rangle \in \alpha$ with $a = f(x_0), a' = f(x_0')$, and take a β-invariant Borel set $E \subseteq Y$. Then $\pi_{1,Y}^{-1}[E] = (E \times Y) \cap \beta$.

Putting all this together, we obtain

$$
\begin{aligned}
M(a, a')(\pi_{1,Y}^{-1}[E]) &= M(f(x_0), f(x_0'))((E \times Y) \cap \beta) \\
&= M_0(x_0, x_0')((g^{-1}[E] \times Y) \cap \beta_g) \\
&= K_0(x_0)(g^{-1}[E]) \\
&= K(f(x))(E) \\
&= K(a)(E).
\end{aligned}
$$

$\square$

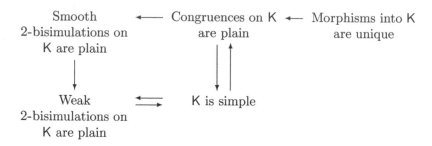

Figure 5.3:   Simple systems: the analytic case

As a consequence the analogue to Theorem 5.54 is obtained for analytic objects. Figure 5.3 suggests a pictorial summary.

THEOREM 5.57
Consider these statements for the analytic object K

(a). K *is simple.*

(b). *Each smooth 2-bisimulation on* K *is plain.*

(c). *Each weak 2-bisimulation on* K *is plain.*

(d). *Let* $f_1, f_2 : M \to K$ *be morphisms, where* M *is an analytic object, then* $f_1 = f_2$.

(e). *Each congruence on* K *is plain.*

Then these implications hold: $(d) \Rightarrow (a) \Leftrightarrow (e) \Rightarrow (b) \Rightarrow (c) \Leftarrow (e)$.

We are now in a position to characterize simple systems over analytic spaces completely. Let $\mathbb{1} := \{*\}$ be the one-element space with the discrete topology (which is Polish) and $\mathfrak{Pow}(\mathbb{1})$ as its Borel sets. This space plays a distinguished rôle:

PROPOSITION 5.58
The analytic objects $(X, \mathbb{1}, K)$ *such that* $x \mapsto K(x)(\mathbb{1})$ *is injective are exactly the simple analytic objects.*

PROOF   1. Let $K = (X, \mathbb{1}, K)$ be such an object, and assume that

$$f = (\phi, \psi) : K \to L = (A, B, L)$$

is a morphism. Then B can have only one element. Since $x \mapsto K(x)(\mathbb{1})$ is one-to-one, we see that $x \neq x'$ implies $L(\phi(x))(B) \neq L(\phi(x'))(B)$, hence $\phi(x) \neq \phi(x')$. Consequently f is an isomorphism. Thus K is simple.

2. Let conversely $K = (X, Y, K)$ be a simple stochastic relation, and define the smooth equivalence relation $\widetilde{\alpha}$ through $\widetilde{\alpha} := \ker(K(\cdot)(Y))$. Put $\widetilde{\omega} := Y \times Y$, then $\widetilde{\omega}$ is also smooth. It is not difficult to see that $c := (\widetilde{\alpha}, \widetilde{\omega})$ is a congruence on K, since $\{\emptyset, Y\}$ is the σ-algebra of $\widetilde{\omega}$-invariant subsets. By Theorem 5.57, c is plain, thus $\widetilde{\alpha} = \Delta_X$ and $\widetilde{\omega} = \Delta_Y$. Hence Y can only have one element, and α is the kernel of an injective map. ⬚

Thus the simple objects in the category of stochastic relations over analytic spaces are in one-to-one correspondence with the injective Borel maps from analytic spaces to the unit interval. Proposition 5.58 is the stochastic counterpart to the coalgebraic characterization of simple systems which says that a system S is simple iff it is isomorphic to $S/\equiv$, where $\equiv$ is the greatest bisimulation on S (Rutten, 2000, Theorem 8.1). This construction is not directly applicable in the present context since the notion of a greatest bisimulation is not available here.

Call finally an object F *final* iff given another object M there exists exactly one morphism $f : M \to F$. In view of Theorem 5.54, a final object is simple. The category of stochastic relations does not have final objects: Being simple, a final object would have the shape $F = (X, \mathbb{1}, F)$ according to Proposition 5.58. But X cannot have more than one element, thus $F = (\mathbb{1}, \mathbb{1}, F)$ with $F(*)(\mathbb{1}) = r$ for some $r, 0 \leq r \leq 1$. But then there would be a unique morphism $(\mathbb{1}, \mathbb{1}, K) \to (\mathbb{1}, \mathbb{1}, K')$ with $K'(*)(\mathbb{1}) = r' \neq r$. This is evidently impossible.

We have, however, the following positive result:

COROLLARY 5.59

The full subcategory of stochastic relations (X, Y, K) with $K(x)(Y) = 1$ for all $x \in X$ has a final object $(\mathbb{1}, \mathbb{1}, F)$.

5.8 Case Study: The Converse of a Stochastic Relation

Bisimilarity is quite robust a relation; this will be demonstrated for the converse of a stochastic relation. Quite apart from this observation, the problem is interesting in its own right, because it suggests an occasion for investigating some similarities between forming the converse for set-theoretic relations and their stochastic cousins. It is shown how the converse is constructed through a disintegration argument (in marked contrast to the set-theoretic case, where merely the order of the pairs needs to be reversed).

For introducing the problem, let R be a set-relation on a set of states. If $\langle x, y \rangle \in R$, then this can be written as $x \to_R y$ and interpreted as a state

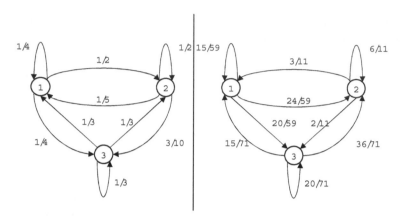

Figure 5.4: A Stochastic Relation and Its Converse

transition from x to y. The converse $R^{\smile}$ shifts attention to the goal of the transition: $y \to_{R^{\smile}} x$ is interpreted as y being the goal of a transition from x.

Now let $p(x, y)$ be the probability that there is a transition from x to y, and the question arises with which probability state y is the goal of a transition from x. This question cannot be answered unless we know the initial probability μ for the states. Then we can calculate $p_{\mu}^{\smile}(y, x)$ as the probability to make a transition from x to y weighted by the probability to start from x conditional to the event to reach y at all, i.e.,

$$p_{\mu}^{\smile}(y, x) := \frac{\mu(x) \cdot p(x, y)}{\sum_{t} \mu(t) \cdot p(t, y)}.$$

Consider as an example the simple transition system p on three states given in the left hand side of Fig. 5.4. The converse $p_{\mu}^{\smile}$ for the initial probability $\mu := [1/2 \ 1/4 \ 1/4]$ is given on the right hand side.

The transition probabilities p are given through

$$\begin{bmatrix} 1/4 & 1/2 & 1/4 \\ 1/5 & 1/2 & 3/10 \\ 1/3 & 1/3 & 1/3 \end{bmatrix}$$

with initial probabilities according to the vector $\mu := [1/2, \ 1/4 \ 1/4,]$. The converse $p_{\mu}^{\smile}$ is then computed as

$$\begin{bmatrix} \frac{15}{31} & \frac{6}{31} & \frac{10}{31} \\ \frac{6}{11} & \frac{3}{11} & \frac{2}{11} \\ \frac{15}{34} & \frac{9}{34} & \frac{5}{17} \end{bmatrix}.$$

The situation is of course more complicated in the nonfinite case. We assume as usual that we work in Polish spaces. A definition of the converse $K_\mu^\smile$ of a stochastic relation K given an initial distribution μ is proposed in terms of disintegration. An interpretation of the converse in terms of random variables is given, and it is shown that the converse behaves with respect to composition like its set-theoretic counterpart, viz., $(K*L)_\mu^\smile = L_{K^\bullet(\mu)}^\smile * K_\mu^\smile$, where $K^\bullet(\mu)$ denotes the image distribution of μ under K (Lemma 5.60), and the composition is the Kleisli composition for the Giry monad (Section 2.3). This is of course the probabilistic counterpart to the corresponding law for relations R and S, which reads $(R*S)^\smile = S^\smile * R^\smile$.

Before entering the discussion on the converse, we study briefly the interplay between stochastic relations and the measures on the codomain.

LEMMA 5.60

Let X and Y be Polish spaces, $K : X \leadsto Y$ be a stochastic relation, and put for $\mu \in \mathfrak{S}(X), B \in \mathcal{B}(B)$

$$K^\bullet(\mu)(B) := \int_X K(x)(B) \ \mu(dx).$$

Then

a. *$K^\bullet$ defines a map $K^\bullet : \mathfrak{S}(X) \to \mathfrak{S}(Y)$ such that*

$$\int_Y g \ dK^\bullet(\mu) = \int_X \int_Y g(y) \ K(x)(dy) \ \mu(dx)$$

holds for each $g \in \mathcal{F}(Y)$.

b. *$(\mu \otimes K)(D) := \int_X K(x)(D_x) \ \mu(dx)$ assigns $\mu \in \mathfrak{S}(X)$ and K a subprobability on $X \times Y$ such that*

$$\int_{X \times Y} g \ d(\mu \otimes K) = \int_Y \int_X g(x,y) \ K(x)(dy) \ \mu(dx)$$

is true whenever $g \in \mathcal{F}(X \times Y)$.

PROOF The proofs work along the following pattern, so often encountered here already: One first shows that the claim is correct for the case of indicator functions, then establishes that things work as expected for step functions as the linear combinations of indicator functions. Using a monotone approximation for nonnegative bounded and measurable functions, the integral's monotone continuity shows that the claim is justified for these functions; finally, a decomposition of a map into the difference of nonnegative functions yields the claim for general measurable and bounded maps. The reader is invited to fill in the details. $\Box$

The map $K^{\bullet} : \mathfrak{S}(X) \to \mathfrak{S}(Y)$ defined in Lemma 5.60 is usually called the *Kleisli extension* to $K : X \rightsquigarrow Y$. This construction is a helpful tool when investigating Kleisli morphisms.

Let us illustrate these constructions for the discrete case.

EXAMPLE 5.61

Assume that $p : \{1,\ldots,n\} \rightsquigarrow \{1,\ldots,m\}$ is a stochastic relation, and let $\mu \in \mathfrak{S}(\{1,\ldots,n\})$ be an initial distribution. Then

1. $p^{\bullet}(\mu)(j) = \sum_{i=1}^{n} \mu(i) \cdot p(i,j)$ is the probability that response j is produced, given the initial probability μ.

2. $(\mu \otimes p)(\langle i,j \rangle) = \mu(i) \cdot p(i,j)$ gives the probability for the input/output pair $\langle i,j \rangle$ to occur, given the initial probability μ (which is responsible for input i), and the probability $p(i,j)$ for output j after input i.

These properties are easily established using elementary computations.

It is remarkable that the construction in part *b* of Definition 5.60 can be reversed, and this is in fact the cornerstone for constructing the converse of a stochastic relation. *Reversing the construction* means that each measure on the product of two Polish spaces can be represented as a product of a stochastic relation with a measure.

PROPOSITION 5.62

Given $\nu \in \mathfrak{S}(X \times Y)$ there exists $\mu \in \mathfrak{S}(X)$ and $K : X \rightsquigarrow Y$ with $\nu = \mu \otimes K$.

PROOF   This is but a reformulation of Proposition 1.85.   ⬚

Recall that the stochastic relation K is known as the *regular conditional distribution of π_Y given π_X*; see Section 1.5.3. Relation K is sometimes called *a version of the disintegration of ζ w.r.t.* $\mathfrak{S}(\pi_{X \times Y,X})(\zeta)$.

EXAMPLE 5.63

Let $\zeta \in \mathfrak{S}(\{1,\ldots,n\} \times \{1,\ldots,m\})$, then the probability $p(i,j)$ for input i generating output j is the probability $\zeta(\langle i,j \rangle)$ for the pair $\langle i,j \rangle$ to occur conditioned on the probability $\sum_{t=1}^{m} \zeta(\langle i,t \rangle)$ that input i is produced at all. Thus relation p satisfies the equation

$$\zeta(\langle i,j \rangle) = \left(\sum_{t=1}^{m} \zeta(\langle i,t \rangle) \right) \cdot p(i,j).$$

This is the discrete version of Proposition 1.85. In contrast to the discrete case, however, the version of the disintegration of ζ with respect to its projection usually cannot be computed explicitly in the general case.

There is a rather helpful interplay between the projection of $\mu \otimes K$ to the second component and $K^\bullet(\mu)$ which will be exploited later on.

LEMMA 5.64
If $\mu \in \mathfrak{S}(X)$ is a subprobability measure, and $K : X \rightsquigarrow Y$ is a stochastic relation, then
$$\mathfrak{S}(\pi_{X \times Y, Y})(\mu \otimes K) = K^\bullet(\mu).$$

PROOF Let $B \subseteq Y$ be a Borel set, then
$$\begin{aligned}
\mathfrak{S}(\pi_{X \times Y, X})(\mu \otimes K)(B) &= (\mu \otimes K)(X \times B) \\
&= \int_X K(x)((X \times B)_x) \; \mu(dx) \\
&= \int_X K(x)(B) \; \mu(dx) \\
&= K^\bullet(\mu)(B).
\end{aligned}$$

$\square$

5.8.1 Converse Relations

Given a substochastic matrix
$$(p(i,j))_{1 \leq i \leq n, 1 \leq j \leq m}$$
representing a stochastic relation
$$\{1, \ldots, n\} \rightsquigarrow \{1, \ldots, m\}$$
and an initial distribution, we saw above that the probability $p_\mu^\smile(j)(i)$ of responding with $j \in \{1, \ldots, m\}$ on a stimulus $i \in \{1, \ldots, n\}$ is calculated as
$$p_\mu^\smile(j)(i) = \frac{\mu(i) \cdot p(i,j)}{\sum_t \mu(t) \cdot p(t,j)}.$$

The probability $p_\mu^\smile$ under consideration reverses p given an initial distribution, so is regarded as the converse of p (*inverse* might at first sight be considered a better name, but this seems to suggest invertibility of the matrix associated with p).

In view of Examples 5.63 and 5.61, this amounts to the disintegration of $\mu \otimes p$ with respect to the distribution $p^\bullet(\mu) = \mathfrak{S}(\pi_{X \times Y, Y})(\mu \otimes p)$.

This observation guides the way for the definition of the converse for a general stochastic relation. Fix a stochastic relation $K : X \rightsquigarrow Y$, and a subprobability measure $\mu \in \mathfrak{S}(X)$. Then $\mu \otimes K \in \mathfrak{S}(X \times Y)$ has a kind of natural converse: define $\tau := \mathfrak{S}(r)(\mu \otimes K)$, where $r : X \times Y \to Y \times X$ switches components. Thus

$$r[R] = R^{\smile} := \{\langle y, x \rangle \mid \langle x, y \rangle \in R\},$$

whenever $R \subseteq X \times Y$ is a relation, so r produces the converse.

Because $\tau \in \mathfrak{S}(Y \times X)$, this measure is (according to Proposition 1.85) representable through a stochastic relation $K_\mu^{\smile} : Y \rightsquigarrow X$ and its projection $\mathfrak{S}(\pi_{Y \times X, Y})(\tau)$ upon writing

$$\tau = \mathfrak{S}(\pi_{Y \times X, Y})(\tau) \otimes K_\mu^{\smile}.$$

Since $\mathfrak{S}(\pi_{Y \times X, Y})(\tau) = K^{\bullet}(\mu)$ by Lemma 5.64, the definition of the converse of a stochastic relation now reads as follows.

DEFINITION 5.65   *The μ-converse $K_\mu^{\smile}$ of the stochastic relation K with respect to the input probability μ is defined by the equation*

$$\mathfrak{S}(r)(\mu \otimes K) = K^{\bullet}(\mu) \otimes K_\mu^{\smile},$$

where $r : X \times Y \ni \langle x, y \rangle \mapsto \langle y, x \rangle \in Y \times X$ switches components.

It is remarked that by Proposition 1.85 the converse $K_\mu^{\smile}$ always exists, and that it is unique μ-almost everywhere. Since

$$\mu(A) = (\mu \otimes K)(A \times Y) = (K^{\bullet}(\mu) \otimes K_\mu^{\smile})((Y \times A)^{\smile})$$

is true for the Borel set $A \subseteq X$,

$$\mu(A) = \int_X \int_Y K_\mu^{\smile}(A) \, K(x)(dy) \, \mu(dx) = \int_Y K_\mu^{\smile}(A) \, K^{\bullet}(\mu)(dy),$$

we infer that

$$\mu = (K_\mu^{\smile})^{\bullet}(K^{\bullet}(\mu)) = (K * K_\mu^{\smile})^{\bullet}(\mu)$$

holds. Hence the converse $K_\mu^{\smile}$ solves the equation $\mu = (K * T)^{\bullet}(\mu)$ for T. This equation does, however, not determine the converse uniquely. This is so because it is an equation in terms of the Borel sets of X, hence may only be carried over to the "strip" $\{A \times Y \mid A \in \mathcal{B}(X)\}$ on the product $X \times Y$. This is not enough to determine a measure on the entire product.

A *probabilistic interpretation* using regular conditional distributions may be given as follows: Let $(\Omega, \mathcal{A}, \mathbb{P})$ be a probability space, $\zeta_i : \Omega \to X_i$ random variables with values in the Polish spaces X_i ($i = 1, 2$). Let μ be the joint distribution of $\langle \zeta_1, \zeta_2 \rangle$, and let μ_i be the marginal distribution of ζ_i. If $\pi_i : X_1 \times X_2 \to X_i$ are the projections, then clearly $\mu_i = \mathfrak{S}(\pi_i)(\mu)$. K denotes the regular conditional distribution of ζ_2 given ζ_1, thus we have for the Borel sets $A_i \subseteq X_i$

$$\mathbb{P}(\{\omega \in \Omega \mid \zeta_1(\omega) \in A_1, \zeta_2(\omega) \in A_2\}) = \mu(A_1 \times A_2)$$
$$= \int_{A_1} K(x_1)(A_2) \, \mu_1(dx_1).$$

We will show now that $K_{\mu_1}^{\smile}$ is the regular conditional distribution of ζ_1 given ζ_2. In fact, let L be the latter distribution, then the definitions of K and L, resp., imply

$$K^{\bullet}(\mu_1) = \mu_2 \text{ and } L^{\bullet}(\mu_2) = \mu_1.$$

Let $A_i \subseteq X_i$ be Borel sets, then

$$(K^{\bullet}(\mu_1) \otimes L)(A_2 \times A_1) = \int_{A_2} L(x_2)(A_1) \, K^{\bullet}(\mu_1)(dx_2)$$
$$= \int_{A_2} L(x_2)(A_1) \, \mu_2(dx_2)$$
$$= \int_{A_1} K(x_1)(A_2) \, \mu_1(dx_2)$$
$$= (\mu_1 \otimes K)(A_1 \times A_2).$$

Interpreting a stochastic relation as a regular conditional distribution of a random variable ζ_1 given ζ_2, its converse may be interpreted as the conditional distribution of ζ_2 given ζ_1. The start probability μ in the definition of $K_{\mu}^{\smile}$ is then interpreted as a marginal distribution.

Returning to the general case, the defining equation for the converse is spelled out in terms of an integral:

$$\int_X K(x)(D^x) \, \mu(dx) = \int_Y K_{\mu}^{\smile}(y)(D_y) \, K^{\bullet}(\mu)(dy).$$

This will be generalized and made use of later:

LEMMA 5.66
Let $f \in \mathcal{F}(X \times Y)$, then this identity holds:

$$\int_X \int_Y f(x,y) \, K(x)(dy) \, \mu(dx) = \int_Y \int_X f(x,y) \, K_{\mu}^{\smile}(y)(dx) \, K^{\bullet}(\mu)(dy).$$

Thus the order of integration of f may be interchanged, as in Fubini's Theorem, but in contrast we need to adjust the measures used for integration (nevertheless it could be called *Fubinito's Lemma*).

Some properties of forming the converse will be investigated now. We begin with an analogue of the property $R^{\smile\smile} = R$ which holds for the set theoretic converse. Taking the initial distribution into account, this property is very similar for the probabilistic case.

PROPOSITION 5.67

If $K : X \rightsquigarrow Y$, and if $\mu \in \mathfrak{S}(X)$, then $(K_\mu^\smile)_{K^\bullet(\mu)}^\smile = K$ holds everywhere except possibly on a set of μ-measure zero.

PROOF   The stochastic relation $(K_\mu^\smile)_{K^\bullet(\mu)}^\smile$ is determined by the equation

$$(K^\bullet(\mu) \otimes K_\mu^\smile)^\smile = \eta \otimes (K_\mu^\smile)_{K^\bullet(\mu)}^\smile$$

with $\eta := K_\mu^\smile(K^\bullet(\mu))$. The defining equation implies $\eta = \mu$, consequently $\mu \otimes K$ equals $\mu \otimes (K_\mu^\smile)_{K^\bullet(\mu)}^\smile$, as expected.   ⬛

The question under what condition a stochastic relation may be represented as the converse of another relation is a little more difficult to answer than for the set-valued case. In view of the probabilistic interpretation using conditional distributions, however, the following solution arises naturally.

COROLLARY 5.68

Let $L : Y \rightsquigarrow X$ be a stochastic relation, and $\mu \in \mathfrak{S}(X)$. Then these conditions are equivalent:

a. $\mu = L^\bullet(\nu)$ for some $\nu \in \mathfrak{S}(Y)$,

b. $L = K_\mu^\smile$ for some $K : X \rightsquigarrow Y$.

Thus $L : Y \rightsquigarrow X$ may be written in a variety of ways as the converse of a stochastic relation, viz., $L = (K_\nu)_{L^\bullet(\nu)}^\smile$ for an arbitrary $\nu \in \mathfrak{S}(Y)$ (where the relation $X \rightsquigarrow Y$ depends on ν). This is in marked contrast again to the set-theoretic case, where the converse of the converse of a relation is the relation itself, hence is uniquely determined.

Compatibility of composition and forming the converse is an important property in the world of set-theoretic relations. In that case it is well known that $(R * S)^\smile = S^\smile * R^\smile$ always holds (which might be called an anti-commutative law). The corresponding property for stochastic relations reads

PROPOSITION 5.69

Let $K : X \rightsquigarrow Y, L : Y \rightsquigarrow T$ be stochastic relations, and let $\mu \in \mathfrak{S}(X)$ be an initial distribution. Then $(K * L)_\mu^\smile = L_{K^\bullet(\mu)}^\smile * K_\mu^\smile$ holds.

PROOF   We will make use of Lemma 5.66 by showing that both relations have the same properties on measurable and bounded functions. Let $f \in$

$\mathcal{F}(X \times Z)$, then

$$\int_{X \times Z} f \, d(\mu \otimes (K * L)) \overset{(1)}{=}$$

$$\int_X \int_Z f(x, z) \ (K * L)(x)(dz) \ \mu(dx) \overset{(2)}{=}$$

$$\int_X \int_Y \int_Z f(x, z) \ L(y)(dz) \ K(x)(dy) \ \mu(dx) \overset{(3)}{=}$$

$$\int_Y \int_X \int_Z f(x, z) \ L(y)(dz) \ K_\mu^{\smile}(y)(dx) \ K^\bullet(\mu)(dy) \overset{(4)}{=}$$

$$\int_Y \int_Z \int_X f(x, z) \ K_\mu^{\smile}(y)(dx) \ L(y)(dz) \ K^\bullet(\mu)(dy) \overset{(5)}{=}$$

$$\int_Z \int_Y \int_X f(x, z) \ K_\mu^{\smile}(y)(dx) \ L_{K^\bullet \cdot \mu}^{\smile}(z)(dy) \ L^\bullet(K^\bullet(\mu))(dz) \overset{(6)}{=}$$

$$\int_Z \int_X f(x, z) \ \left(L_{\overline{K}^\bullet(\mu)}^{\smile} * K_\mu^{\smile} \right)(z)(dx) \ L^\bullet(K^\bullet(\mu))(dz)$$

Equation (1) applies the definition of $\mu \otimes (K * L)$ to the first integral. In equation (2) the definition of $K * L$ is expanded, and in equation (3) Lemma 5.66 is applied to the two outermost integrals, similarly for equation (5). Fubini's Theorem is used for interchanging integrals in equations (4) and (6). The latter equation applies the definition of the composition of kernels to $L_{\overline{K}^\bullet(\mu)}^{\smile}$ and $K_\mu^{\smile}$.

On the other hand,

$$\int_{X \times Z} f \, d(\mu \otimes (K * L)) = \int_X \int_Z f(x, z) \ (K * L)(x)(dz) \ \mu(dx)$$

$$= \int_Z \int_X f(x, z) \ (K * L)_\mu^{\smile}(z)(dx) \ L^\bullet(K^\bullet(\mu))(dz)$$

is inferred from Lemma 5.66. Comparing the results established the claim. ∎

This is again a place to note algebraic similarities between set-theoretic and stochastic relations, but also to record exceptions. Take, e.g., *Schröder's Cycle Rule*

$$Q * R \subseteq S \Leftrightarrow Q^{\smile} * \overline{S} \subseteq \overline{R} \Leftrightarrow \overline{S} * R^{\smile} \subseteq \overline{Q},$$

the bar denoting complementation. This rule is very helpful in practical applications, but it does not enjoy a direct counterpart for stochastic relations, since the respective notions of negation, and of containment do not carry over.

REMARK 5.70 It can be shown that the converse has a quite interesting topological property, because it forms essentially a relatively compact subset of $\mathfrak{S}(X)$. To be specific, it can be shown:

Given $K : X \rightsquigarrow Y$ with X, Y Polish, and $\mu \in \mathfrak{S}(X)$, then there exists a Borel set $A \subseteq Y$ with $K(x)(A) = 0$ for μ-almost all $x \in X$, so that the set $\{K_\mu^\smile(y) \mid y \notin A\}$ is a relatively compact subset of $\mathfrak{S}(X)$.

This is established through the observation that the converse as a whole lives essentially on a compact subset of X which is in turn produced through tightness of μ (see Section 1.5.1). Thus $\{K_\mu^\smile(y) \mid y \notin A\}$ is uniformly tight, which by Prohorov's characterization of compactness on spaces on measures (Parthasarathy, 1967, Theorem II.6.7) implies that this set is relatively compact. The reader is referred to (Doberkat, 2004, Proposition 7).   ⬚

5.8.2   Preserving Bisimilarity

We will show that bisimilar relations give rise to bisimilar converses, so that bisimilarity is preserved under forming converses. We have to take into account, however, that forming the converse does not only depend on the relation itself, but that also an initial distribution is needed. Hence we extend the notion of bisimilarity to subprobabilities as well by treating them as constant stochastic relations.

We have discussed different notions of bisimilarity with ties to congruences in this chapter. The variant that fits here best is 2-similarity, because domain and range of the mediating relation are part of the Cartesian product of the domain resp. range of the given relations, rather than being somewhat unrelated, abstractly given spaces. Thus *bisimilar* means in this section always *2-bisimilar*.

DEFINITION 5.71   *Let X_1, X_2 be Polish spaces with $\mu_i \in \mathfrak{S}(X_i)$ ($i = 1, 2$). Then $\langle X_1, \mu_1 \rangle$ is said to be 2-bisimilar to $\langle X_2, \mu_2 \rangle$ iff there exists a subset $Z \subseteq X_1 \times X_2$ and $\zeta \in \mathfrak{S}(Z)$ such that*

a. Z is a Borel subset of $X_1 \times X_2$,

b. $\mu_1 = \mathfrak{S}(\pi_{Z,X_1})(\zeta)$ and $\mu_2 = \mathfrak{S}(\pi_{Z,X_2})(\zeta)$,

c. there exists Borel sets $C_1 \subseteq X_1, C_2 \subseteq X_2$ with

$$\emptyset \neq Z \cap (C_1 \times X_2) = Z \cap (X_1 \times C_2) \neq Z.$$

$\langle Z, \zeta \rangle$ is said to mediate *for $\langle X_1, \mu_1 \rangle$ and $\langle X_2, \mu_2 \rangle$.*

The first condition is quite necessary for otherwise it would be difficult to define a measure on Z, the second one is just a translation of the requirement that the corresponding diagram should be commutative, and the third one postulates that there is a nontrivial common event; see Section 5.6.

EXAMPLE 5.72

In the discrete setting, the mediating subprobability measure may be represented as a matrix. In fact, let $\langle\{1,\ldots,n\},\mu_1\rangle$ and $\langle\{1,\ldots,m\},\mu_2\rangle$ be bisimilar with mediating $\langle Z,\zeta\rangle$. Then ζ is represented as an $n\times m$ matrix $(a_{i,j})_{1\leq i\leq n,1\leq j\leq m}$ such that

a. $0\leq a_{i,j}\leq 1$,

b. for each i, the sum $\sum_{j=1}^m a_{i,j}$ equals $\mu_1(i)$,

c. for each j, the sum $\sum_{i=1}^n a_{i,j}$ equals $\mu_2(j)$.

The set Z is determined as the set of indices $\langle i,j\rangle$ for which $a_{i,j}\neq 0$.

Let $X_1 = \{1,2,3\}, \mu_1 = [1/2,1/4,1/4]$ and $X_2 = \{1,2\}, \mu_2 = [3/8,5/8]$. Then $\langle Z,\zeta\rangle$ mediates between $\langle X_1,\mu_1\rangle$ and $\langle X_2,\mu_2\rangle$, where

$$Z := \{\langle 1,2\rangle, \langle 2,1\rangle, \langle 2,2\rangle, \langle 3,1\rangle\}$$

and ζ is given through the matrix

$$\begin{bmatrix} 0 & 1/2 \\ 1/8 & 1/4 \\ 1/8 & 0 \end{bmatrix}$$

Bisimulations are maintained by forming products, and by transporting a measure through a stochastic relation, as we will see now:

PROPOSITION 5.73

Let $\mathsf{K}_i = \langle X_i, Y_i, K_i\rangle$ be *2-bisimilar Polish objects* $(i = 1,2)$ *for which* $N : U \rightsquigarrow V$ *mediates, and assume that* $\mu_i \in \mathfrak{S}(X_i)$ *such that* $\langle X_1,\mu_1\rangle$ *and* $\langle X_2,\mu_2\rangle$ *are 2-bisimilar with mediating* $\langle Z,\zeta\rangle$. *Assume that* $Z \subseteq U$ *holds, then*

a. $\langle Y_1, K_1{}^\bullet(\mu_1)\rangle$ *is 2-bisimilar to* $\langle Y_2, K_2{}^\bullet(\mu_2)\rangle$ *with mediating* $\langle V, N^\bullet(\zeta)\rangle$,

b. $\langle X_1 \times Y_1, \mu_1 \otimes K_1\rangle$ *is 2-bisimilar to* $\langle X_2 \times Y_2, \mu_2 \otimes K_2\rangle$ *with mediating* $\langle t[E], \mathfrak{S}(t)(\zeta \otimes N)\rangle$, *where* $E := Z \times V$ *and* t *switches components,* $t(x_1, x_2, y_1, y_2) := \langle x_1, y_1, x_2, y_2\rangle$.

PROOF   0. Because $Z \subseteq U$, we know that for $z \in Z$ the equality $\pi_{Z,X_1}(z) = \pi_{U,X_1}(z)$ holds, so that

$$K_1(\pi_{Z,X_1}(z)) = K_1(\pi_{U,X_1}(z)) = \mathfrak{S}(\pi_{V,Y_1})(N(z))$$

is true; similarly for K_2.

1. For establishing a, let $f_1 \in \mathcal{F}(Y_1)$, then

$$\int_{Y_1} f_1 \, dK_1^{\bullet}(\mu_1) = \int_{X_1} \int_{Y_1} f_1(y_1) \, K_1(x_1)(dy_1) \, \mu_1(dx_1)$$

$$= \int_Z \int_{Y_1} f_1 \, dK_1(\pi_{Z,X_1}(z)) \, \zeta(dz)$$

$$= \int_Z \int_B (f_1 \circ \pi_{V,Y_1}) \, dN(z) \, \zeta(dz)$$

$$= \int_B f_1 \circ \pi_{V,Y_1} \, dN^{\bullet}(\zeta).$$

This implies $K_1^{\bullet}(\mu_1) = \mathfrak{S}(\pi_{V,Y_1})(N^{\bullet}(\zeta))$. In the same way, $K_2^{\bullet}(\mu_2) = \mathfrak{S}(\pi_{V,Y_2})(N^{\bullet}(\zeta))$ is established. This proves the first part of the assertion, because the σ-algebra of common events for K_1 and K_2 can be used for the common events of $\langle Y_1, K_1^{\bullet}(\mu_1)\rangle$ and $\langle Y_2, K_2^{\bullet}(\mu_2)\rangle$.

2. An argument very similar to the preceding one shows that for $f_1 \in \mathcal{F}(X_1 \times Y_1)$ these equalities hold:

$$\int_{X_1 \times Y_1} f_1 \, d(\mu_1 \otimes K_1) = \int_{X_1} \int_{Y_1} f_1(x_1, y_1) \, K_1(x_1)(dy_1) \, \mu_1(dx_1)$$

$$= \int_E f_1 \, d(\mathfrak{P}(\pi_{E,X_1 \times Y_1})(\zeta \otimes N)).$$

A similar calculation shows for $f_2 \in \mathcal{F}(X_2 \times Y_2)$ that

$$\int_{X_2 \times Y_2} f_2 \, d(\mu_2 \otimes K_2) = \int_E f_2 \, d(\mathfrak{P}(\pi_{E,X_2 \times Y_2})(\zeta \otimes N)).$$

This implies the assertion, since the isomorphism t only serves to reorder variables. □

The argumentation above shows that bisimilar relations and bisimilar initial distributions lead to bisimilar measures on the product. The process can be reversed: the idea is that disintegrating 2-bisimilar measures on a product leads to 2-bisimilar stochastic relations.

LEMMA 5.74

Let X_i, Y_i be Polish spaces, $\mu_i \in \mathfrak{S}(X_i \times Y_i)$ for $i = 1, 2$. Assume that $\langle X_1 \times Y_1, \mu_1\rangle$ is 2-bisimilar to $\langle X_2 \times Y_2, \mu_2\rangle$. Define the Polish objects $K_i := \langle X_i, Y_i, K_i\rangle$ through the disintegrations of μ_i w.r.t $\mathfrak{S}(\pi_{X_i \times Y_i, X_i})(\mu_i)$. Then there exists a Polish object $M = \langle X_1 \times X_2, Y_1 \times Y_2, M\rangle$ that mediates between K_1 and K_2.

PROOF   1. Assume that $\langle E, \zeta\rangle$ is mediating between $\langle X_1 \times Y_1, \mu_1\rangle$ and $\langle X_2 \times Y_2, \mu_2\rangle$. Put $E_0 := t[E]$, $\zeta_0 := \mathfrak{S}(t)(\zeta)$, where t rearranges components, as in Proposition 5.73. Let $\gamma := \mathfrak{S}\left(\pi_{E_0, X_1 \times X_2}\right)(\zeta_0) \in \mathfrak{S}\left(X_1 \times X_2\right)$, and let M' be the disintegration of ζ_0 with respect to γ.

2. Let $\mathcal{G}_i$ be a countable generator for the σ-algebra on $X_i \times Y_i$, so that $\mathcal{G}_i$ is closed under finite intersections $(i = 1, 2)$. Let $G \in \mathcal{G}_1$, then

$$\mu_1(G) = \zeta_0(G \times X_2 \times Y_2)$$

$$= \int_{X_1 \times X_2} \mathfrak{S}\left(\pi_{Y_1 \times Y_2, Y_1}\right)\left(M'(x_1, x_2)\right)\left(G_{x_1}\right) \gamma(d\langle x_1, x_2\rangle),$$

and

$$\mu_1(G) = \int_{X_1} K_1(x_1)(G_{x_1}) \, \mathfrak{S}\left(\pi_{X_1 \times Y_1, X_1}\right)(\mu_1)$$

by the definition of K_1. Since

$$\mathfrak{S}\left(\pi_{X_1 \times Y_1, X_1}\right)(\mu_1) = \mathfrak{S}\left(\pi_{X_1 \times Y_1, X_1}\right)\left(\mathfrak{S}\left(\pi_{E_0, X_1 \times Y_1}\right)(\zeta_0)\right)$$

$$= \mathfrak{S}\left(\pi_{E_0, X_1}\right)(\zeta_0)$$

$$= \mathfrak{S}\left(\pi_{X_1 \times X_2, X_1}\right)(\gamma),$$

the latter integral may be expressed as

$$\mu_1(G) = \int_{X_1 \times X_2} K_1(x_1)(G_{x_1}) \, \gamma(d\langle x_1, x_2\rangle).$$

Thus

$$A_G := \{\langle x_1, x_2\rangle \in X_1 \times X_2 \mid K_1(x_1)(G_{x_1}) \neq \mathfrak{S}\left(\pi_{Y_1 \times Y_2, Y_1}\right)\left(M'(x_1, x_2)\right)\left(G_{x_1}\right)\}$$

is a measurable subset of $X_1 \times X_2$ which has γ-measure 0. Put

$$A_1 := \bigcup\{A_G \mid G \in \mathcal{G}_1\},$$

then clearly $\gamma(A_1) = 0$, and $K_1(x_1)(G_{x_1}) = \mathfrak{S}\left(\pi_{Y_1 \times Y_2, Y_1}\right)\left(M'(x_1, x_2)\right)\left(G_{x_1}\right)$ holds for all measurable subsets $G \subseteq X_1 \times Y_1$ whenever $\langle x_1, x_2\rangle \notin A_1$. This is so since by the π-λ-Theorem 1.1 a $\cap$-stable generator uniquely determines a finite measure, and since the equation above is true for all $G \in \mathcal{G}_1$. In a similar way a measurable subset A_2 of $X_1 \times X_2$ can be found with $\gamma(A_2) = 0$, so that for $\langle x_1, x_2\rangle \notin A_2$ and for all measurable subsets $G \subseteq X_1 \times Y_2$ the equality

$$K_2(x_2)(G_{x_2}) = \mathfrak{S}\left(\pi_{Y_1 \times Y_2, Y_2}\right)\left(M'(x_1, x_2)\right)\left(G_{x_2}\right)$$

holds.

3. Define M as M' outside $A_1 \cup A_2$, and set $M(x_1, x_2) := K_1(x_1) \otimes K_2(x_2)$, for $\langle x_1, x_2\rangle \in A_1 \cup A_2$, then $M : X_1 \times X_2 \rightsquigarrow Y_1 \times Y_2$ has the desired properties.

$\square$

Showing that bisimilarity is maintained when forming the converse is now an easy consequence:

PROPOSITION 5.75

Let $\mathsf{K}_i = \langle X_i, Y_i, K_i \rangle$ be 2-bisimilar Polish objects ($i = 1, 2$) between which $N : U \rightsquigarrow V$ mediates, and assume that $\mu_i \in \mathfrak{S}(X_i)$ such that $\langle X_1, \mu_1 \rangle$ and $\langle X_2, \mu_2 \rangle$ are 2-bisimilar with mediating $\langle Z, \varsigma \rangle$. Assume that $Z \subseteq U$ holds. Then $\mathsf{K}_{\widetilde{1,\mu_1}}$ is 2-bisimilar to $\mathsf{K}_{\widetilde{2,\mu_2}}$.

PROOF   We know from Proposition 5.73 that $\langle X_1 \times Y_1, \mu_1 \otimes K_1 \rangle$ and $\langle X_2 \times Y_2, \mu_2 \otimes K_2 \rangle$ are 2-bisimilar. Bisimilarity is plainly not destroyed by interchanging coordinates. The assertion follows from Lemma 5.74, because the common events for $\langle X_1, \mu_1 \rangle$ and $\langle X_2, \mu_2 \rangle$ are also common events for the disintegrations.   ☐

5.9   Case Study: Simple Relations for Counting

The characterization of simple systems helps in analyzing the average behavior of algorithms by discussing two examples. The results are not new, the approach, however, is. Rutten (Rutten, 2002; Rutten, 2003) shows how a stream calculus based on coinduction is used for counting, and hence for some aspects of the average case analysis of algorithms. This is made possible through the existence of final systems for the functor considered. By Corollary 5.59, the situation discussed here is different in that for the probabilistic case only a trivial final system exists, and it may be doubted whether this can be put to significant use.

Despite this somewhat restricted situation simple relations may be put to work; we will discuss the average case analysis of two algorithms and show how the continuous and the discrete case interact. This suggests the formal justification for using continuous models in the average case analysis of discrete algorithms. Various relations will be constructed that are simple. Through Proposition 5.58 quantities that otherwise can be obtained only with difficulties are derived. Denote in the discussion that follows by V_n the set of permutations on $\{1, \ldots, n\}$.

5.9.1   Left to Right Maxima

Given an array `a[1..n]` of natural numbers, the following algorithm identifies the index `m` of the maximal element.

ALGORITHM 5.76

```
m := 1;
for i := 2 to n do
    if a[m] < a[i] then m := i; fi;
end for; ♣
```

The expected number of times the variable m changes its value in Algorithm 5.76 is asked for; Knuth discusses this algorithm and arrives at the result that this expectation equals $H_n - 1$, where $H_n := \sum_{i=1}^{n} i^{-1}$ is the nth harmonic number, provided the array has n mutually different components (Knuth, 1973a, Section 1.2.10). To be more specific, he shows that the number $p_{n,k}$ of permutations on $\{1, \ldots, n\}$ for which the step in question is executed exactly k times equals

$$p_{n,k} = \frac{1}{n!} \cdot \begin{bmatrix} n \\ k+1 \end{bmatrix}$$

with $\begin{bmatrix} n \\ k \end{bmatrix}$ as a Stirling number of the first kind (Knuth, 1973a, Section 1.2.10, Equation (9)). These numbers are defined through

$$z \cdot (z+1) \cdot \ldots \cdot (z+n-1) = \sum_k \begin{bmatrix} n \\ k \end{bmatrix} \cdot z^k.$$

They are interpreted combinatorially through cycles: $\begin{bmatrix} n \\ k \end{bmatrix}$ is the number of ways to arrange n objects into k cycles; see (Graham et al., 1989, Section 6.1).

Define the stochastic relation $\mathsf{K}_n := (\{1, \ldots, n\}, \mathbf{1}, K_n)$ with $K_n(k)(\mathbf{1}) := p_{n,k}$, and assume that the values $p_{n,0}, \ldots, p_{n,n}$ are mutually different (if they are not, factor). Then K_n is a simple relation, thus for each other relation K there exists at most one morphism into it.

We want to compute the expected value when we have continuous data. Let $(\Omega, \mathcal{A}, \mathbb{P})$ be a probability space, $\zeta : \Omega \to [0,1]^n$ be a uniformly distributed random variable, and $\tau : \Omega \to \mathbb{N}$ the number of times the value corresponding to m is changed. Thus if $\omega \in \Omega$ is observed, the vector $\zeta(\omega)$ is the input to the algorithm; $\tau(\omega) = Z(\zeta(\omega))$ counts the corresponding number, where Z is the function for counting. We are looking for the expected value $\mathbb{E}(\tau)$. Since τ takes only discrete values, and since

$$\mathbb{E}(\tau) = \sum_{k=1}^{\infty} k \cdot \mathbb{P}(\tau = k),$$

it is sufficient to compute the probability $\mathbb{P}(\tau = k)$ that the random variable τ has the value k. Now put $K(x)(\mathbf{1}) := \mathbb{P}(\{\omega \in \Omega \mid \zeta(\omega) = Z(x)\})$. Then the simplicity of K_n implies that defining $K(x)(\mathbf{1}) := p_{n,Z(x)}$ is the *only* way to define the stochastic relation $\mathsf{K} = ([0,1]^n, \mathbf{1}, K)$ making $Z : \mathsf{K} \to \mathsf{K}_n$ a morphism. Consequently, $\mathbb{P}(\tau = k) = p_{n,k}$, and $\mathbb{E}(\tau) = H_n - 1$, as in the discrete case.

This illustrates how the transfer between discrete and continuous stochastic systems works: the behavior is known in the discrete case, and defining an appropriate simple system helps in transporting that knowledge to the continuous case. The example gives insight into the relationship between discrete and continuous systems. The quantitative result is not new, however, and the central recurrence from which Knuth derives the expected value, viz.,

$$p_{n,k} = \frac{1}{n} \cdot p_{n-1,k-1} + \frac{n-1}{n} \cdot p_{n-1,k}, \text{ and } p_{1,k} = \delta_{0,k}$$

can be easily derived directly for the continuous case.

5.9.2   Williams' Algorithm to Construct Heaps

Recall that a permutation $p \in V_n$ is a heap iff $p_{\lfloor i/2 \rfloor} < p_i$ holds for each index i with $2 \leq i \leq n$. Heaps are usually represented through binary trees with node 1 as the root and node $\lfloor i/2 \rfloor$ as the father of node i, so that the heap condition entails that each node has a label p_i which is larger than the label $p_{\lfloor i/2 \rfloor}$ of its father. Denote by H_n all elements of V_n that are heaps.

Let $x \in [0,1]^n$ be a vector of n components taken from $[0,1]$ which are mutually different, then $\wp_n(x) \in V_n$ is the permutation that arises from x by order statistics, i.e., if $\wp_n(x) = p$, then $p_i = k$ iff x_i is the kth-largest component of x. We will deal in the sequel with uniformly distributed elements of $[0,1]^n$. Since equality of components happens only on a set of measure zero, those elements can be neglected, so that $\wp_n$ is defined almost everywhere on $[0,1]^n$.

Assume that n has the binary representation $/1b_{\nu-1} \ldots b_0/_2$, then the node

$$t(n,\kappa) := /1b_{\nu-1} \ldots b_{\nu-\kappa}/_2$$

is called the *special node* on level κ (Knuth, 1973b, Exercise 5.2.3.20).

Now let

$$S_{n+1,0} := \{p \in V_{n+1} \mid \wp_n(p_1, \ldots, p_n) \in H_n, p_{n+1} > p_{t(n+1,1)}\},$$
$$S_{n+1,\kappa} := \{p \in V_{n+1} \mid \wp_n(p_1, \ldots, p_n) \in H_n, p_{t(n+1,\kappa)} > p_{n+1} > p_{t(n+1,\kappa+1)}\}$$
$$(1 \leq \kappa \leq \nu := \lfloor \log_2(n+1) \rfloor),$$
$$S_{n+1,\nu} := \{p \in V_{n+1} \mid \wp_n(p_1, \ldots, p_n) \in H_n, p_{t(n+1,\nu)} > p_{n+1}\}.$$

The task at hand is to count the number of elements in $S_{n+1,0}, \ldots, S_{n+1,\nu}$. This is important for determining the average complexity of Williams' algorithm to insert an element into a heap: Suppose $p \in V_{n+1}$ is a permutation with $n+1$ elements such that $p_1, \ldots, p_n$ forms a heap, then p_{n+1} is inserted into this heap according to Williams' algorithm, which searches the path $n+1, \lfloor (n+1)/2 \rfloor, .., 1$ that goes from node $n+1$ to the root for the correct position of p_{n+1} and inserts it there, specifically:

ALGORITHM 5.77

```
j := n+1; i := ⌊j/2⌋; q := pₙ₊₁
while (i > 0) && (q < pᵢ) do
        pⱼ := pᵢ; j := i; i := ⌊i/2⌋;
od;
pⱼ := q; ♣
```

This algorithm may be used for iteratively building up a heap, it is one of the classics (Williams, 1964; Knuth, 1973b). Nevertheless, the average case analysis is surprisingly complicated (Doberkat, 1981). We will show here through simple stochastic relations that probabilistic arguments help in counting permutations. Put

$$W_n := \{x \in [0,1]^n \mid x \text{ is a heap}\},$$
$$G := W_n \times [0,1].$$

We will assume the inputs from W_n and from $[0,1]$ are uniformly distributed with λ^n as Lebesgue measure on (the Borel sets of) $[0,1]^n$. It is claimed for later use that

$$\lambda^{n+1}(\{x \in G \mid \wp_{n+1}(x) = \mathsf{p}\}) = \frac{\lambda^n(W_n)}{n+1}$$

independently of $\mathsf{p} \in V_{n+1}$. One first notes that the Change of Variable formula (Proposition 1.95) implies that

$$\lambda^{n+1}(\{x \in G \mid \wp_{n+1}(x) = \mathsf{p}\}) = \lambda^{n+1}(\{x \in G \mid \wp_{n+1}(x) = \mathsf{p}'\})$$

with p' as the element of V_{n+1} the first n components of which form a heap, and $\mathsf{p}'_{n+1} < \mathsf{p}'_1$. This is so since the Jacobian of a permutation for the coordinates equals 1. Thus we obtain

$$\lambda^{n+1}(\{x \in G \mid \wp_{n+1}(x) = \mathsf{p}\}) = \lambda^{n+1}(\{x \in G \mid \wp_{n+1}(x) = \mathsf{p}'\})$$
$$\overset{(\dagger)}{=} \int_0^1 \lambda^n(\{y \in [0,1]^n \mid y \text{ is a heap}, x < y_1\}) \, dx$$
$$\overset{(\ddagger)}{=} \lambda^n(W_n) \cdot \int_0^1 (1-x)^n \, dx$$
$$= \frac{\lambda^n(W_n)}{n+1}.$$

Equation $(\dagger)$ is Fubini's Theorem on product integration. Equation $(\ddagger)$ is again the Change of Variable formula, since the transformation

$$(y_1, \ldots, y_n) \mapsto ((1-x) \cdot y_1 + x, \ldots, (1-x) \cdot y_n + x)$$

which maps all heaps with n elements

$$\{y \in [0,1]^n \mid y \text{ is a heap}\}$$

bijectively to those heaps the root of which is greater than x,

$$\{y \in [0,1]^n \mid y \text{ is a heap}, x < y_1\}$$

has the Jacobian $(1-x)^n$.

Now let $g(n,i)$ be the number of nodes in the subtree rooted at node i, then Knuth (Knuth, 1973b, Equation 5.2.3-14) shows that

$$g(n, /b_{\nu-1} \ldots b_\kappa/2) = /1b_{\kappa-1} \ldots b_0/2,$$

and that, if node i is on level κ,

$$g(n,i) = \begin{cases} 2^{\nu-\kappa} - 1 & i \text{ is a right node}, \\ 2^{\nu-\kappa+1} - 1 & i \text{ is a left node} \end{cases}$$

where *right* and *left* are relative to the node's position with respect to the special node on that level.

It follows from (Knuth, 1973b, Equation 5.2.3-16, Exercise 5.1.4-20) that for the number h_n of heaps on n elements

$$h_n = \prod_{i=1}^{n} g(n,i)$$

holds; from this the equality

$$\lambda^n(W_n) = \prod_{i=1}^{n} g(n,i)^{-1} =: \chi_n$$

is easily established (Doberkat, 1981). Now let for $0 \le \kappa \le \lfloor \log_2(n+1) \rfloor$

$$\begin{aligned} S(\kappa) &:= \{x \in G \mid \wp_{n+1}(x) \in S_{n+1,\kappa}\} \\ &= Z^{-1}[\{\kappa\}]. \end{aligned}$$

Here

$$Z : \begin{cases} G & \to Q_{n+1} \\ x & \mapsto \kappa, \text{ if } \wp_{n+1}(x) \in S_{n+1,\kappa} \end{cases}$$

(with $Q_{n+1} := \{p \in V_{n+1} \mid (p_1, \ldots, p_n) \text{ is a heap}\}$) has the rôle of an index map, since $\wp_{n+1}(x) \in S_{n+1,Z(x)}$ holds for all $x \in G$. Thus the membership of x in $S(\kappa)$ determines the number of levels the new element climbs up the heap, hence from it the complexity of Algorithm 5.77 can be computed. Note that $|Q_{n+1}| = (n+1) \cdot h_n$ holds. It can be shown (Doberkat, 1981, Proposition 3)

PROPOSITION 5.78

If $n+1$ has the binary representation $/1c_{\nu-1} \ldots c_0/2$, then

$$\lambda^{n+1}(S(\kappa)) = \frac{\chi_n}{/1c_{\kappa-1} \ldots c_0/2} \cdot \prod_{j=\kappa+1}^{\nu} \frac{/1c_{j-1} \ldots c_0/2 - 1}{/1c_{j-1} \ldots c_0/2}.$$

Now define the stochastic relation $\mathsf{K} = (G, \mathbb{1}, K)$ through $K(x)(\mathbb{1}) := \lambda^{n+1}(S(Z(x)))$, and put $\mathsf{K}' := (\{0, \ldots, \lfloor \log_2(n+1) \rfloor\}, \mathbb{1}, K')$ with

$$K'(\kappa)(\mathbb{1}) := \frac{\chi_n}{n+1} \cdot \mid S_{n+1,\kappa} \mid .$$

We claim that $K(x)(\mathbb{1}) = K'(Z(x))(\mathbb{1})$ holds for each $x \in G$, rendering Z a morphism $\mathsf{K} \to \mathsf{K}'$. In fact,

$$\mid S_{n+1,\kappa} \mid = \sum_{\mathsf{p} \in S_{n+1,\kappa}} 1 =$$

$$\frac{n+1}{\chi_n} \cdot \sum_{\mathsf{p} \in S_{n+1,\kappa}} \lambda^{n+1}(\{x \in S(\kappa) \mid \wp_{n+1}(x) = \mathsf{p}\}) =$$

$$\frac{n+1}{\chi_n} \cdot \mathsf{h}_n \cdot \lambda^{n+1}(\{x \in G \mid \wp_{n+1}(x) \in S_{n+1,\kappa}\}) =$$

$$\frac{n+1}{\chi_n} \cdot \mathsf{h}_n \cdot \lambda^{n+1}(S(\kappa)).$$

This is a cornerstone for establishing

PROPOSITION 5.79

The number $\mid S_{n+1,\kappa} \mid$ *of permutations* p *on* $\{1, \ldots, n+1\}$ *such that* $\mathsf{p}_1, \ldots \mathsf{p}_n$ *is a heap, and* p_{n+1} *climbs up* κ *levels during Williams' algorithm equals*

$$\frac{(n+1) \cdot \mathsf{h}_n}{/1c_{\kappa-1} \ldots c_0/_2} \cdot \prod_{j=\kappa+1}^{\nu} \frac{/1c_{j-1} \ldots c_0/_2 - 1}{/1c_{j-1} \ldots c_0/_2},$$

assuming that $n+1$ *has the binary representation* $/1c_{\nu-1} \ldots c_0/_2$.

PROOF We assume again without loss of generality that the numbers in question are mutually different. Then they represent the only way of defining the simple system K' in such a way that Z becomes a morphism. ⬜

This explicit result is apparently new. It should be noted, however, that results of this kind cannot be used for exploiting the average complexity of Williams' algorithm. It makes essential use of the fact that the underlying probability is uniform, but it is well known that uniform distribution is destroyed by inserting an element into a heap through this algorithm (or removing an element from it using Floyd's).

5.10   Bibliographic Notes

Bisimilarity. Bisimilarity is introduced in its coalgebraic version as a span of morphisms (Joyal et al., 1996; Rutten, 2000). For coalgebras based on the category of sets, this definition agrees with the one through relations, originally given by Milner (see (Rutten, 2000); (van Breugel et al., 2005) work with a relationally oriented definition of bisimulation). Given the broad interest in this subject, it is of course impossible to indicate the history of the subject. Behavioral equivalence was studied by Kurz (Kurz, 2000); see Pattinson (Pattinson, 2004). In (Desharnais et al., 2000) the authors call a bisimulation what we introduced as a congruence, albeit that paper restricts itself to labeled Markov transition systems, thus technically to families of stochastic relations $S \rightsquigarrow S$ for some state space S. It seems conceptually to be clearer to distinguish spans of morphisms from equivalence relations, thus we make this distinction here.

The close relationships between bisimulations and certain equivalence relations have been known since at least the Hennessy-Milner Theorem (Hennessy and Milner, 1980). Having a closer look at the relation that is defined there, it is evident that the relation defined through

$$s \equiv s' \Longleftrightarrow \forall \phi : [s \models \phi \Leftrightarrow s' \models \phi]$$

is countably generated. Once one can show in a probabilistic interpretation of modal logics that the sets of states for which a formula is valid is measurable, the relation is recognized as smooth, so the tools from the theory of Borel sets (Kechris, 1994; Srivastava, 1998; Arveson, 1976) developed for countably generated equivalence relations become available; this is exactly how we will proceed in Section 6.2.

The observation that factoring an analytic space through a smooth equivalence relation will yield an analytic space again is the reason that analytic spaces figure prominently in this development; on the other hand it shows upon factoring that these relations will not lead to bizarre spaces that are technically difficult to grasp.

The Converse. Abramsky, Blute, and Panangaden (Abramsky et al., 1999) investigate the category **PRel** of probability spaces, hereby introducing the converse of a probabilistic relation as we do through the product measure (Abramsky et al., 1999, Section 7). The process by which they arrive at this construction is quite similar to disintegration, as proposed here but makes heavier use of absolute continuity (in fact, morphisms in **PRel** use absolute continuity in a crucial way). The argumentation that has been used in the present discussion seems to be closer to the set-theoretic case by looking at what happens when we compute the probability for a converse relation. Further investigations of the converse do not include the anti-commutative law Proposition 5.69. This is probably due to the fact that integration techniques are directly used in the present paper, whereas (Abramsky et al., 1999) prefers arguing using absolute continuity, and consequently, with the Radon-Nikodym Theorem.

The analogy between set-theoretic and stochastic relations is like a central thread to many investigations in this area, since it is sometimes annoying, sometimes exciting to see that constructions that can be carried out without great difficulties for relations can be done only with great effort for the probabilistic case. An example is given through the converse, another one is apparent when studying pullbacks. The monograph (Freyd and Ščedrov, 1990) is a general approach to fit relations under a single roof.

Simple Systems. Simple systems are of course a topic in algebra (Lang, 1965). The work in coalgebras for which (Rutten, 2000) stands as a representative has given a tight relationship between simplicity and certain forms of bisimulations. This is not only for reasons to better explore the structure of coalgebras but also because final systems are at the basis of the proof principle of coinduction. It is based on the observation that a final system has exactly one morphism going into it, so it constructs its argument usually by constructing scenarios from which the desired properties are inferred through uniqueness. This requires the existence of a final system, which in turn is sometimes a rather complicated business, and usually the underlying functor must at least preserve weak pullbacks. This situation is analyzed in a paper by Gumm and Schröder for the category of sets, and it is shown that the functor governing the coalgebra preserves kernel pairs iff every congruence is a bisimulation (Gumm and Schröder, 2005, Theorem 5.7). It is interesting to compare this statement with Proposition 4.14 and Proposition 5.51 for the subprobability functor.

In (Moss and Viglizzo, 2004) the probability functor $\mathfrak{P}$ is considered on measurable spaces that are endowed with a initial σ-algebra related to the weak-*-σ-algebra. Given a discrete space I, final coalgebras for a functor derived from $\mathfrak{P}$ yielding a type space over category $\mathfrak{Meas}^I$ are discussed. Besides establishing the existence of a final coalgebra for the functor through satisfied theories, the main result states that functors polynomial in the original type functor have final coalgebras as well. This result is extended in (Viglizzo, 2005) using the final sequence of the functor under consideration, where the method is shown to work for other, set-valued functors as well. In (Cîrstea, 2004) coalgebraic simulation is considered, and one of the application areas for the discussion is probabilistic transition systems. They are modeled as coalgebras for the functor that assigns each set its discrete probability measures. All this shows that special assumptions and constructions are needed for securing the existence of a final system that is related with the probability functor even over finite spaces.

Chapter 6

Interpreting Modal and Temporal Logics

6.1   Introduction

Consider the simple modal logic the formulas of which are given through

$$\varphi ::= \top \mid p \mid \varphi_1 \wedge \varphi_2 \mid \Diamond\varphi,$$

where $p \in P$ with P a set of atomic propositions. A Kripke model $\mathcal{R} = (S, R, V)$ is given by a set S of states (the possible worlds), a relation $R \subseteq S \times S$ and a map $V : P \to \mathfrak{Pow}(S)$. $V(p)$ indicates for an atomic proposition in which worlds it holds. Given $s \in S$, we say that formula $\Diamond\varphi$ holds in s iff we can find $s' \in S$ with $\langle s, s'\rangle \in R$ such that φ holds in s'. Thus $[\![\Diamond\varphi]\!] = \exists R([\![\varphi]\!])$, see Example 1.55, where as usual $[\![\varphi]\!]$ is the set of all states in which formula φ holds.

A probabilistic counterpart for Kripke models cannot restrict itself to stating that a formula holds (or that it doesn't) but should provide quantitative arguments: we want to know the probability with which a formula is true. Thus we want to know the probability for $\Diamond\varphi$ to hold, subject to φ being true. Hence we replace the relation $R \subseteq S \times S$ by a stochastic relation $K : S \rightsquigarrow S$, and the single diamond $\Diamond$ by a whole family $(\Diamond_q)_{0 \le q \le 1}$ indicating the degree to which $\Diamond\varphi$ holds, and state formally that in world $s \in S$ the formula $\Diamond_q\varphi$ is true iff $K(s)([\![\varphi]\!]) \ge q$ holds. This is the faithful probabilistic counterpart to the model above.

This chapter studies probabilistic interpretations of modal and temporal logics. It does not cost much more to replace the very simple modal logic from above by a more comfortable one which commands a collection of modal operators of arbitrary arity, and which includes negation. We will define in Section 6.2 nondeterministic and stochastic Kripke models, give examples for some well-known logics, and we will compare the nondeterministic and the stochastic approach. A stochastic Kripke model is seen as a quantitative model, while a nondeterministic one obviously stresses the qualitative character. Consequently it is sensible to relate nondeterministic to stochastic models through a refinement relation: if a formula holds in the stochastic model with probability one, then it should hold as well in the nondeterministic one. This relation between models is investigated, and we show under which conditions a nondeterministic model can be refined stochastically. Not surprisingly, selection theorems for set-valued maps enter the argumentation.

But the really interesting topic is bisimilarity in its relation to behavioral and logical equivalence — under what conditions are two stochastic Kripke models bisimilar? We show that the Hennessy-Milner Theorem provides an answer in this scenario as well: bisimilarity and logical equivalence are the same. This requires, however, a careful discussion of morphisms and the bisimulations associated with them. At this point we harvest from the investment we undertook in discussing bisimulations from a very general and abstract point of view, since we can just stretch out our hands and pluck the results, once the scenario is set up properly. This, then, gives a rather general result for stochastic Kripke models, and the Bibliographic Notes at the end of the present chapter will put things into the proper context.

We will turn our attention then to temporal logics and propose interpretations for the logics **CSL** and its closed relative μ**CSL**; the latter logic includes a fixed point operator. The important point to be made is to show first how to interpret path formulas, i.e., formulas the validity of which depends on an infinite path. This requires some measure-theoretic preparations, since we have to build up a probability that works on infinite paths from the probabilities for just one step being performed. The main tool to use will be the projective limit of a suitable projective system.

Characteristic features of both **CSL** and μ**CSL** include the explicit incorporation of residence times, and we will show that the tools we collected from smooth equivalence relations and from congruences can be put to good use for investigating subsets of all formulas that govern the behavior of a model on all formulas. Putting it less cryptically, we take a subset $F \subseteq \mathfrak{L}_P$ of all formulas, and ask the question, under what conditions the equivalence relation

$$s \equiv_F s' \Leftrightarrow \forall \phi \in F : [s \models \phi \Leftrightarrow s' \models \phi]$$

equals the equivalence relation $\equiv_{\mathfrak{L}_P}$ addressing the set $\mathfrak{L}_P$ of all formulas (with P again as the set of atomic formulas). It is clear that the answer is interesting for model checking, since such a set F of formulas eases the task of a model checker tremendously. We will look into the case that $F = P$, identifying under which conditions the simplest possible case will do. Surprisingly, it turns out that the invariant Borel sets for the smooth relation $\equiv_F$ will play a leading rôle.

The stochastic interpretation of both logics proposed here is new. In contrast to the traditional approach that starts from a rate function, this approach assumes only a stochastic law governing state changes and residence times. This may be specialized to assuming the stochastic independence of both, but for the development of the theory this is not vital. We use rather an interpretation as a stochastic relation, enabling the use of the tools developed for investigating these relations.

The main contributions of this approach to **CSL** and the understanding of stochastic logics in general lie in investigation of bisimulations for these logics in terms of congruences, together with the development of criteria for the equivalence of different notions of bisimulations. Another point worth emphasizing is that this approach permits the formulation of a general approach for the investigation of bisimulations for this type of logic through the theory of congruences for stochastic relations (which, in turn, is developed further). These logics all have their roots in approaches to model checking, thus they have a pronounced practical side. For the logic at hand this means that computational issues are important, and it becomes evident that structural properties need to be looked at not only for their own interest. The results for **CSL** entail such practical considerations.

The investigations into μ**CSL** take up a trail that was laid for stochastic relations in general and for Kripke models for modal logics in particular: we want to characterize the relationship of bisimilarity and logical equivalence. But with this logic — as with **CSL** — we are in a multi-layered scenario by having to discuss both state formulas and path formulas. This will affect the discussion, because congruences as the tool of choice for these investigations become somewhat awkward to handle. Nevertheless we will find that all these ways of describing the behavior of models are closely related. The approach emphasizes the usefulness of congruences for these investigations, and it shows that criterion for bisimilarity given in Section 5.4 is quite general and versatile, opening up the road to investigations of coalgebraic stochastic logics.

6.2   Modal Logics

We have established a criterion for bisimilarity through simulation equivalent congruences and discussed bisimilarity in terms of isomorphic factor spaces; see in particular Section 5.4. We will now apply this to modal logic. This section defines the logic we will be working with, and Kripke models are defined in their usual nondeterministic and their stochastic versions, together with their satisfaction relation. In Section 6.2.1 some examples are given in order to exhibit probabilistic models for specific logics, and we relate in Section 6.2.2 nondeterministic to stochastic interpretations by introducing probabilistic refinements.

Let P be a countable set of propositional letters which is fixed throughout,

$O \neq \emptyset$ is a set of modal operators. Following (Blackburn et al., 2001), $\tau = (O, \rho)$ is called a *modal similarity type* iff $O \neq \emptyset$, and if $\rho : O \to \mathbb{N}$ is a map, assigning each modal operator $\triangle$ its arity $\rho(\triangle) \geq 1$. We will not deal with modal operators of arity zero, since they do not have to be dealt with as modal constants in an interpretation. The similarity type τ will be fixed.

We define three modal languages based on τ and P. The formulas of the *basic modal language* $\mathfrak{Mod}_b(\tau, P)$ are given by the syntax

$$\varphi ::= p \mid \top \mid \varphi_1 \wedge \varphi_2 \mid \neg\varphi \mid \triangle(\varphi_1, \ldots \varphi_{\rho(\triangle)}),$$

where $p \in P$. If we have $O' = \{\Diamond\}$ with $\rho(\Diamond) = 1$, we obtain the formulas of the basic modal language with negation. Omitting negation in $\mathfrak{Mod}_b(\tau, P)$ defines the formulas in the *negation free basic modal language* $\mathfrak{Mod}_1(\tau, P)$. Finally the *extended modal language* $\mathfrak{Mod}_s(\tau, P)$ is defined through the syntax

$$\varphi ::= p \mid \top \mid \varphi_1 \wedge \varphi_2 \mid \neg\varphi \mid \triangle_q(\varphi_1, \ldots \varphi_{\rho(\triangle)}),$$

where $q \in \mathbb{Q} \cap [0, 1]$ is a rational number, and $p \in P$ is a propositional letter. Again, if we deal with $O = O'$, then we get an entire line of new formulas through $(\Diamond_q)_{q \in \mathbb{Q} \cap [0,1]}$.

A *nondeterministic τ- Kripke model* $\mathcal{R} = (S, R_\tau, V)$ consists of a state space S, a family $R_\tau = ((R_\triangle)_{\triangle \in O})$ of set valued maps $R_\triangle : S \to \mathfrak{Pow}(S^{\rho(\triangle)})$ and a set valued map $V : P \to \mathfrak{Pow}(S)$.

The satisfaction relation $\models$ for a nondeterministic τ-Kripke model $\mathcal{R}$ is defined as usual for $\mathfrak{Mod}_b(\tau, P)$:

- $\mathcal{R}, s \models p \Leftrightarrow s \in V(p)$

- $\mathcal{R}, s \models \neg\varphi \Leftrightarrow \mathcal{R}, s \not\models \varphi$

- $\mathcal{R}, s \models \varphi_1 \wedge \varphi_2 \Leftrightarrow \mathcal{R}, s \models \varphi_1$ and $\mathcal{R}, s \models \varphi_2$

- $\mathcal{R}, s \models \triangle(\varphi_1, \ldots, \varphi_{\rho(\triangle)}) \Leftrightarrow$ there exists $\langle s_1, \ldots, s_{\rho(\triangle)} \rangle \in R_\triangle(s)$ with $\mathcal{R}, s_i \models \varphi_i$ for $1 \leq i \leq \rho(\triangle)$.

Denote by

$$[\![\varphi]\!]_\mathcal{R} := \{s \in S \mid \mathcal{R}, s \models \varphi\}$$

the set of states for which formula φ is valid, and by

$$Th_\mathcal{R}(s) := \{\varphi \in \mathfrak{Mod}_b(\tau, P) \mid \mathcal{R}, s \models \varphi\}$$

the theory of state s in $\mathcal{R}$.

An easy calculation shows that

$$\mathcal{R}, s \models \triangle(\varphi_1, \ldots, \varphi_{\rho(\triangle)}) \Leftrightarrow R_\triangle(s) \cap [\![\varphi_1]\!]_\mathcal{R} \times \ldots \times [\![\varphi_{\rho(\triangle)}]\!]_\mathcal{R} \neq \emptyset$$
$$\Leftrightarrow s \in \exists R_\triangle([\![\varphi_1]\!]_\mathcal{R} \times \ldots \times [\![\varphi_{\rho(\triangle)}]\!]_\mathcal{R}).$$

In analogy, a *stochastic τ-Kripke model* $\mathcal{K} = (S, K_\tau, V)$ has a state space S which is endowed with a σ-algebra $\mathcal{A}$, a family $K_\tau = (K_\triangle)_{\triangle \in O}$ of stochastic relations $K_\triangle : S \rightsquigarrow S^{\rho(\triangle)}$ and a set valued map $V : P \to \mathcal{A}$. We will always assume that S is a Polish space, and that the σ-algebra are the Borel sets.

The interpretation of formulas in $\mathfrak{Mod}_s(\tau, P)$ for a stochastic τ-Kripke model $\mathcal{K}$ is fairly straightforward, the interesting case arising when a modal operator is involved:

$$\mathcal{K}, s \models \triangle_q(\varphi_1, \ldots, \varphi_{\rho(\triangle)})$$

holds iff there exists measurable subsets $\Lambda_1, \ldots, A_{\rho(\triangle)}$ of S such that $\mathcal{K}, s_i \models \varphi_i$ holds for all $s_i \in A_i$ for $1 \le i \le \rho(\triangle)$, and

$$K_\triangle(s)(A_1 \times \ldots \times A_{\rho(\triangle)}) \ge q.$$

Arguing from the point of view of state transition systems, this interpretation of validity reflects that upon the move indicated by $\triangle_q$, a state s satisfies $\triangle_q(\varphi_1, \ldots, \varphi_{\rho(\triangle)})$ iff we can find states s_i satisfying φ_i with a $K_\triangle$-probability exceeding q. Note that the usual operators $\triangle$ and ∇ are replaced by a whole spectrum of operators $\triangle_q$ which permit a finer and probabilistically more adequate notion of satisfaction.

Again, let $[\![\varphi]\!]_\mathcal{K}$ be the set of all states for which $\varphi \in \mathfrak{Mod}_s(\tau, P)$ is satisfied under $\mathcal{K}$, and

$$Th_\mathcal{K}(s) := \{\varphi \in \mathfrak{Mod}_s(\tau, P) \mid \mathcal{K}, s \models \varphi\}$$

the theory for state $s \in S$.

It turns out that the sets $[\![\varphi]\!]_\mathcal{K}$ are measurable, so that they may be used as arguments for the stochastic relations we are working with:

LEMMA 6.1
$[\![\varphi]\!]_\mathcal{K}$ is a measurable subset of S for each $\varphi \in \mathfrak{Mod}_s(\tau, P)$.

PROOF The proof proceeds by induction on φ. If $\varphi = p \in P$, then $[\![\varphi]\!]_\mathcal{K} = V(p)$ holds, which is measurable by assumption. Since the measurable sets are closed under complementation and intersection, the only interesting case is again the one in which a modal operator is involved. Because

$$[\![\triangle_q(\varphi_1, \ldots, \varphi_{\rho(\triangle)})]\!]_\mathcal{K} = \{s \in S \mid K_\triangle(s)([\![\varphi_1]\!]_\mathcal{K} \times \ldots \times [\![\varphi_{\rho(\triangle)}]\!]_\mathcal{K}) \ge q\},$$

the assertion follows from the induction hypothesis and the fact that $K_\triangle$ is a stochastic relation. ◻

As in the case of stochastic relations we need to exclude trivial cases:

DEFINITION 6.2 *A τ-Kripke model $\mathcal{K}$ with state space S is called degenerate iff $[\![\varphi]\!]_\mathcal{K} = S$ or $[\![\varphi]\!]_\mathcal{K} = \emptyset$ holds for each formula $\varphi \in \mathfrak{Mod}_s(\tau, P)$.*

Hence a degenerate model does usually not carry useful information. The restriction is quite similar to not permitting the universal relation as a part of a congruence, and of requesting the existence of nontrivial common events for bisimulations. We will see that these constraints are closely related.

6.2.1   Examples

We show how some popular logics may be interpreted through Kripke models, indicating that specific logics require specific probabilistic arguments. But first we indicate that each stochastic relation may be "trained" to interpret a modal logic simply by interpreting the subformulas of a compound formula as stochastically independent. Then we introduce the well-known logic associated with labeled transition systems. This example is of historic significance, given the seminal work of Larsen and Skou (Larsen and Skou, 1991). It is shown also that the basic temporal language may be interpreted stochastically by reversing a relation. Arrow logic as a popular logic modeling simple programming constructs is interpreted through a simple transformation of a distribution.

In presenting these examples we follow essentially the representation of the respective logics in (Blackburn et al., 2001).

A stochastic relation on the state space induces a stochastic τ-Kripke model. This is illustrated through the following example.

EXAMPLE 6.3
Let $K : S \rightsquigarrow S$ be a stochastic relation on the state space S, and define for $s \in S$ and for the modal operator $\triangle$

$$K_\triangle(s) := \bigotimes_{i=1}^{\rho(\triangle)} K(s),$$

then $K_\triangle : S \rightsquigarrow S^{\rho(\triangle)}$ is a stochastic relation. Let $V : P \to \mathcal{B}(S)$, then

$$\mathcal{K}_{K,V} := (S, (K_\triangle)_{\triangle \in O}, V)$$

is a stochastic τ-Kripke model such that

$$\mathcal{K}_{K,V}, s \models \triangle_q(\varphi_1, \ldots, \varphi_{\rho(\triangle)}) \Leftrightarrow K(s)(\llbracket \varphi_1 \rrbracket_{\mathcal{K}_{K,V}}) \cdot \ldots \cdot K(s)(\llbracket \varphi_{\rho(\triangle)} \rrbracket_{\mathcal{K}_{K,V}}) \geq q.$$

Thus the arguments to each modal operator are stochastically independent. Consequently, $\mathcal{K}_{K,V}, s \models \triangle_q(\varphi_1, \ldots, \varphi_{\rho(\triangle)})$ holds if we can make a move to states $s_1, \ldots, s_{\rho(\triangle)}$ so that $\mathcal{K}_{K,V}, s_i \models \varphi_i$ holds independently from each other.

EXAMPLE 6.4
Suppose that L is a countable alphabet of actions. Each action $a \in \mathsf{L}$ is

associated with a binary modal operator $\langle a \rangle$, so put $\tau := (O, \rho)$ with $O :=$ $\{\langle a \rangle \mid a \in \mathsf{L}\}$ and $\rho(\langle a \rangle) := 1$.

A nondeterministic τ-Kripke model is based on a labeled transition system $(S, (\rightarrow_a)_{a \in \mathsf{L}})$ which associates a binary relation $\rightarrow_a \subseteq S \times S$ with each action a. Thus

$$s \models \langle a \rangle \varphi \Leftrightarrow \exists s' : s \rightarrow_a s' \wedge s' \models \varphi.$$

A stochastic τ-model is based on a labeled Markov transition system, say $(S, (k_a)_{a \in \mathsf{L}})$, which associates with each action a a stochastic relation $k_a : S \rightsquigarrow S$. Thus

$$s \models \langle a \rangle_q \varphi \Leftrightarrow k_a(s)([\![\varphi]\!]) \geq q;$$

hence making a transition is replaced by a probability with which a transition can happen.

Variants of the logic $\mathfrak{Mod}_s(\tau, P)$ with $P = \emptyset$ were investigated in the literature by Larsen and Skou, and by Desharnais, Edalat and Panangaden with a reference to the logic investigated by Hennessy and Milner; we refer to them also as *Hennessy-Milner logic*.

EXAMPLE 6.5
The basic temporal language has two unary modal operators $\mathbf{F}$ (forward) and $\mathbf{B}$ (backward), so that $O = \{\mathbf{F}, \mathbf{B}\}$. A nondeterministic τ-Kripke model interprets the forward operator $\mathbf{F}$ through a relation $R \subseteq S \times S$ and the backward operator $\mathbf{B}$ through the converse $R^{\smile}$ of relation R, thus $R^{\smile} := \{\langle s', s \rangle \mid \langle s, s' \rangle \in R\}$. Consequently, we have

$$s \models \mathbf{B}\varphi \Leftrightarrow \exists t \in S : \langle t, s \rangle \in R \wedge t \models \varphi.$$

A probabilistic interpretation interprets $\mathbf{F}$ through a stochastic relation $K : S \rightsquigarrow S$, so that

$$s \models \mathbf{F}_q \varphi \Leftrightarrow K(s)([\![\varphi]\!]) \geq q.$$

The backward operator $\mathbf{B}$ is interpreted through the converse $K_\mu^{\smile} : S \rightsquigarrow S$, provided the state space S is Polish and an initial probability μ is given. It was shown in Section 5.8 that the converse $K_\mu^{\smile}$ of stochastic relation K given μ is the stochastic relation $L : S \rightsquigarrow S$ such that

$$\int_S K(s)(B_s) \, \mu(ds) = \int_S L(s')(B^{s'}) \, \mu(ds')$$

holds for each Borel set $B \subseteq S \times S$. We know from Lemma 5.64 that for Polish S the converse relation exists. Thus

$$s \models \mathbf{B}_q \varphi \Leftrightarrow K_\mu^{\smile}(s)([\![\varphi]\!]) \geq q.$$

An easy calculation shows that

$$s \models \mathbf{B}_1 \mathbf{F}_1 \varphi \Leftrightarrow K_\mu^{\smile}(s) \left(\{s' \mid K(s')([\![\varphi]\!]) = 1\}\right) = 1$$

$$\Leftrightarrow \int_S K(s')([\![\varphi]\!]) \, K_\mu^{\smile}(s)(ds') = 1.$$

Note that the definition of the converse requires an initial probability (this is intuitively clear: if the probability for a backward running process is described, one has to say where to start). It is also noteworthy that a topological assumption has been made; if the state space is not a Polish space, then the technical arguments permitting the definition of the converse are not available.

EXAMPLE 6.6

Arrow logic has three modal operators modeling reversal, composition and skip, resp., thus $O = \{\mathbf{1}, \otimes, \circ\}$. with respective arities $\rho(\mathbf{1}) = 0, \rho(\otimes) = 1, \rho(\circ) = 2$. The usual interpretation of arrow logic is done over a world of pairs, so the base state space is $S \times S$ for some S, with associated relations

$$R_1 = \{\langle s, s \rangle \mid s \in S\},$$
$$R_\otimes = \{\langle\langle s_0, s_1 \rangle, \langle s_1, s_0 \rangle\rangle \mid s_0, s_1 \in S\},$$
$$R_\circ = \{\langle\langle s_0, s_1 \rangle, \langle s_0, s \rangle, \langle s, s_1 \rangle\rangle \mid s, s_0, s_1 \in S\}.$$

Thus, e.g.,

$$\langle s, s' \rangle \models \phi \circ \psi \Leftrightarrow \exists s_0 : \langle s, s_0 \rangle \models \phi \wedge \langle s_0, s' \rangle \models \psi$$

and

$$\langle s, s' \rangle \models \otimes\phi \Leftrightarrow \langle s', s \rangle \models \phi.$$

Now assume again that S is a Polish space, and let $\mu \in \mathfrak{S}(S)$ be a sub-probability. Put for $A \in \mathcal{B}(S \times S)$

$$\hat{\mu}(A) := \mu(\{s \in S \mid \langle s, s \rangle \in A\}),$$

thus $\hat{\mu}$ transports a Borel set in S to a Borel set in the diagonal of $S \times S$.

Interpret the composition operator $\circ_q$ through the stochastic relation

$$K_\circ(s, s') := \delta_s \otimes \hat{\mu} \otimes \delta_{s'}.$$

Note that the operator $\otimes$ is somewhat overloaded: it denotes the modal operator for reversal, and the product operator for measures. The context should make it clear which version is meant.

We obtain then

$$K_\circ(s, s')(\llbracket\phi\rrbracket \times \llbracket\psi\rrbracket) = (\delta_s \otimes \hat{\mu} \otimes \delta_{s'})(\llbracket\phi\rrbracket \times \llbracket\psi\rrbracket)$$
$$= \hat{\mu}(\{\langle s_1, s_2 \rangle \mid \langle s, s_1 \rangle \in \llbracket\phi\rrbracket, \langle s_2, s' \rangle \in \llbracket\psi\rrbracket\})$$
$$= \mu(\{s_1 \mid \langle s, s_1 \rangle \in \llbracket\phi\rrbracket, \langle s_1, s' \rangle \in \llbracket\psi\rrbracket\}).$$

Consequently,

$$\langle s, s' \rangle \models \phi \circ_1 \psi \Leftrightarrow \langle s, s_1 \rangle \models \phi \wedge \langle s_1, s' \rangle \models \psi \text{ for } \mu\text{-almost all } s_1$$

(here μ-*almost all* s_1 means as usual that the set of all s_1 for which the property does not hold has μ-measure 0). More generally, $\langle s, s' \rangle \models \phi \circ_q \psi$ iff $\langle s, s_1 \rangle \models \phi \wedge \langle s_1, s' \rangle \models \psi$ for all s_1 from a Borel set S_0 with $\mu(S_0) \geq q$. Finally, put $K_\otimes(s, s') := \delta_{\langle s, s' \rangle}$, then $\langle s, s' \rangle \models \otimes_q \phi \Leftrightarrow \langle s', s \rangle \models \phi$, for all rational q with $0 \leq q \leq 1$ (which is evidently independent of q), and let

$$K_1(s, s') := \begin{cases} 0, & s \neq s' \\ \delta_{\langle s, s \rangle}, & s = s' \end{cases}$$

(here 0 is the null measure), then

$$\langle s, s' \rangle \models 1 \Leftrightarrow s = s'.$$

Note that in general we did exclude modal constants, i.e., modal operators of arity 0, when defining modal similarity types. The example shows that it is possible to include them nevertheless without much ado.

6.2.2 Refinements

Given a nondeterministic and a stochastic interpretation, we want to compare both. Intuitively, the stochastic interpretation is more precise than its nondeterministic cousin: whereas nondeterministically we can only talk about possibilities, we can assign weights to these possibilities using probabilities. To say that after a certain input the output will be a, b or c conveys certainly less information than saying that the probabilities for these outputs will be, respectively, $p(a) = 1/100$, $p(b) = 1/50$ and $p(c) = 97/100$.

Since negation has its own problems, we will restrict the discussion to the negation free logic $\mathfrak{Mod}_1(\tau, P)$, and we will deal in the present Section 6.2.2 exclusively with stochastic relations which assign always the whole space probability one.

DEFINITION 6.7 *Let $\mathcal{R}$ and $\mathcal{K}$ be a nondeterministic and a stochastic τ-Kripke model, and assume that $K_\triangle(s)(S \times \ldots \times S^{\rho(\triangle)}) = 1$ holds for each $s \in S$ (we will call these models* probabilistic*). $\mathcal{K}$ is said to* refine $\mathcal{R}$ *(abbreviated as $\mathcal{K} \circledS \mathcal{R}$) iff $[\![\varphi]\!]_\mathcal{K} \subseteq [\![\varphi]\!]_\mathcal{R}$ holds for all $\varphi \in \mathfrak{Mod}_1(\tau, P)$.*

Consequently, given the interpretations $\mathcal{K}$ and $\mathcal{R}$, we have $\mathcal{K} \circledS \mathcal{R}$ if $\mathcal{R}, s \models \varphi$ holds only if $\mathcal{K}, s \models \varphi$ is true for each formula φ in the negation free fragment of the logic.

We will investigate the relationship between nondeterministic and stochastic satisfaction by showing that for each stochastic interpretation $\mathcal{K}$ we can find a nondeterministic one $\mathcal{R}$ with $\mathcal{K} \circledS \mathcal{R}$ by simply taking all possible state changes and making it into a Kripke model. Conversely, we will look into the possibility of refining a given nondeterministic Kripke model into a stochastic model. This requires some topological assumptions (for otherwise the notion *all possible states* cannot be made precise). Thus from now on the state space S is a Polish space with its Borel sets as σ-algebra.

The set of all states possible for a probability μ on a Polish space is captured through the support of a probability μ.

LEMMA 6.8

Let $\mu \in \mathfrak{S}(X), \mu \neq 0$ for the Polish space X. Then there exists a unique closed set $C_\mu \subseteq X$ with the following properties

a. $\mu(C_\mu) = \mu(X)$,

b. if $D \subseteq X$ is a closed set with $\mu(D) = \mu(X)$, then $C_\mu \subseteq D$,

c. $x \in C_\mu$ iff $\mu(U) > 0$ for all open neighborhoods U of x.

PROOF   1. Let $U_\mu := \bigcup \{U \mid U \text{ open with } \mu(U) = 0\}$, then U_μ is open. There exist countably many open sets $(U_n)_{n \in \mathbb{N}}$ with $U_\mu = \bigcup_{n \in \mathbb{N}} U_n$. This is so since X is Polish, hence has a countable base for its topology. Thus

$$\mu(U_\mu) \leq \sum_{n \in \mathbb{N}} \mu(U_n) = 0.$$

Put $C_\mu := X \setminus U_\mu$, then plainly C_μ is closed with $\mu(C_\mu) = \mu(X)$. If $D \subseteq X$ is closed with $\mu(D) = \mu(X)$, the construction of U_μ yields $X \setminus D \subseteq U_\mu$, thus $C_\mu \subseteq D$.

2. If $x \in C_\mu$ and U is an open neighborhood of x with $\mu(U) = 0$, we have $U \subseteq U_\mu$, which contradicts $x \in C_\mu$. If $x \notin C_\mu$, the set U_μ is an open neighborhood if x with $\mu(U_\mu) = 0$.   □

DEFINITION 6.9   Let $0 \neq \mu \in \mathfrak{S}(X)$ for the Polish space X. The set C_μ constructed in Lemma 6.8 is denoted by $\mathsf{supp}(\mu)$ and is called the **support** of μ.

Having a look at the properties of the support in Lemma 6.8, we see that this is exactly what we want: a set in which all points have the property that all neighborhoods have positive measure.

PROPOSITION 6.10

Let $\mathcal{K} = \left(S, (K_\triangle)_{\triangle \in O}, V \right)$ be a probabilistic τ-Kripke model. Define for the modal operator $\triangle \in O$ the set valued map

$$R_\triangle^{\mathcal{K}}(s) := \mathsf{supp}(K_\triangle(s)).$$

Put

$$\mathcal{R}_{\mathcal{K}} := \left(S, (R_\triangle^{\mathcal{K}})_{\triangle \in O}, V \right),$$

then $\mathcal{K}$ ⑤ $\mathcal{R}_\mathcal{K}$.

PROOF   The proof proceeds by induction on the structure of the formulas. Assume that $\triangle$ is a modal operator, and that we know $[\![\varphi_i]\!]_\mathcal{K} \subseteq [\![\varphi_i]\!]_{\mathcal{R}^\mathcal{K}}$ for $1 \leq i \leq \rho(\triangle)$. Now suppose $R^\mathcal{K}_\triangle, s \not\models \triangle_1(\varphi_1, \ldots, \varphi_{\rho(\triangle)})$ for some state s. Thus $R^\mathcal{K}_\triangle(s) \cap [\![\varphi_1]\!]_{\mathcal{R}^\mathcal{K}} \times \ldots \times [\![\varphi_{\rho(\triangle)}]\!]_{\mathcal{R}^\mathcal{K}} = \emptyset$, and, consequently, by the hypothesis, $R^\mathcal{K}_\triangle(s) \cap [\![\varphi_1]\!]_\mathcal{R} \times \ldots \times [\![\varphi_{\rho(\triangle)}]\!]_\mathcal{R} = \emptyset$. But this means $K_\triangle(s)([\![\varphi_1]\!]_\mathcal{R} \times \ldots \times [\![\varphi_{\rho(\triangle)}]\!]_\mathcal{R}) < 1$, hence $\mathcal{K}, s \not\models \triangle_1(\varphi_1, \ldots, \varphi_{\rho(\triangle)})$.   □

Thus each probabilistic Kripke model carries a nondeterministic one with it, and it refines this companion (one is tempted to perceive this as a *nondeterministic shadow*: a shadow as a coarser, black-and-white image of a probably more colorful and picturesque original).

It will be shown now that the converse of Proposition 6.10 is also true: Given a nondeterministic Kripke model, there exists a stochastic one refining it. Intuitively, and in the finite case, one simply assigns a uniform weight as a probability to all possible outcomes. This observation is the starting point for the nonstandard approach to probability; see (Lindstrøm, 1988, Example II.2.1).

Actually, this is basically what we will do here, too, but we have to be a bit more careful since in an uncountable setting this idea requires some additional underpinning. It is provided by the structure of the support map when combined with a stochastic relation, yielding a set-valued map with favorable properties.

It is immediate that the support produces a measurable relation for a probabilistic relation $K : Y \rightsquigarrow Z$: put

$$R_K := \{\langle y, z \rangle \in Y \times Z \mid z \in \mathsf{supp}(K(y))\},$$

then

$$\forall R_K(F) = \{y \in Y \mid K(y)(F) = 1\}$$

for the closed set $F \subseteq Z$, and

$$\exists R_K(G) = \{y \in Y \mid K(y)(G) > 0\}$$

for the open set $G \subseteq Z$. Both sets are measurable.

It is also plain that a representation of R through a stochastic relation K which is given by

$$(*) \ \forall y \in Y : R(y) = \mathsf{supp}(K(y))$$

implies that R has to be a measurable relation.

Given a set-valued relation R, a probabilistic relation K with $(*)$ can be found. For this, R has to take closed values, and a condition of measurability is imposed. We will obtain the existence of such a relation from Proposition 1.57.

LEMMA 6.11

Let $R \subseteq Y \times Z$ be a measurable relation for Y, Z Polish. There exists a probabilistic relation $K : Y \rightsquigarrow Z$ such that $R(y) = \mathsf{supp}(K(y))$ holds for each $y \in Y$.

PROOF   Because R is measurable we obtain from Proposition 1.57, part b, a sequence $(f_n)_{n \in \mathbb{N}}$ of measurable maps $f_n : Y \to Z$ such that $\{f_n(y) \mid n \in \mathbb{N}\}$ is dense in $R(y)$ for each $y \in Y$. Define $K_n(y) := \sum_{i=1}^{n} 2^{-i} \cdot \delta_{f_i(y)}$, then $K_n : Y \rightsquigarrow Z$, with

$$\mathsf{supp}(K_n(y)) = \{f_i(y) \mid 1 \le i \le n\} \subseteq R(y).$$

It is not difficult to see that

$$K_n(y) \to_w K(y) := \sum_{j \in \mathbb{N}} 2^{-j} \cdot \delta_{f_j(y)},$$

that $K : Y \rightsquigarrow Z$, and that

$$\mathsf{supp}(K(y)) = \mathsf{cl}\left(\{f_j(y) \mid j \in \mathbb{N}\}\right) = R(y).$$

$\square$

Thus we can find a probabilistic Kripke structure refining a given nondeterministic one, provided we impose a measurability condition:

PROPOSITION 6.12

Suppose $\mathcal{R} := \left(S, (R_\triangle)_{\triangle \in O}, V\right)$ is a nondeterministic τ-Kripke model such that

a. $V(p) \in \mathcal{B}(S)$ for all $p \in P$,

b. $R_\triangle$ is a measurable relation on $S \times S^{\rho(\triangle)}$ for each $\triangle \in O$.

Then there exists a probabilistic τ-Kripke model $\mathcal{K} = \left(S, (K_\triangle)_{\triangle \in O}, V\right)$ with $\mathcal{K} \circledS \mathcal{R}$.

PROOF   Applying Lemma 6.11, one finds for each modal operator $\triangle \in O$ a transition probability $K_\triangle : S \rightsquigarrow S^{\rho(\triangle)}$ such that $R_\triangle(s) = \mathsf{supp}(K_\triangle(s))$ holds for all $s \in S$. The argumentation in the proof of Lemma 6.10 establishes the claim. $\square$

It is clear that the probabilistic τ-Kripke model is underspecified by merely requiring it to be a refinement to a nondeterministic one. This is supported through the following observation:

COROLLARY 6.13

Let $\mathcal{R}$ be a nondeterministic τ-Kripke model satisfying the conditions of Proposition 6.12. Assume that $\mathcal{K}_i = \left(S, (K_{\triangle,i})_{\triangle \in O}, V \right)$ is a probabilistic τ-Kripke model with $\mathcal{K}_i \textcircled{S} \mathcal{R}$ for each $i \in \mathbb{N}$. Let $(\alpha_i)_{i \in \mathbb{N}}$ be a sequence of positive real numbers such that $\sum_{i \in \mathbb{N}} \alpha_i = 1$, and define for $\triangle \in O$ the stochastic relation

$$K_{\triangle}(s) := \sum_{i \in \mathbb{N}} \alpha_i \cdot K_{\triangle,i}(s).$$

Then $\left(S, (K_{\triangle})_{\triangle \in O}, V \right) \textcircled{S} \mathcal{R}$.

PROOF   Let $(\mu_i)_{i \in \mathbb{N}}$ be a sequence of probability measures. Since all α_i are positive, the definition of the support function yields that

$$\mathsf{supp}(\sum_{i \in \mathbb{N}} \alpha_i \cdot \mu_i) = \mathsf{cl}\left(\bigcup_{i \in \mathbb{N}} \mathsf{supp}(\mu_i) \right)$$

holds. Thus $R_{\triangle}$ equals $\mathsf{supp}(K_{\triangle})$. The assertion now follows from Proposition 6.12. $\square$

Thus we know that not only a probabilistic τ-Kripke model is the refinement of a probabilistic one, but also that refinements offer a considerable degree of freedom, because they are closed under countable convex combinations (in fact, it can also be shown that it is closed under integration as the generalization of convex combinations). This supports the intuitive feeling that a probabilistic model conveys much more information than a nondeterministic one, but that it is also much harder to obtain.

6.2.3   Bisimulations for Kripke Models

This section investigates morphisms for stochastic τ-Kripke models; we want to relate bisimilarity and logical equivalence to each other. To this end we first discuss morphisms that are based on morphisms for stochastic relations — a τ-Kripke model is built from a family of stochastic relations, after all — and indicate that this notion of morphism is not adequate for our purposes and propose the notion of a strong morphism. We show that strong morphisms are suitable for our purposes.

Assume first that the set P of propositional letters is empty, rendering the initial discussion a bit less technical. Then a stochastic τ-Kripke model $\mathcal{K} := (S, (K_{\triangle})_{\triangle \in O})$ is determined through the Polish state space S and the family $K_{\triangle} : S \rightsquigarrow S^{\rho(\triangle)}$ of stochastic relations. A morphism

$$\Phi : (S, (K_{\triangle})_{\triangle \in O})) \rightarrow \left(S', (K'_{\triangle})_{\triangle \in O} \right)$$

for stochastic τ-Kripke models is then a family $\Phi = ((\phi_\triangle, \psi_\triangle)_{\triangle \in O})$ of morphisms

$$(\phi_\triangle, \psi_\triangle) : (S, S^{\rho(\triangle)}, K_\triangle) \to (S', (S')^{\rho(\triangle)}, K'_\triangle)$$

for the associated relations.

Consider a modal operator $\triangle$. The σ-algebra $\mathcal{A}_\triangle$ generated by

$$\{[\![\varphi_1]\!]_\mathcal{K} \times \ldots \times [\![\varphi_{\rho(\triangle)}]\!]_\mathcal{K} \mid \varphi_1, \ldots, \varphi_{\rho(\triangle)} \in \mathfrak{Mod}_s(\tau, P)\}$$

is evidently countably generated, thus gives rise to a smooth equivalence relation $\beta_\triangle$ on $S^{\rho(\triangle)}$, and the relation

$$s\alpha_\triangle s' \Leftrightarrow \forall B \in \mathcal{A}_\triangle : K_\triangle(s)(B) = K_\triangle(s')(B)$$

is smooth due to $\mathcal{A}_\triangle$ being countably generated. Consequently, $(\alpha_\triangle, \beta_\triangle)$ is a congruence for $K_\triangle : S \rightsquigarrow S^{\rho(\triangle)}$, and if $\mathcal{K}$ is nondegenerate, this congruence is nontrivial.

Let $\mathcal{K}' = \left(S', (K'_\triangle)_{\triangle \in O}\right)$ be another τ-Kripke model which is logical equivalent to the first one in the sense that for the states the corresponding theories mutually coincide. To be more precise:

DEFINITION 6.14   *The stochastic τ-Kripke models $\mathcal{K}$ and $\mathcal{K}'$ are said to be* logical equivalent *(abbreviated as $\mathcal{K} \sim \mathcal{K}'$) iff $\{Th_\mathcal{K}(s) \mid s \in S\} = \{Th_{\mathcal{K}'}(s') \mid s' \in S'\}$.*

Thus $\mathcal{K} \sim \mathcal{K}'$ iff given $s \in S$ there exists $s' \in S'$ such that $Th_\mathcal{K}(s) = Th_{\mathcal{K}'}(s')$, and vice versa.

Assume both $\mathcal{K}$ and $\mathcal{K}'$ are nondegenerate. Construct for $\mathcal{K}'$ the congruence $(\alpha'_\triangle, \beta'_\triangle)$ for each modal operator $\triangle$ as above, then it can be shown that $\mathcal{K} \sim \mathcal{K}'$ implies that the congruences $(\alpha_\triangle, \beta_\triangle)$ and $(\alpha'_\triangle, \beta'_\triangle)$ are simulation equivalent. From Proposition 5.39 we see that $K_\triangle$ and $K'_\triangle$ are bisimilar for each modal operator$\triangle$, so that there exists a span of morphisms

$$(S, S^{\rho(\triangle)}, K_\triangle) \xleftarrow{(\phi_\triangle, \psi_\triangle)} (A_\triangle, B_\triangle, M_\triangle) \xrightarrow{(\phi'_\triangle, \psi'_\triangle)} (S', (S')^{\rho(\triangle)}, K'_\triangle).$$

This is rather satisfying from the point of view of stochastic relations, but not when considering stochastic τ-Kripke models. This is so since in general $((A_\triangle, B_\triangle, M_\triangle)_{\triangle \in O})$ fails to be such a model, because there is no way to guarantee that all $A_\triangle$ coincide with, say, a Polish space T, and so that $B_\triangle$ equals $T^{\rho(\triangle)}$.

Consequently, we have to strengthen the requirements for a morphism in order to achieve some uniformity. This will be done now, and we admit propositional letters again. The basic idea is to have just one map ϕ between the state spaces so that

$$K'_\triangle(\phi(s))(A) = K_\triangle(s)(\{\langle s_1, \ldots, s_{\rho(\triangle)}\rangle \mid \langle \phi(s_1), \ldots, \phi(s_{\rho(\triangle)})\rangle \in A\})$$

holds for each state $s \in S$ and each Borel set $A \subseteq (S')^{\rho(\triangle)}$, making the diagram

$$
\begin{array}{ccc}
S & \xrightarrow{\ \phi\ } & S' \\
\left\downarrow{\scriptstyle K_\triangle}\right. & & \left\downarrow{\scriptstyle K'_\triangle}\right. \\
\mathfrak{S}\left(S^{\rho(\triangle)}\right) & \xrightarrow[\mathfrak{S}\left(\phi^{\rho(\triangle)}\right)]{} & \mathfrak{S}\left((S')^{\rho(\triangle)}\right)
\end{array}
$$

commutative (where $\phi^n : \langle x_1, \ldots, x_n \rangle \mapsto \langle \phi(x_1), \ldots, \phi(x_n) \rangle$ distributes ϕ into the components), and we want to have $s \in V(p)$ iff $\phi(s) \in V'(p)$ for each propositional letter p. This leads to

DEFINITION 6.15   *Let $\mathcal{K}$ and $\mathcal{K}'$ be stochastic τ-Kripke models with $\mathcal{K} = (S, (K_\triangle)_{\triangle \in O}), V)$ and $\mathcal{K}' = \left(S', (K'_\triangle)_{\triangle \in O}), V'\right)$. A strong morphism $\phi : \mathcal{K} \to \mathcal{K}'$ is determined through a measurable and surjective map $\phi : S \to S'$ so that these conditions are satisfied:*

a. $\forall p \in P : V(p) = \phi^{-1}\left[V'(p)\right]$,

b. for each modal operator $\triangle$,

$$
K'_\triangle \circ \phi = \mathfrak{S}\left(\phi^{\rho(\triangle)}\right) \circ K_\triangle
$$

holds.

Thus, if $\phi : \mathcal{K} \to \mathcal{K}'$ is a strong morphism, then

$$
(\phi, \phi^{\rho(\triangle)}) : (S, S^{\rho(\triangle)}, K_\triangle) \to (S', (S')^{\rho(\triangle)}, K'_\triangle)
$$

is a morphism between the corresponding stochastic relations for each modal operator $\triangle \in O$. Note that we take also the propositional letters into account.

It is clear that stochastic τ-Kripke models over general measurable spaces form a category p.$\mathfrak{Kripke}$ with this notion of morphism, because the composition of strong morphisms is again a strong morphism, and because the identity is a strong morphism, too. Furthermore, each modal operator $\triangle$ induces a functor $F_\triangle : $ p.$\mathfrak{Kripke} \to \mathfrak{Stoch}$ which forgets all but $K_\triangle$. We will below make (rather informal) use of this functor.

Because we work on the safe grounds of a category, we have bisimulations at our disposal, which can be defined again as spans of strong morphisms. Similarly, we define behavioral equivalence through a cospan of morphisms, essentially mimicking the corresponding definition for stochastic relations.

DEFINITION 6.16
Let $\mathcal{K}_1$ and $\mathcal{K}_2$ be stochastic τ-Kripke models.

a. $\mathcal{K}_1$ and $\mathcal{K}_2$ are called strongly bisimilar *iff there exists a stochastic τ-Kripke model $\mathcal{M}$ and strong morphisms* $\mathcal{K}_1 \xleftarrow{\phi_1} \mathcal{M} \xrightarrow{\phi_2} \mathcal{K}_2$, *such that the σ-algebra of common events* $\phi_1^{-1}[\mathcal{B}(S_1)] \cap \phi_2^{-1}[\mathcal{B}(S_2)]$ *is nontrivial (here S_i is the state space of $\mathcal{K}_i, i = 1, 2$).*

b. $\mathcal{K}_1$ and $\mathcal{K}_2$ are called behavioral equivalent *iff there exists a stochastic τ-Kripke model $\mathcal{L}$ and strong morphisms* $\mathcal{K}_1 \xrightarrow{\phi_1} \mathcal{L} \xleftarrow{\phi_2} \mathcal{K}_2$.

Since the product σ-algebra is the smallest σ-algebra which contains all the measurable rectangles, it is not difficult to see that $\phi_1^{-1}[\mathcal{B}(S_1)] \cap \phi_2^{-1}[\mathcal{B}(S_2)]$ is nontrivial iff for each modal operator $\triangle \in O$ the σ-algebra

$$\bigotimes_{i=1}^{\rho(\triangle)} \phi_1^{-1}[\mathcal{B}(S_1)] \cap \bigotimes_{i=1}^{\rho(\triangle)} \phi_2^{-1}[\mathcal{B}(S_2)]$$

is nontrivial. Thus the second part of condition a in Definition 6.16 will imply that this notion of bisimilarity is compatible to the one used for stochastic relations in general. Similarly, behavioral equivalence is adapted to Kripke models.

We will relate logical equivalence, strong bisimilarity and behavioral equivalence of Kripke models $\mathcal{K}$ and $\mathcal{K}'$, provided the models are based on Polish spaces. Fix the stochastic τ-Kripke models $\mathcal{K} := (S, (K_\triangle)_{\triangle \in O}), V)$ and $\mathcal{K}' := \left(S', (K'_\triangle)_{\triangle \in O}, V' \right)$.

It is well known that morphisms preserve theories for the Hennessy-Milner logic (Desharnais et al., 2002). This is also true for stochastic relations:

LEMMA 6.17

If $\phi : \mathcal{K} \to \mathcal{K}'$ is a strong morphism, then $Th_\mathcal{K}(s) = Th_{\mathcal{K}'}(\phi(s))$ holds for all states $s \in S$.

PROOF   1. We show by induction on the formula $\varphi \in \mathfrak{Mod}_s(\tau, P)$ that

$$\mathcal{K}, s \models \varphi \Leftrightarrow \mathcal{K}', \phi(s) \models \varphi$$

holds; putting it slightly different, we want to show

$$(*) \quad [\![\varphi]\!]_\mathcal{K} = [\![\varphi]\!]_{\mathcal{K}'}$$

for all these φ.

2. If $\varphi = p \in P$, this follows from $V(p) = \phi^{-1}[V'(p)]$. The interesting case in the induction step is the application of a n-ary modal operator $\triangle_q$ with

rational q. Suppose the assertion is true for $[\![\varphi_1]\!]_{\mathcal{K}}, \ldots, [\![\varphi_n]\!]_{\mathcal{K}}$, then

$$
\begin{aligned}
\mathcal{K}, s \models \triangle_q(\varphi_1, \ldots, \varphi_n) &\Leftrightarrow K_\triangle(s)([\![\varphi_1]\!]_{\mathcal{K}} \times \ldots \times [\![\varphi_n]\!]_{\mathcal{K}}) \geq q \\
&\overset{(\dagger)}{\Leftrightarrow} K_\triangle(s)((\phi^n)^{-1}[[\![\varphi_1]\!]_{\mathcal{K}'} \times \ldots \times [\![\varphi_n]\!]_{\mathcal{K}'}]) \geq q \\
&\Leftrightarrow (\mathfrak{S}(\phi^n) \circ K_\triangle)(s)([\![\varphi_1]\!]_{\mathcal{K}'} \times \ldots \times [\![\varphi_n]\!]_{\mathcal{K}'}) \geq q \\
&\overset{(\ddagger)}{\Leftrightarrow} K'_\triangle(\phi(s))([\![\varphi_1]\!]_{\mathcal{K}'} \times \ldots \times [\![\varphi_n]\!]_{\mathcal{K}'}) \geq q \\
&\Leftrightarrow \mathcal{K}', \phi(s) \models \triangle_q(\varphi_1, \ldots, \varphi_n).
\end{aligned}
$$

In ($\dagger$) we use reformulation ($*$) for the induction hypothesis; in ($\ddagger$) we make use of the defining equation of a (strong) morphism. ☐

Define the equivalence relation α on state space S through

$$
s_1 \; \alpha \; s_2 \Leftrightarrow Th_{\mathcal{K}}(s_1) = Th_{\mathcal{K}}(s_2),
$$

thus two states are α-equivalent iff they satisfy exactly the same formulas in $\mathfrak{Mod}_s(\tau, P)$; in a similar way α' is defined on S'. Because we have at most countably many formulas, α and α' are smooth equivalence relations. Define the equivalence relation $\beta_\triangle$ on $S^{\rho(\triangle)}$ through

$$
\langle s_1, \ldots, s_{\rho(\triangle)} \rangle \; \beta_\triangle \; \langle t_1, \ldots, t_{\rho(\triangle)} \rangle \Leftrightarrow s_1 \; \alpha \; t_1 \wedge \cdots \wedge s_{\rho(\triangle)} \; \alpha \; t_{\rho(\triangle)},
$$

then $\beta_\wedge$ is smooth, and we know that the σ-algebra of β-invariant sets can be written in terms of the α-invariant sets, viz., $\mathcal{INV}\left(\mathcal{B}(S^{\rho(\triangle)}), \beta_\triangle\right) = \bigotimes_{i=1}^{\rho(\triangle)} \mathcal{INV}\left(\mathcal{B}(S), \alpha\right)$ (see Lemma 5.14). The relation $\beta'_\triangle$ is defined in the same way for α'.

The equivalence of $\mathcal{K}$ and $\mathcal{K}'$ makes these relations into simulation equivalent congruences:

LEMMA 6.18

If $\mathcal{K} \sim \mathcal{K}'$ for the nondegenerate Kripke models $\mathcal{K}$ and $\mathcal{K}'$, then $(\alpha, \beta_\triangle)$ and $(\alpha', \beta'_\triangle)$ are simulation equivalent and nontrivial congruences for the stochastic relations $F_\triangle(\mathcal{K})$ and $F_\triangle(\mathcal{K}')$.

PROOF 1. The equivalence relations involved are all smooth, so it first has to be demonstrated that each pair forms indeed a congruence. Assume that $s_1 \alpha s_2$ holds, then

$$
K_\triangle(s_1)([\![\varphi_1]\!]_{\mathcal{K}} \times \ldots \times [\![\varphi_{\rho(\triangle)}]\!]_{\mathcal{K}}) = K_\triangle(s_2)([\![\varphi_1]\!]_{\mathcal{K}} \times \ldots \times [\![\varphi_{\rho(\triangle)}]\!]_{\mathcal{K}})
$$

follows (otherwise we could find a rational number q with

$$
\mathcal{K}, s_1 \models \triangle_q(\varphi_1, \ldots, \varphi_{\rho(\triangle)})
$$

but

$$\mathcal{K}, s_2 \not\models \Delta_q(\varphi_1, \ldots, \varphi_{\rho(\Delta)}),$$

or vice versa). Because

$$\mathcal{B}_0 := \{ [\![\varphi_1]\!]_\mathcal{K} \times \cdots \times [\![\varphi_{\rho(\Delta)}]\!]_\mathcal{K} \mid \varphi_1, \ldots, \varphi_{\rho(\Delta)} \in \mathfrak{Mod}_\mathfrak{s}(\tau, P) \}$$

forms a generator for $\mathcal{INV}\left(\mathcal{B}(S^{\rho(\Delta)}), \beta_\Delta\right)$, we infer that (α, β_Δ) is a congruence for $F_\Delta(\mathcal{K})$. The same arguments show that also (α', β'_Δ) is a congruence for $F_\Delta(\mathcal{K}')$.

2. $\mathcal{A}_0 := \{ [\![\varphi]\!]_\mathcal{K} \mid \varphi \in \mathfrak{Mod}_\mathfrak{s}(\tau, P) \}$ is a countable generator of the σ-algebra $\mathcal{INV}\left(\mathcal{B}(S), \alpha\right)$, and since the logic is closed under conjunction, $\mathcal{A}_0$ is closed under finite intersections. Given $s \in S$ there exists $s' \in S'$ such that $Th_\mathcal{K}(s) = Th_{\mathcal{K}'}(s')$ holds; define $\Upsilon([s]_\alpha) := [s']_{\alpha'}$, then $\Upsilon : S/\alpha \to S'/\alpha'$ is well defined, and $\Upsilon_{[\varphi]_\mathcal{K}} = [\varphi]_{\mathcal{K}'}$ holds. Consequently, $\{ \Upsilon_A \mid A \in \mathcal{A}_0 \}$ generates $\mathcal{INV}\left(\mathcal{B}(S'), \alpha'\right)$. Hence α spawns α' via $(\Upsilon, \mathcal{A}_0)$.

3. The construction of β_Δ yields

$$\left[\langle s_1, \ldots, s_{\rho(\Delta)} \rangle \right]_{\beta_\Delta} = [s_1]_\alpha \times \ldots \times \left[s_{\rho(\Delta)} \right]_\alpha.$$

An argument very similar to that used above shows that β_Δ spawns β'_Δ via $(\Theta, \mathcal{B}_0)$, where

$$\Theta : \left[\langle s_1, \ldots, s_{\rho(\Delta)} \rangle \right]_{\beta_\Delta} \mapsto \Upsilon([s_1]_\alpha) \times \ldots \times \Upsilon([s_{\rho(\Delta)}]_\alpha),$$

and $\mathcal{B}_0$ is defined above.

4. An argumentation very close to the first part of the proof shows that $Th_\mathcal{K}(s) = Th_{\mathcal{K}'}(s')$ for $s \in S, s' \in S'$ implies for all formulas $\varphi_1, \ldots, \varphi_{\rho(\Delta)}$ that

$$K_\Delta(s)([\![\varphi_1]\!]_\mathcal{K} \times \ldots \times [\![\varphi_{\rho(\Delta)}]\!]_\mathcal{K}) = K'_\Delta(s')([\![\varphi_1]\!]_{\mathcal{K}'} \times \ldots \times [\![\varphi_{\rho(\Delta)}]\!]_{\mathcal{K}'})$$

(cp. part 2 of the proof of Lemma 6.17). Thus (α, β_Δ) simulates (α', β'_Δ), and in the same way, interchanging the roles of $\mathcal{K}$ and $\mathcal{K}'$, we infer that (α', β'_Δ) simulates (α, β_Δ).

5. Because $\mathcal{K}$ is nondegenerate, the σ-algebra

$$\mathcal{INV}\left(\mathcal{B}(S), \alpha\right) = \sigma(\{ [\![\varphi]\!]_\mathcal{K} \mid \varphi \in \mathfrak{Mod}_\mathfrak{s}(\tau, P) \})$$

is nontrivial; because

$$\mathcal{INV}\left(\mathcal{B}(S^{\rho(\Delta)}), \beta_\Delta\right) = \bigotimes_{i=1}^{\rho(\Delta)} \mathcal{INV}\left(\mathcal{B}(S), \alpha\right),$$

we see that $\mathcal{INV}\left(\mathcal{B}(S^{\rho(\Delta)}), \beta_\Delta\right)$ contains a set of the form $B^{\rho(\Delta)}$ for some B with $\emptyset \neq B \neq S$, thus we may conclude that β_Δ is not the universal relation. Thus $(\alpha, \beta_\triangle)$ is a nontrivial congruence. Replacing $\mathcal{K}$ by $\mathcal{K}'$, this is also established for the congruence $(\alpha', \beta'_\triangle)$. This completes the proof. $\quad\square$

Accordingly, we know from Proposition 5.39 that for logical equivalent Kripke models $\mathcal{K}$ and $\mathcal{K}'$ and for each modal operator $\triangle$ the stochastic relations $F_\triangle(\mathcal{K})$ and $F_\triangle(\mathcal{K}')$ are bisimilar. All the mediating relations can be collected to form a mediating Kripke model. This requires, however, that we know a wee bit about the internal structure of the semi-pullback which is constructed along the way. We will see this in the proof of the following result, the extended Hennessy-Milner Theorem for stochastic τ-Kripke models:

THEOREM 6.19
Assume that $\mathcal{K}$ and $\mathcal{K}'$ are nondegenerate stochastic τ-Kripke models over Polish spaces, then the following statements are equivalent:

a. $\mathcal{K}$ and $\mathcal{K}'$ are strongly bisimilar,

b. $\mathcal{K}$ and $\mathcal{K}'$ are logical equivalent.

c. $\mathcal{K}$ and $\mathcal{K}'$ are behavioral equivalent.

PROOF　1. Because both $a \Rightarrow b$ and $c \Rightarrow b$ follow from Lemma 6.17, we may concentrate on the respective proofs for $b \nrightarrow a$ and $b \Rightarrow c$.

2. We deal with $b \Rightarrow a$ first. Since $\mathcal{K} \sim \mathcal{K}'$, we know from Lemma 6.18 that the congruences $c_\triangle := (\alpha, \beta_\triangle)$ and $c'_\triangle := (\alpha', \beta'_\triangle)$ are simulation equivalent for each modal operator $\triangle$. Let $\mathcal{M}_\triangle = (M_\triangle, N_\triangle, L_\triangle)$ be the mediating stochastic relation, which exists by Proposition 5.39. The proof of Theorem 4.10 shows that $(n := \rho(\triangle))$

$$M_\triangle = \{\langle s, s'\rangle \in S \times S' \mid s\,(\alpha \diamond \alpha')\,s'\},$$
$$N_\triangle = \{\langle s_1, s'_1, \ldots, s_n, s'_n\rangle \in (S \times S')^n \mid s_i\,(\alpha \diamond \alpha')\,s'_i \text{ for } 1 \leq i \leq n\},$$

which may be rendered Polish spaces. Note that $S'' := M_\triangle$ does not depend at all on the modal operator, and that $N_\triangle$ depends only on its arity. Furthermore, we may infer for the $\mathfrak{P} - \mathfrak{Stoch}$-morphisms

$$F_\triangle(\mathcal{K}) \xleftarrow{\ f_\triangle\ } \mathcal{M}_\triangle \xrightarrow{\ f'_\triangle\ } F_\triangle(\mathcal{K}')$$

that $f_\triangle = (\pi_{1,S}, \pi^n_{1,S}), f'_\triangle = (\pi_{2,S'}, \pi^n_{2,S'})$ holds, where the π denote the projections.

Now define for the propositional letter $p \in P$

$$W(p) := \{\langle s, s'\rangle \in M_\triangle \mid s \in V(p), s' \in V'(p)\},$$

then it is immediate that the equations $W(p) = \pi_1^{-1}[V(p)] = \pi_2^{-1}[V'(p)]$ hold. Consequently, $\mathcal{M} := (S'', (L_\triangle)_{\triangle \in O}, W)$ is a stochastic τ-Kripke model with

$$\mathcal{K} \xleftarrow{\quad \pi_{1,S} \quad} \mathcal{M} \xrightarrow{\quad \pi_{2,S'} \quad} \mathcal{K}'$$

in p.Kriple.

3. We need additionally to show that the σ-algebra $\pi_{1,S}^{-1}[\mathcal{B}(S)] \cap \pi_{2,S'}^{-1}[\mathcal{B}(S')]$ is nontrivial. This is essentially the same argument as the one used in the third part of the proof to Proposition 5.39. Since $\mathcal{K}$ is nondegenerate, we can find a formula φ with $\emptyset \neq [\![\varphi]\!]_{\mathcal{K}} \neq S$. The set $[\![\varphi]\!]_{\mathcal{K}}$ is an α-invariant Borel subset of S. We know from the proof of Lemma 6.18 that α spawns α' via $(\Upsilon, \{[\![\phi]\!]_{\mathcal{K}} \mid \phi \in \mathfrak{Mod}_s(\tau, P)\})$ for some suitably chosen Υ. Thus

$$\pi_{1,S}^{-1}[[\![\varphi]\!]_{\mathcal{K}}] = \pi_{2,S'}^{-1}[\Upsilon_{[\![\varphi]\!]_{\mathcal{K}}}],$$

consequently we see that

$$\pi_{1,S}^{-1}[[\![\varphi]\!]_{\mathcal{K}}] \in \pi_{1,S}^{-1}[\mathcal{B}(S)] \cap \pi_{2,S'}^{-1}[\mathcal{B}(S')].$$

Since $\emptyset \neq [\![\varphi]\!]_{\mathcal{K}} \neq S$ we conclude that $\emptyset \neq \pi_{1,S}^{-1}[[\![\varphi]\!]_{\mathcal{K}}] \neq M_\triangle$, hence the σ-algebra in question is indeed not trivial.

4. Turning to $b \Rightarrow c$, we know that $(K_\triangle + K'_\triangle)/(\mathsf{c}_\triangle \diamond \mathsf{c}'_\triangle)$ is isomorphic to $K_\triangle/\mathsf{c}_\triangle$ and to $K'_\triangle/\mathsf{c}'_\triangle$ by Corollary 5.40. From the construction of the factor relation it is inferred that $K_\triangle/\mathsf{c}_\triangle = (S/\alpha, S^{\rho(\triangle)}/\beta_\triangle, K_{\triangle,\mathsf{c}_\triangle})$. Now it is easy to see that $S^{\rho(\triangle)}/\beta_\triangle$ is Borel isomorphic to $(S/\alpha)^{\rho(\triangle)}$. Let

$$\widetilde{K_\triangle} : S/\alpha \rightsquigarrow (S/\alpha)^{\rho(\triangle)}$$

be the corresponding stochastic relation. Because we have for $s \in S, s' \in S'$ with $[s]_{\alpha \diamond \alpha'} = [s']_{\alpha \diamond \alpha'}$ that $s \in V(p)$ iff $s' \in V'(p)$, whenever $p \in P$ is a propositional letter, we put $\widetilde{V}(p) := \eta_\alpha[V(p)]$ and note that $\widetilde{V}(p)$ is a Borel set in S/α by Lemma 5.7. Thus

$$\widetilde{\mathcal{K}} := \left(S/\alpha, \left(\widetilde{K_\triangle}\right)_{\triangle \in O}, \widetilde{V}\right)$$

defines a stochastic Kripke model for which we can find strong morphisms

$$\mathcal{K} \xrightarrow{\quad \phi_1 \quad} \widetilde{\mathcal{K}} \xleftarrow{\quad \phi_2 \quad} \mathcal{K}'.$$

$\square$

Looking back at the development, it is noted that Theorem 6.19 is derived from Proposition 5.39, hence from a condition that arose from the consideration of stochastic relations alone. This is in marked contrast to the proofs proposed in (Desharnais et al., 2002; Doberkat, 2003) which start from the logic and develop the properties of simulation equivalent congruences implicitly. It is also clear that the model constructed in the last part of the proof will usually not be defined over a Polish space. This is so since factoring destroys Polishness.

The following example gives a brief illumination.

EXAMPLE 6.20
Consider Kripke models for the basic modal language that has just one modal operator, traditionally denoted by $\Diamond$, which is unary. Assume that there are at least two propositional letters. Let $\mathcal{K} = (S, K_\Diamond, V)$ and $\mathcal{L} = (S, L_\Diamond, W)$ be stochastic Kripke models such that $\{\langle V(p), W(p)\rangle \mid p \in P\}$ is a block for $K_\Diamond, L_\Diamond$ (see Definition 5.42). Then $\mathcal{K} \sim \mathcal{L}$.

This is so since the Kripke models are strongly bisimilar by Corollary 5.43 and by Theorem 6.19.

Logical equivalence appears here as a catalyst which permits proving that bisimilarity and behavioral equivalence describe the same phenomenon, a link that is missing in the general development of stochastic relations; see Section 5.4. There we have simulation equivalent congruences at our disposal, which are always tied to a relation, while the logic serves here as an arbitrator which is completely independent of a Kripke model interpreting it.

6.3   Projective Limits for Interpreting Temporal Logics

The interpretation of modal logics rests on relational properties, e.g., we say that

$$s \models \triangle_q(\varphi_1, \ldots, \varphi_{\rho(\triangle)}) \Leftrightarrow K_\triangle(s)(\llbracket \varphi_1 \rrbracket \times \llbracket \varphi_{\rho(\triangle)} \rrbracket) \geq q,$$

so we use the relation associated with the modal operator $\triangle$ to associate a probability to the finite path of those words of length $\rho(\triangle)$ that satisfy $\varphi_1, \ldots, \varphi_{\rho(\triangle)}$. Changing to infinite paths, we basically could do the same: assign a probability to that set of paths that we want to consider. This requires relations that work on those paths which unfortunately are usually not given offhand. Modeling a reactive system with possibly nonterminating computations, we have to piece together the probabilities for these infinite paths from their finite components. They are usually given through relations that describe what happens in a single step. This is what we will discuss next. Since finding an adequate probability is sometimes a bit intricate, we will first have a look at the measure-theoretic mechanisms. Then we will apply this by showing how the machinery developed so far may be put to use for discussing the continuous time stochastic logics **CSL** and μ**CSL**, which incorporate time explicitly. They offer a very powerful approach to describing systems, and an additional challenge for the treatment of its probabilistic properties, in particular to the important problem of bisimilar states.

6.3.1   Setting the Stage: Infinite Paths

Fix a Polish state space S over which the logics will be interpreted. A path σ is an element of the set $(S \times \mathbb{R}_+)^\infty$. Path $\sigma = \langle s_0, t_0, s_1, t_1, \ldots \rangle$ may be written as $s_0 \xrightarrow{t_0} s_1 \xrightarrow{t_1} \ldots$ with the interpretation that t_i is the time spent in state s_i. Given $i \in \mathbb{N}$, denote s_i by $\sigma[i]$ as the $(i+1)$st state of σ, and let $\delta(x, i) := t_i$. Let for $t \in \mathbb{R}_+$ the index i be the smallest index k such that $t < \sum_{i=0}^{k} t_i$, and put $\sigma @ t := \sigma[i]$, if i is defined; set $\sigma @ t := \#$, otherwise (here $\#$ is a new symbol not in $S \cup \mathbb{R}_+$). $S_\#$ denotes $S \cup \{\#\}$; this is a Polish space when endowed with the sum σ-algebra. The definition of $\sigma @ t$ makes sure that for any time t we can find a rational time t' with $\sigma @ t = \sigma @ t'$.

We will deal only with infinite paths. This is no loss of generality because events that happen at a certain time with probability 0 will have the effect that the corresponding infinite paths occur only with probability 0. Thus we do not prune the path; this makes the notation somewhat easier to handle without losing any substance.

The Borel sets $\mathcal{B}((S \times \mathbb{R}_+)^\infty)$ are the smallest σ-algebra which contains all the cylinder sets

$$\{\prod_{j=1}^{n}(B_j \times I_j) \times \prod_{j>n}(S \times \mathbb{R}_+) \mid n \in \mathbb{N}, I_1, \ldots, I_n \text{ rational intervals},$$
$$B_1, \ldots, B_n \in \mathcal{B}(S)\}.$$

Thus a cylinder set is an infinite product that is determined through the finite product of an interval with a Borel set in S; see Section 1.2. It will be helpful to remember that the intersection of two cylinder sets is again a cylinder set.

The following Lemma looks innocent, but will turn out to be an important device:

LEMMA 6.21
$\langle \sigma, t \rangle \mapsto \sigma @ t$ *is a Borel measurable map from* $(S \times \mathbb{R}_+)^\infty \times \mathbb{R}_+$ *to* $S_\#$. *In particular, the set* $\{\langle \sigma, t \rangle \mid \sigma @ t \in S\}$ *is a measurable subset of* $(S \times \mathbb{R}_+)^\infty \times \mathbb{R}_+$.

Before we prove it, we need a simple auxiliary statement

LEMMA 6.22
Let $(N, \mathcal{N})$ *be a measurable space,* $f : N \to \mathbb{R}$ *be a Borel measurable map. Then*

$$\{\langle n, x \rangle \mid f(n) > x\} \in \mathcal{N} \otimes \mathcal{B}(\mathbb{R}).$$

PROOF   Put $f_0(n, x) := \langle f(n), x \rangle$, then $f_0 : N \times \mathbb{R} \to \mathbb{R} \times \mathbb{R}$ is $\mathcal{N} \otimes \mathcal{B}(\mathbb{R})$-$\mathcal{B}(\mathbb{R} \times \mathbb{R})$-measurable. This is so since

$$\mathcal{D} := \{B \in \mathcal{B}(\mathbb{R} \times \mathbb{R}) \mid f_0^{-1}[B] \in \mathcal{N} \otimes \mathcal{B}(\mathbb{R})\}$$

is a σ-algebra, and since $f_0^{-1}[B \times E] = f^{-1}[B] \times E$, hence we know that $\mathcal{D}$ contains all measurable rectangles, thus $\mathcal{D} = \mathcal{B}(\mathbb{R} \times \mathbb{R})$.

Since $\{\langle n, x \rangle \mid f(n) > x\} = f_0^{-1}[L]$ with $L := \{\langle u, v \rangle \mid u > v\} \in \mathcal{B}(\mathbb{R} \times \mathbb{R})$, the assertion is established.   ⬚

The set $\{\langle n, x \rangle \mid f(n) > x\}$ may be visualized for $N = \mathbb{R}$ as the area below the graph of f.

PROOF   of Lemma 6.21

0. Note that we claim joint measurability in both components (which is strictly stronger than measurability in each component). Thus we have to show that $\{\langle \sigma, t \rangle \mid \sigma@t \in A\}$ is a measurable subset of $(S \times \mathbb{R}_+)^\infty \times \mathbb{R}_+$, whenever $A \subseteq S_\#$ is Borel.

1. Because for fixed $i \in \mathbb{N}$ the map $\sigma \mapsto \delta(\sigma, i)$ is a projection, $\delta(\cdot, i)$ is measurable, hence $\sigma \mapsto \sum_{i=0}^{j} \delta(\sigma, i)$ is. Consequently,

$$\{\langle \sigma, t \rangle \mid \sigma@t = \#\} = \{\langle \sigma, t \rangle \mid \forall j : t > \sum_{i=0}^{j} \delta(\sigma, i)\}$$

$$= \bigcap_{j \geq 0} \{\langle \sigma, t \rangle \mid t \geq \sum_{i=0}^{j} \delta(\sigma, i)\}.$$

This is clearly a measurable set.

2. Put

$$\mathsf{Stop}(\sigma, t) := \inf\{k \geq 0 \mid t < \sum_{i=0}^{k} \delta(\sigma, i)\},$$

thus

$$X_k := \{\langle \sigma, t \rangle \mid \mathsf{Stop}(\sigma, t) = k\} = \{\langle \sigma, t \rangle \mid \sum_{i=0}^{k-1} \delta(\sigma, i) \leq t < \sum_{i=0}^{k} \delta(\sigma, i)\}$$

is a measurable set by Lemma 6.22. Now let $B \in \mathcal{B}(S)$ be a Borel set, then

$$\{\langle \sigma, t \rangle \mid \sigma@t \in B\} = \bigcup_{k \geq 0} \{\langle \sigma, t \rangle \mid \sigma@t \in B, \mathsf{Stop}(\sigma, t) = k\}$$

$$= \bigcup_{k \geq 0} \{\langle \sigma, t \rangle \mid \sigma[k] \in B, \mathsf{Stop}(\sigma, t) = k\}$$

$$= \bigcup_{k \in \mathbb{N}} \left(X_k \cap \left(\prod_{i<k} (S \times \mathbb{R}_+) \times (B \times \mathbb{R}_+) \times \prod_{i>k} (S \times \mathbb{R}_+) \right) \right).$$

Because X_k is measurable, the latter set is measurable. This establishes measurability of the @-map.   ⬚

As a consequence, we obtain the measurability of some sets and maps which will be important for the later development. A notational convention for improving readability is met: the letter σ will always denote a generic element of $(S \times \mathbb{R}_+)^\infty$, and the letter τ always a generic element of $\mathbb{R}_+ \times (S \times \mathbb{R}_+)^\infty$.

PROPOSITION 6.23

We observe the following properties:

a. *The set of Zeno paths $\{\sigma \mid \sum_{i \geq 0} \delta(\sigma, i)$ exists and is finite$\}$ is a measurable subset of $(S \times \mathbb{R}_+)^\infty$,*

b. *$\{\langle \sigma, t \rangle \mid \lim_{i \to \infty} \delta(\sigma, i) = t\}$ is a measurable subset of $(S \times \mathbb{R}_+)^\infty \times \mathbb{R}_+$,*

c. *both*

$$s \mapsto \liminf_{t \to \infty} N_\infty(s)(\{\tau \mid \langle s, \tau \rangle @ t \in A\})$$

and

$$s \mapsto \limsup_{t \to \infty} N_\infty(s)(\{\tau \mid \langle s, \tau \rangle @ t \in A\})$$

are measurable maps $X \to \mathbb{R}_+$ for each Borel set $A \subseteq S$, provided $N_\infty : S \leadsto (\mathbb{R}_+ \times S)^\infty$ is a stochastic relation.

PROOF   0. The proof makes crucial use of the fact that the real line is a complete metric space (so each Cauchy sequence converges), and that the rational numbers are dense, forming a countable set.

1. Since $\sum_{i \geq 0} \delta(\sigma, i)$ exists and is finite iff given $\epsilon > 0$ there exists $n \in \mathbb{N}$ such that $| \sum_{i=n_1}^{n_2} \delta(\sigma, i) | < \epsilon$ whenever $n_1, n_2 \geq n$, we see that

$$\{\sigma \mid \sum_{i \geq 0} \delta(\sigma, i) \text{ exists and is finite}\} =$$

$$\bigcap_{\mathbb{Q} \ni \epsilon > 0} \bigcup_{n \in \mathbb{N}} \bigcap_{n_1, n_2 \geq n} \{\sigma \mid \left| \sum_{i=n_1}^{n_2} \delta(\sigma, i) \right| < \epsilon\}.$$

Measurability of $\sigma \mapsto \delta(\sigma, i)$ for each i follows from Lemma 6.21. This implies measurability of the set in part a, since the union and the intersections are defined over countable index sets.

2. The same argument applies basically to set of all paths and those times the timing labels converge to in part b:

$$\{\langle \sigma, t \rangle \mid \lim_{i \to \infty} \delta(\sigma, i) = t\} = \bigcap_{\mathbb{Q} \ni \epsilon > 0} \bigcup_{n \in \mathbb{N}} \bigcap_{m \geq n} \{\langle \sigma, t \rangle \mid | \delta(\sigma, m) - t | < \epsilon\}.$$

By Lemma 6.21, the set

$$\{\langle \sigma, t \rangle \mid\mid \delta(\sigma, m) - t \mid < \epsilon\} = \{\langle \sigma, t \rangle \mid \delta(\sigma, m) > t - \epsilon\} \cap \{\langle \sigma, t \rangle \mid \delta(\sigma, m) < t + \epsilon\}$$

is a measurable subset of $(S \times \mathbb{R}_+)^\infty \times \mathbb{R}_+$, and since the union and the intersections are countable, measurability is inferred.

3. From the definition of the @-operator it is immediate that given an infinite path σ and a time $t \in \mathbb{R}_+$, there exists a rational t' with $\sigma @ t = \sigma @ t'$. Thus we obtain for an arbitrary real number x, an arbitrary Borel set $A \subset S$ and $s \in S$

$$\liminf_{t \to \infty} N_\infty(s)(\{\tau \mid \langle s, \tau \rangle @ t \in A\}) \leq x \Leftrightarrow$$

$$\sup_{t \geq 0} \inf_{r \geq t} N_\infty(s)(\{\tau \mid \langle s, \tau \rangle @ r \in A\}) \leq x \Leftrightarrow$$

$$\sup_{\mathbb{Q} \ni t \geq 0} \inf_{\mathbb{Q} \ni r \geq t} N_\infty(s)(\{\tau \mid \langle s, \tau \rangle @ r \in A\}) \leq x \Leftrightarrow$$

$$s \in \bigcap_{\mathbb{Q} \ni t \geq 0} \bigcup_{\mathbb{Q} \ni r \geq t} A_{r,x}$$

with

$$A_{r,x} := \{s' \mid N_\infty(s')(\{\tau \mid \langle s', \tau \rangle @ r \in A\}) \leq x\}.$$

We infer that $A_{r,x}$ is a measurable subset of S from the fact that N_∞ is a stochastic relation and from Lemma 1.86. Since a map $f : W \to \mathbb{R}$ is measurable iff each of the sets $\{w \in W \mid f(w) \leq s\}$ is a measurable subset of W, the assertion follows for the first map in part a. The second part is established in exactly the same way, using that $f : W \to \mathbb{R}$ is measurable iff $\{w \in W \mid f(w) \geq s\}$ is a measurable subset of W, and observing

$$\limsup_{t \to \infty} N_\infty(s)(\{\tau \mid \langle x, \tau \rangle @ t \in A\}) \geq x \Leftrightarrow$$

$$\inf_{\mathbb{Q} \ni t \geq 0} \sup_{\mathbb{Q} \ni r \geq t} N_\infty(x)(\{\tau \mid \langle s, \tau \rangle @ r \in A\}) \geq x.$$

$$\square$$

As a consequence we obtain that the set on which the asymptotic behavior of the transition times is reasonable in the sense that it tends probabilistically to a limit is well behaved in terms of measurability:

COROLLARY 6.24

Let $A \subseteq X$ be a Borel set, and assume that $N_\infty : S \rightsquigarrow (\mathbb{R}_+ \times S)^\infty$ is a stochastic relation. Then

a. *the set $Q_A := \{s \in S \mid \lim_{t \to \infty} N_\infty(s)(\{\tau \mid \langle s, \tau \rangle @ t \in A\})$ exists$\}$ on which the limit exists is a Borel subset of S,* *b.* $s \mapsto \lim_{t \to \infty} N_\infty(s)(\{\tau \mid \langle s, \tau \rangle @t \in A\}$ *is a measurable map* $Q_A \to \mathbb{R}_+$.

PROOF   Since $s \in Q_A$ iff

$$\liminf_{t \to \infty} N_\infty(x)(\{\tau \mid \langle s, \tau \rangle @t \in A\}) = \limsup_{t \to \infty} N_\infty(x)(\{\tau \mid \langle s, \tau \rangle @t \in A\}),$$

and since the set on which two Borel measurable maps coincide is a Borel set itself, the first assertion follows from Proposition 6.23, part *c*. This implies the second assertion.   ⬜

When dealing with the semantics of the until operator later, we will also need to establish measurability of certain sets. Preparing for that, we state:

LEMMA 6.25
Assume that A_1 and A_2 are Borel subsets of S, and let $I \subseteq \mathbb{R}_+$ be an interval, then

$$U(I, A_1, A_2) := \{\sigma \mid \exists t \in I : \sigma @t \in A_2 \wedge \forall t' \in [0, t[: \sigma @t' \in A_1\}$$

is a measurable set of paths, thus $U(I, A_1, A_2) \in \mathcal{B}((S \times \mathbb{R}_+)^\infty)$.

PROOF   0. Remember that, given a path σ and a time $t \in \mathbb{R}_+$, there exists a rational time $t_r \leq t$ with $\sigma @t = \sigma @t_r$. Consequently,

$$U(I, A_1, A_2) = \bigcup_{t \in \mathbb{Q} \cap I} \left(\{\sigma \mid \sigma @t \in A_2\} \cap \bigcap_{t' \in \mathbb{Q} \cap [0,t]} \{\sigma \mid \sigma @t' \in A_1\} \right).$$

The inner intersection is countable and is performed over measurable sets by Lemma 6.21, thus forming a measurable set of paths. Intersecting it with a measurable set and forming a countable union yields a measurable set again. ⬜

We want to capture paths with our probabilistic model as well, so we need to compute probabilities for sets of paths. Let $M : S \rightsquigarrow \mathbb{R}_+ \times S$ be a stochastic relation. Fix a state $s \in S$, and proceed inductively: Put $M_1(s) := M(s)$, and set in the inductive step for the Borel set $D \subseteq (\mathbb{R}_+ \times S)^{n+1}$

$$M_{n+1}(s)(D) :=$$
$$\int_{(\mathbb{R}_+ \times S)^n} M(s_n)(\{\langle t, s \rangle \mid \langle t_0, s_1, \ldots, t_{n-1}, s_n, t, s \rangle \in D\}) \times$$
$$\times M_n(s)(d\langle t_0, s_1, \ldots, t_{n-1}, s_n \rangle) =$$
$$\int_{(\mathbb{R}_+ \times S)^n} M(\hbar_S(\mathbf{w}))(D_{\mathbf{w}}) \, M_n(s)(d\mathbf{w}),$$

where we have set $\hbar_S(t_0, s_1, \ldots, t_{n-1}, s_n) := s_n$ for simplifying the notation. Thus the argument to $M(s_n) = M(\hbar_S(\mathbf{w}))$ is the set of all times and states $\langle t, s \rangle$ such that $\langle \mathbf{w}, t, s \rangle = \langle t_0, s_1, \ldots, t_{n-1}, s_n, t, s \rangle$ is a member of D. Analyzing the expression further, we see that at step $n+1$ the probability for the pair that consists of timing a transition and changing a state is an element of $\{\langle t, s \rangle \mid \langle t_0, s_1, \ldots, t_{n-1}, s_n, t, s \rangle \in D\}$ equals

$$M(\hbar_S(\mathbf{w}))(D_{\mathbf{w}}) = M(s_n)\left(\{\langle t, s \rangle \mid \langle t_0, s_1, \ldots, t_{n-1}, s_n, t, s \rangle \in D\}\right),$$

provided the corresponding times and states that have been run through during steps $1, \ldots, n$ are given by $\mathbf{w} = \langle t_0, s_1, \ldots, t_{n-1}, s_n \rangle$ which in turn is captured through $M_n(s)(d\mathbf{w})$.

Standard arguments show that $M_n : S \rightsquigarrow (\mathbb{R}_+ \times S)^n$ is a stochastic relation. For each state $s \in S$ the sequence $(M_n(s))_{n \in \mathbb{N}}$ forms a projective system (Definition 1.87), provided $M(s)(\mathbb{R}_+ \times S) = 1$ holds for each $s \in S$: for each Borel set $B \subseteq (\mathbb{R}_+ \times S)^n$ the equality

$$M_{n+1}(B \times (\mathbb{R}_+ \times S)) = M_n(s)(B)$$

holds. Consistency of this family has as a consequence that the measures can be extended to Borel sets of infinite sequences. We obtain from Corollary 1.90 the existence of the projective limit.

PROPOSITION 6.26

Given a stochastic relation $M : S \rightsquigarrow \mathbb{R}_+ \times S$ *with*

$$\forall s \in S : M(s)(\mathbb{R}_+ \times S) = 1,$$

there exists a unique stochastic relation $M_\infty : S \rightsquigarrow (\mathbb{R}_+ \times S)^\infty$ *such that*

$$M_\infty(s)\left(B \times \prod_{j > n}(\mathbb{R}_+ \times S)\right) = M_n(s)(B)$$

for each Borel set $B \in \mathcal{B}((\mathbb{R}_+ \times S)^n)$ *and each state* $s \in S$. M_∞ *is called the projective limit of* $(M_n)_{n \in \mathbb{N}}$.

The projective limit displays indeed limiting behavior: suppose B is an infinite measurable cube $\prod_{n \in \mathbb{N}} B_n$ with $B_n \in \mathcal{B}(\mathbb{R}_+ \times S)$ as Borel sets. Because

$$B = \bigcap_{n \in \mathbb{N}}\left(\prod_{1 \leq j \leq n} B_j \times \prod_{j > n}(\mathbb{R}_+ \times S)\right),$$

is represented as the intersection of a monotonically decreasing sequence, we have for $s \in S$

$$M_\infty(s)(B) = \lim_{n \to \infty} M_\infty(s)(\prod_{1 \leq j \leq n} B_j \times \prod_{j > n} (\mathbb{R}_+ \times S))$$

$$= \lim_{n \to \infty} M_n(s)(\prod_{1 \leq j \leq n} B_j).$$

Hence we obtain $M_\infty(s)(B)$ as the limit of the probabilities $M_n(s)(B_n)$ at step n.

In this way models based on a Polish state space S yield stochastic relations $S \rightsquigarrow (\mathbb{R}_+ \times S)^\infty$ through projective limits. Without this limit it would be difficult to model the transition behavior on infinite paths; the assumption that we work in Polish spaces makes sure that these limits in fact do exist. We need to assume that given a state $s \in S$, there is always a state to change into after a finite amount of time. Thus we make this assumption for the rest of this chapter.

> All stochastic relations assign probability one to their target space.

As a first consequence of the construction for the projective limit we obtain a recursive formulation for the transition law $M : X \rightsquigarrow (\mathbb{R}_+ \times S)^\infty$ that reflects the domain equation $(\mathbb{R}_+ \times S)^\infty = (\mathbb{R}_+ \times S) \times (\mathbb{R}_+ \times X)^\infty$.

LEMMA 6.27
If $D \in \mathcal{B}((\mathbb{R}_+ \times S)^\infty)$, then

$$M_\infty(s)(D) = \int_{\mathbb{R}_+ \times S} M_\infty(s')(D_{\langle t,s' \rangle}) \, M_1(s)(d\langle t, s' \rangle)$$

holds for all $s \in S$.

PROOF   Recall that $D_{\langle t,s' \rangle} = \{ \tau \mid \langle t, s', \tau \rangle \in D \}$. Let

$$D = (H_1 \times \ldots \times H_{n+1}) \times \prod_{j > n} (\mathbb{R}_+ \times X)$$

be a cylinder set with $H_i \in \mathcal{B}(\mathbb{R}_+ \times S), 1 \leq i \leq n + 1$. The equation in question in this case boils down to

$$M_{n+1}(s)(H_1 \times \ldots \times H_{n+1}) = \int_{H_1} M_n(s')(H_2 \times \ldots \times H_{n+1}) M_1(s)(d\langle t, x' \rangle).$$

This may easily be derived from Lemma 6.28. Consequently, the equation in question holds for all cylinder sets, thus the π-λ-Theorem (Proposition 1.1) implies that it holds for all Borel subsets of $(\mathbb{R}_+ \times S)^\infty$.   ⏹

This decomposition indicates that we may first select in state s a new state and a transition time; with these data the system then works just as if the selected new state would have been the initial state. New states and transition times are being averaged over, since we select these items according to a probability law. Lemma 6.27 may accordingly be interpreted as a Markov property for a process the behavior of which is independent of the specific step that is undertaken.

6.3.2 Independence and Zeno Paths

Assume that the state transitions are given through a stochastic relation $K :$ $S \rightsquigarrow S$ and the times are triggered through another relation $L : S \rightsquigarrow \mathbb{R}_+$, so that the probability for jumping from a state s to another state occurs within the time interval $[t_1, t_2]$ is given through $L(s)([t_1, t_2])$. If state transitions and times are stochastically independent, then $M(s) := L(s) \otimes K(s)$ governs the system's behavior.

We state for later use:

LEMMA 6.28
Let $f : (\mathbb{R}_+ \times S)^n \to \mathbb{R}$ be measurable and bounded, and assume that $M(s) = K(s) \otimes L(s)$ holds for each $s \in S$. Then

$$\int_{(\mathbb{R}_+ \times S)^n} f \ dM_n(s) = \int_S \cdots \int_S \int_0^\infty \cdots \int_0^\infty f(t_0, s_1, \ldots, t_{n-1}, s_n) \times$$
$$\times L(s_{n-1})(dt_{n-1})L(s_{n-2})(dt_{n-2}) \ldots L(s)(dt_0) \times$$
$$\times K(s_{n-1})(ds_n)K(s_{n-2})(ds_{n-1}) \ldots K(s)(ds_1).$$

PROOF   (Sketch) One first shows that the representation on the right hand side is true if $f = \chi_A$ for some Borel set $A \subseteq (\mathbb{R}_+ \times S)^n$. Correctness follows in this case from the definition, since

$$\int_{(\mathbb{R}_+ \times S)^n} \chi_A \ dM_n(s) = M_n(s)(A).$$

The validity for indicator functions implies the validity for step functions, i.e., functions f of the form $f = \sum_{i=0}^n \alpha_i \cdot \chi_{A_i}$ with Borel sets A_i and real α_i through the integral's additivity. One then observes that each nonnegative bounded Borel map can be approximated by an increasing sequence of step functions from below, thus the equality is true for nonnegative f by the monotone convergence theorem (Proposition 1.60). The general case writes $f = f^+ + f^-$ with $f^+(s) := \max(f(s), 0), f^-(s) := \min(f(s), 0)$ and applies the previous case.

The last step rearranges the integrals according to the dependencies of their integration variables. This is admissible through Fubini's Theorem on product integration, which permits interchanging the order of integration for product measures, in this case over the domains S and $\mathbb{R}_+$. ▯

Observation 1 The usual approach to interpreting Markov chains with continuous time runs via a rate function. Assume that R represents the rate, then

a. $R(s)$ is for every $s \in S$ a finite measure on S such that $E(s) := R(s) > 0$ is always strictly positive,

b. $s \mapsto R(s)(B)$ is for every $B \in \mathcal{B}(S)$ a measurable function $S \to \mathbb{R}_+$.

The rate function models the transition rate: if the system is in state s, then the transition rate for jumping to a new state that is a member of the Borel set $D \subseteq S$ is given by $R(s)(D)$. This transition rate is assumed to be finite. We also assume in the rate model that there is no blind state, so transitions are assumed to be possible from all states, thus $E(s) > 0$.

Put

$$K(s)(D) := \frac{R(s)(D)}{E(s)}$$

and set for the probability of making a transition from state s within t time units

$$L(s)([0, t]) := 1 - e^{-E(s) \cdot t},$$

then

$$L(s)(F) = \frac{1}{E(s)} \cdot \int_F e^{-E(s) \cdot t} \, dt.$$

Consequently, the approach discussed here fits into the usual set up to model continuous time Markov processes, and generalizes it.

6.3.2.1   Zeno Paths

We have seen in Proposition 6.23 that the set

$$Z := \{\sigma \mid \sum_{i \geq 0} \delta(\sigma, i) \text{ exists and is finite}\}$$

of all *Zeno paths* is measurable in the universe of our paths. *Math 101* tells us that $Z \subseteq C$ with C as the set of all paths the transition times of which tend to zero, thus

$$C := \{\sigma \mid \lim_{i \to \infty} \delta(\sigma, i) = 0\}.$$

By Proposition 6.23 this is also a measurable set. We will show now that a Zeno path will only occur with probability 0, provided the probability L that governs the transitions does not concentrate its mass close to zero (this means that very short transition times will occur quite infrequently).

We establish this property of Zeno paths under the assumption that the relation L is uniformly bounded at the origin, but we treat the problem a wee bit more generally.

DEFINITION 6.29   *The stochastic relation $L : S \rightsquigarrow \mathbb{R}_+$ is called bounded at time t iff given $\epsilon > 0$ there is $\delta > 0$ such that*

$$\sup_{s \in S} L(s)([t - \delta, t + \delta] \cap \mathbb{R}_+) < \epsilon.$$

Thus *bounded at t* means for L that no positive mass is associated with t, uniformly for all states. Consequently t is a time in which state changes cannot occur with positive probability for any state.

Observation 2 Assume that we work in the rate model. Let as in Observation 1

$$L(s)(F) := \frac{1}{E(s)} \cdot \int_F e^{-E(s) \cdot t} \, dt,$$

and assume that $\rho := \sup_{s \in S} R(s)(S)$ is finite. Then the relation L is bounded at each time t. This is so since

$$L(s)([t_1, t_2]) = e^{-E(s) \cdot t_1} \cdot \left(1 - e^{-E(s) \cdot (t_2 - t_1)}\right) \leq 1 - e^{-\rho \cdot (t_2 - t_1)}.$$

This difference, which is independent of state x, can be brought arbitrarily close to 0.

Boundedness has as a consequence that the Zeno paths can be neglected, when we are working under stochastic independence of state changes and residence times.

PROPOSITION 6.30
   Assume that $M(s) = L(s) \otimes K(s)$ for each $s \in S$, where $K : S \rightsquigarrow S$ and $L : S \rightsquigarrow \mathbb{R}_+$. Then the following holds for each $s \in S$

a. *$M_\infty(s)(\{\tau \mid \lim_{i \to \infty} \delta(\sigma, i) = t\}) = 0$, provided L is bounded at t,*

b. *$M_\infty(s)(\{\tau \mid \sum_{i \geq 0} \delta(\langle s, \tau \rangle, i)$ exists and is finite$\}) = 0$, provided L is bounded at the origin. Consequently, the set of all Zeno paths is negligible.*

PROOF   0. The remarks preceding Definition 6.29 imply that we only have to establish part a; part b will then follow immediately for $t = 0$. Fix $t \in \mathbb{R}_+$ and assume that L is bounded at t.
   1. Since

$$\lim_{i \to \infty} \delta(\sigma, i) = t \Leftrightarrow \forall \epsilon > 0 \exists n \in \mathbb{N} \forall k \geq n : \mid \delta(\sigma, j) - t \mid < \epsilon,$$

we can represent C as

$$C = \bigcap_{\mathbb{Q} \ni \epsilon > 0} \bigcup_{n \in \mathbb{N}} N_{\epsilon,n}$$

with

$$N_{r,n} := \{\sigma \mid\mid \delta(\sigma, n+j) - t \mid < r \text{ for all } j \in \mathbb{N}\}.$$

It is clear from the definition that

$$N_{r,n} = \bigcap_{k \geq 0} N'_{r,n,k},$$

where

$$N'_{r,n,k} := \{\sigma \mid\mid \delta(\sigma, n+j) - t \mid < r \text{ for } 0 \leq j \leq k\}$$
$$= N''_{r,n,k} \times \prod_{j > n+k} (\mathbb{R}_+ \times X).$$

The sequence $(N'_{r,n,k})_{k \in \mathbb{N}}$ is monotonically decreasing, hence

$$M_\infty(s)(\{\tau \mid \langle s, \tau \rangle \in N_{r,n}\}) = \inf_{k \in \mathbb{N}} M_\infty(s)(\{\tau \mid \langle s, \tau \rangle \in N'_{r,n,k}\}).$$

From the construction of $M_\infty(s)$ as projective limit of $(M_n)_{n \in \mathbb{N}}$ we see that

$$M_\infty(s)(\{\tau \mid \langle s, \tau \rangle \in N'_{r,n,k}\}) = M_{n+k}(s)(\{\tau \mid \langle x, \tau \rangle \in N''_{r,n,k}\}),$$

which by Lemma 6.28 may be evaluated as

$$M_{n+k}(s)(\{\tau \mid \langle s, \tau \rangle \in N''_{r,n,k}\}) =$$
$$\int_S \cdots \int_S L(s_{n+k-1})([t-r, t+r] \cap \mathbb{R}_+) \cdot \cdots \cdot L(s_{n-1})([t-r, t+r] \cap \mathbb{R}_+) \times$$
$$\times K(s_{n+k-2})(ds_{n+k-1}) \ldots K(s_1)(dx_2) K(s)(ds_1).$$

Well, that's not too bad.

2. Now if $0 < \epsilon < 1$ is given, we can find $\eta > 0$ such that

$$L(s)([t-\eta, t+\eta] \cap \mathbb{R}_+) < \epsilon$$

is true for all $s \in S$ due to L being bounded at the origin. Consequently,

$$M_{n+k}(s)(\{\tau \mid \langle s, \tau \rangle \in N''_{\eta,n,k}\}) \leq \epsilon^{k+1},$$

which implies

$$M_\infty(s)(\{\tau \mid \langle x, \tau \rangle \in N_{\eta,n}\}) \leq \inf_{k \in \mathbb{N}} \epsilon^k = 0.$$

But this trivially implies

$$M_\infty(s)(\{\tau \mid \lim_{i \to \infty} \delta(\sigma, i) = t\}) = 0.$$

$\Box$

If we assume that L is bounded at each point in time, then the mass associated with each $L(s)$ is not concentrated at any time t, for otherwise the probability of hitting arbitrary small intervals $[t - \delta, t + \delta]$ around t cannot be made arbitrary small. Thus there is no preferred, pointed timing behavior.

6.4   F-Bisimulations for CSL

The continuous time stochastic logic **CSL** will be analyzed in greater detail now. In particular we will look at equivalence relations that are given through subsets of formulas.

Fix P as a countable set of atomic propositions. We define recursively state formulas and path formulas for **CSL**:

State formulas are defined through the syntax

$$\varphi ::= \top \mid a \mid \neg \varphi \mid \varphi \wedge \varphi' \mid \mathcal{S}_{\bowtie p}(\varphi) \mid \mathcal{P}_{\bowtie p}(\psi).$$

Here $a \in P$ is an atomic proposition, ψ is a path formula, $\bowtie$ is one of the relational operators $<, \leq, \geq, >$, and $p \in [0, 1]$ is a rational number.

Path formulas are defined through

$$\psi ::= \mathcal{X}^I \, \varphi \mid \varphi \, \mathcal{U}^I \, \varphi'$$

with φ, φ' as state formulas, $I \subseteq \mathbb{R}_+$ a closed interval of the real numbers with rational bounds (including $I = \mathbb{R}_+$).

We denote the set of all state formulas by $\mathfrak{L}_P$.

The operator $\mathcal{S}_{\bowtie p}(\varphi)$ gives the *steady-state probability* for φ to hold with the boundary condition $\bowtie p$; the formula $\mathcal{P}$ replaces quantification: the *path-quantifier* formula $\mathcal{P}_{\bowtie p}(\psi)$ holds in a state s iff the probability of all paths starting in s and satisfying ψ is specified by $\bowtie p$. Thus ψ holds on almost all paths starting from s iff s satisfies $\mathcal{P}_{\geq 1}(\psi)$, a path being an alternating infinite sequence $\sigma = \langle s_0, t_0, s_1, t_1, \ldots \rangle$ of states x_i and of times t_i. Note that the time is being made explicit here. The *next-operator* $\mathcal{X}^I \, \varphi$ is assumed to hold on path σ iff s_1 satisfies φ, and $t_0 \in I$ holds. Finally, the *until-operator* $\varphi_1 \, \mathcal{U}^I \, \varphi_2$ holds on path σ iff we can find a point in time $t \in I$ such that the state $\sigma@t$ which σ occupies at time t satisfies φ_2, and for all times t' before that, $\sigma@t'$ satisfies φ_1.

The basic operators are introduced now more formally. We will also have a look at issues of measurability: the basic operators will be shown to represent measurable functions. This will help in establishing that many of the sets of paths and, derived from them, sets of states that occur in conjunction with set and path formulas are measurable, thus lie in the domain of the probabilities that we will be working with (suppose in the contrary that an important set is not measurable, then we cannot measure it, we cannot assign a probability to it. This would jeopardize the programme of a stochastic interpretation of **CSL**). As a consequence the sets of states resp. paths on which a formula is valid are measurable.

6.4.1   Interpreting the Logic

Now that we know how to probabilistically describe the behavior of paths, we are ready for a probabilistic interpretation of **CSL**. We assume that we have a stochastic relation $M : S \rightsquigarrow \mathbb{R}_+ \times S$ with $M(s)(\mathbb{R}_+ \times S) = 1$ for all $s \in S$ according to the general assumption from page 278 from which the stochastic relation $M_\infty : S \rightsquigarrow \mathbb{R}_+ \times (S \times \mathbb{R}_+)^\infty$ has been constructed. The interpretations for the formulas are established, and we show that the sets of states resp. paths on which formulas are valid are Borel measurable.

To get started on the formal definition of the semantics, we assume that we know for each atomic proposition which state it is satisfied in, so we fix a map $\mathbf{L}$ that maps P to $\mathcal{B}(S)$, assigning each atomic proposition a Borel set of states.

The semantics is described as usual recursively through relation $\models$ between states resp. paths, and formulas as follows:

a. $s \models \top$ is true for all $s \in S$.

b. $s \models a$ iff $s \in \mathbf{L}(a)$.

c. $s \models \varphi_1 \wedge \varphi_2$ iff $s \models \varphi_1$ and $s \models \varphi_2$.

d. $s \models \neg\varphi$ iff $s \models \varphi$ is false.

e. $s \models \mathcal{S}_{\bowtie p}(\varphi)$ iff $\lim_{t\to\infty} M_\infty(s)(\{\tau \mid \langle s, \tau \rangle @t \models \varphi\})$ exists and is $\bowtie p$.

f. $s \models \mathcal{P}_{\bowtie p}(\psi)$ iff $M_\infty(s)(\{\tau \mid \langle s, \tau \rangle \models \psi\}) \bowtie p$.

g. $\sigma \models \mathcal{X}^I \varphi$ iff $\sigma[1] \models \varphi$ and $\delta(\sigma, 0) \in I$.

h. $\sigma \models \varphi_1 \mathcal{U}^I \varphi_2$ iff $\exists t \in I : \sigma @t \models \varphi_2$ and $\forall t' \in [0, t[: \sigma @t' \models \varphi_1$.

Denote by $[\![\varphi]\!]$ again the set of all states for which the state formula φ holds, resp. the set of all paths for which the path formula φ is valid. We do not distinguish notationally between these sets, as far as the basic domains are concerned, since it should always be clear whether we describe a state formula or a path formula.

We show that we are dealing with measurable sets. Most of the work for establishing this has been done already, so we have to fit in the patterns that we have set up in Proposition 6.23 and its Corollaries.

PROPOSITION 6.31
The set $[\![\varphi]\!]$ is Borel, whenever φ is a state formula or a path formula.

PROOF 0. The proof proceeds by induction on the structure of the formula φ. The induction starts with the formula $\top$, for which the assertion is true, and with the atomic propositions, for which the assertion follows from the assumption on **L**: $[\![a]\!] = \mathbf{L}(a) \in \mathcal{B}(S)$. We assume for the induction step that we have established that $[\![\varphi]\!]$, $[\![\varphi_1]\!]$ and $[\![\varphi_2]\!]$ are Borel measurable.

1. For the next-operator we write

$$[\![\mathcal{X}^I \varphi]\!] = \{\sigma \mid \sigma[1] \in [\![\varphi]\!] \text{ and } \delta(\sigma, 0) \in I\}.$$

This is the cylinder set $(S \times I \times [\![\varphi]\!] \times \mathbb{R}_+) \times (S \times \mathbb{R}_+)^\infty$, hence is a Borel set.

2. The until-operator may be represented through

$$[\![\varphi_1 \, \mathcal{U}^I \, \varphi_2]\!] = U(I, [\![\varphi_1]\!], [\![\varphi_2]\!]),$$

which is a Borel set by Lemma 6.25.

3. Since $M_\infty : S \rightsquigarrow (\mathbb{R}_+ \times S)^\infty$ is a stochastic relation, we know that

$$[\![\mathcal{P}_{\bowtie p}(\psi)]\!] = \{s \in S \mid M_\infty(s)(\{\tau \mid \langle s, \tau \rangle \in [\![\varphi]\!]\}) \bowtie p\}$$

is a Borel set.

4. We know from Corollary 6.24 that the set

$$Q_{[\varphi]} := \{s \in S \mid \lim_{t \to \infty} M_\infty(s)(\{\tau \mid \langle s, \tau \rangle @ t \in [\![\varphi]\!]\}) \text{ exists}\}$$

is a Borel set, and that

$$\ell_\varphi : Q_{[\varphi]} \ni s \mapsto \lim_{t \to \infty} M_\infty(x)(\{\tau \mid \langle s, \tau \rangle @ t \in [\![\varphi]\!]\}) \in [0, 1]$$

is a Borel measurable function. Consequently,

$$[\![\mathcal{S}_{\bowtie p}(\varphi)]\!] = \{s \in Q_{[\varphi]} \mid \ell_\varphi(s) \bowtie p\}$$

is a Borel set. ∎

Measurability of the sets on which a given formula is valid is of course a prerequisite for computing interesting properties. So we can compute, e.g.,

$$\mathcal{P}_{\geq 0.5}((\neg down) \, \mathcal{U}^{[10,20]} \, \mathcal{S}_{\geq 0.8}(up_2 \vee up_3)))$$

as the set of all states that with probability at least 0.5 will reach a state between 10 and 20 time units so that the system is operational ($up_2, up_3 \in P$) in a steady state with a probability of at least 0.8; prior to reaching this state, the system must be operational continuously ($down \in P$).

6.4.2   Definition and Properties of ρ_F

Returning to the logic, fix a set F of state formulas, and define the central equivalence relation

$$s \; \rho_F \; s' \Leftrightarrow \forall \varphi \in F : [s \models \varphi \Leftrightarrow s' \models \varphi],$$

then ρ_F is smooth due to F being countable. We will investigate in this section the equivalence ρ_F. First, the closure $\mathsf{wrap}\,(F)$ of F will be defined as the smallest set of formulas containing F and being closed under the logic's operators, and it will be investigated under which conditions $\rho_{\mathsf{wrap}(F)} = \rho_F$ holds. An answer to this question makes life easier, since testing satisfaction only on F is presumably much easier than testing on $\mathsf{wrap}\,(F)$, in particular when $F = P$ (so that $\mathsf{wrap}\,(F) = \mathfrak{L}_P$). We will examine an enabling condition, using smooth equivalence relations and congruences as the decisive tool. This leads to a discussion of bisimulations; the results obtained for congruences will be transported for an investigation of bisimilar states. Conditions under which P-bisimilarity and the satisfaction of the same formulas are related will be formulated at the end of this section.

The closure $\mathsf{wrap}\,(F)$ of F is defined as the smallest set of formulas in $\mathfrak{L}_P$ which contains F and which is closed under the defining operations for the logic. Formally, $\mathsf{wrap}\,(F)$ is the set of all F-state formulas which are defined through the following rules:

F-state formulas are defined through the syntax

$$\varphi ::= \top \mid \Phi \mid \neg\varphi \mid \varphi \wedge \varphi' \mid \mathcal{S}_{\bowtie p}(\varphi) \mid \mathcal{P}_{\bowtie p}(\psi).$$

Here $\Phi \in F$ is a formula in F, ψ is an F-path formula, $\bowtie$ is one of the relational operators $<, \leq, \geq, >$, and $p \in [0,1]$ is a rational number.

F-path formulas are defined through

$$\psi ::= \mathcal{X}^I \varphi \mid \varphi \, \mathcal{U}^I \varphi'$$

with φ, φ' as F-state formulas, $I \subseteq \mathbb{R}_+$ a closed interval of the real numbers with rational bounds.

Thus we start in building up F-formulas from elements of F as the base, just as we started building up $\mathfrak{L}_P$ from the set P of atomic propositions. Observe that $\mathsf{wrap}\,(P) = \mathfrak{L}_P$. We will investigate the smooth relations ρ_F and $\rho_{\mathsf{wrap}(F)}$ and will establish that under a mildly restrictive condition $\rho_F = \rho_{\mathsf{wrap}(F)}$ holds. This result looks rather modest, but it has some interesting consequences in terms of bisimulations. They will be discussed after the proof.

The mild condition that will enable us to establish the relations' equality was detected by Desharnais and Panangaden in (Desharnais and Panangaden, 2003) for the fragment of **CSL** investigated there.

DEFINITION 6.32   *A set F of formulas is said to satisfy the DP-condition iff F has these properties: F is closed under conjunctions, and $\mathcal{P}_{\bowtie p}(\mathcal{X}^I \varphi) \in F$ whenever $\varphi \in F, p \in [0,1]$ is rational, and $I \subseteq \mathbb{R}_+$ is a closed interval with rational endpoints.*

We will see that the closedness under conjunction will later enable us to apply the π-λ-Theorem for making sure that a condition carries over from the set of generators (in this case $\{[\![\varphi]\!] \mid \varphi \in F\}$) to the σ-algebra generated from it. Closedness under the next operator will have a special consequence, as we will see in a moment.

The DP-condition makes sure that the probabilities for a transition of ρ_F-equivalent states into a state in which a formula in F is valid are identical. This is quite comparable to the observation one makes for stochastic Kripke models for modal logics: there it is well known that the probabilities for making a move into a state in which the same formula is satisfied after an action coincide for equivalent states as well; see Lemma 6.18.

DEFINITION 6.33   *Define for a set $B \in \mathcal{B}(S)$ the probability that a move is made from state $s \in S$ into B by $K(s)(B) := M(s)(\mathbb{R}_+ \times B)$.*

It is apparent that $K : S \rightsquigarrow S$ characterizes the transition behavior in just one step. This is interesting when investigating the probabilities for which state s moves into another state that satisfies a formula in F.

LEMMA 6.34
If $s \, \rho_F \, s'$ and $\varphi \in F$, then $K(s)([\![\varphi]\!]) = K(s')([\![\varphi]\!])$, provided F satisfies the DP-condition.

PROOF   Suppose that we find for $s \, \rho_F \, s'$ a formula $\varphi' \in F$ such that

$$K_1(s)([\![\varphi']\!]) < r \leq K_1(s')([\![\varphi']\!]),$$

where r may be assumed to be rational. Since

$$\{\tau \mid \langle s, \tau \rangle \models \mathcal{X}^{\mathbb{R}_+} \varphi'\} = (\mathbb{R}_+ \times [\![\varphi']\!]) \times (\mathbb{R}_+ \times S)^\infty,$$

we conclude that

$$K(s)([\![\varphi']\!]) = M_\infty(s)(\{\tau \mid \langle s, \tau \rangle \models \mathcal{X}^{\mathbb{R}_+} \varphi'\}).$$

But this implies that $s \models \mathcal{P}_{<r}(\mathcal{X}^{\mathbb{R}_+} \varphi')$, similarly, $s' \not\models \mathcal{P}_{<r}(\mathcal{X}^{\mathbb{R}_+} \varphi')$. But the DP-condition implies that $\mathcal{P}_{<r}(\mathcal{X}^{\mathbb{R}_+} \varphi') \in F$, which is a contradiction. ☐

This Lemma is actually a first step towards establishing that ρ_F generates a congruence for M. It requires an extension of the equivalence relation ρ_F

on S to $(\mathbb{R}_+ \times S)^\infty$. The basic idea is to relate the alternating states in such a sequence through ρ_F, and to leave the residence times alone, which means to relate them through the identity relation $\Delta_{\mathbb{R}_+}$. Thus $\langle t_0, s_1, t_1, \ldots \rangle$ will be related to $\langle t_0', s_1', t_1', \ldots \rangle$ iff $s_i \; \rho_F \; s_i'$ and $t_i = t_i'$ for all indices i. In view of Lemma 5.14 we form the product relation $\times_{n \in \mathbb{N}} (\Delta_{\mathbb{R}_+} \times \rho_F) = (\Delta_{\mathbb{R}_+} \times \rho_F)^\infty$.

To alleviate the heavy notation somewhat, we abbreviate

$$\rho_F^{(n)} := (\Delta_{\mathbb{R}_+} \times \rho_F)^n,$$
$$\rho_F^{(\infty)} := (\Delta_{\mathbb{R}_+} \times \rho_F)^\infty.$$

PROPOSITION 6.35

Assume that F satisfies the DP-condition, then

$$c_F := (\rho_F, \rho_F^{(\infty)})$$

is a congruence for $M : S \rightsquigarrow (\mathbb{R}_+ \times S)^\infty$.

PROOF   0. We need to show that

$$(\dagger) \; M_\infty(s)(D) = M_\infty(s')(D)$$

holds for each $\rho_F^{(\infty)}$-invariant Borel set D, provided $s \; \rho_F \; s'$ holds. We know that

$$\mathcal{INV}\left(\mathcal{B}((\mathbb{R}_+ \times S)^\infty), \rho_F^{(\infty)}\right) = \bigotimes_{n \in \mathbb{N}} \mathcal{INV}\left(\mathcal{B}(\mathbb{R}_+ \times S), \Delta_{\mathbb{R}_+} \times \rho_F\right)$$

holds (Lemma 5.14), and from the construction of the infinite product of measurable spaces we see that we may restrict our attention to cylinder sets the factors of which are $\Delta_{\mathbb{R}_+} \times \rho_F$-invariant. But since the σ-algebra $\mathcal{INV}\left(\mathcal{B}(\mathbb{R}_+ \times S), \Delta_{\mathbb{R}_+} \times \rho_F\right)$ is generated by

$$\{I \times [\![\varphi]\!] \mid I \subseteq \mathbb{R}_+ \text{ is an interval}, \varphi \in F\},$$

it is sufficient for establishing Eq. ($\dagger$) that the equation

$$(\ddagger) \; M_n(s)\left((I_1 \times [\![\varphi_1]\!]) \times \ldots \times (I_n \times [\![\varphi_n]\!])\right) =$$
$$M_n(s')\left((I_1 \times [\![\varphi_1]\!]) \times \ldots \times (I_n \times [\![\varphi_n]\!])\right)$$

holds, whenever $s \; \rho_F \; s'$, where $I_1, \ldots, I_n$ are intervals in $\mathbb{R}_+$ with rational endpoints and $\varphi_1, \ldots, \varphi_n$ are formulas in F. This is done by induction on n.

Fix s, s' with $s \; \rho_F s'$, intervals $(I_n)_{n \in \mathbb{N}}$ in $\mathbb{R}_+$ with rational endpoints, and formulas $(\varphi_n)_{n \in \mathbb{N}}$ in F, and put $B_n := [\![\varphi_n]\!]$ as the set of states in which φ_n is valid.

1. The induction starts at $n = 1$ with the observation that

$$M_1(s)(I_1 \times B_n) = M_\infty(s)(\{\tau \mid \langle s, \tau \rangle \in I_1 \times B_1\}) =$$
$$M_\infty(s)(\{\tau \mid \langle s, \tau \rangle \models \mathcal{X}^{I_1} \varphi_1\}).$$

Thus we have for an arbitrary rational p

$$M_1(s)(I_1 \times B_1) \leq p \Leftrightarrow s \models \mathcal{P}_{\bowtie p}(\mathcal{X}^{I_1} \varphi_1)$$
$$\Leftrightarrow s' \models \mathcal{P}_{\bowtie p}(\mathcal{X}^{I_1} \varphi_1) \quad \text{(since } s \, \rho_F \, s')$$
$$\Leftrightarrow M_1(s')(I_1 \times B_1) \leq p.$$

Consequently, $M_1(s)(I_1 \times B_1) = M_1(s')(I_1 \times B_1)$ is established.

2. Assume for the induction step that the assertion is true for n. This implies in particular that $(\rho_F, \rho_F^{(n)})$ is a congruence for $M_n : S \rightsquigarrow (\mathbb{R}_+ \times S)^n$. From the Markov property in Lemma 6.27 we infer that

$$M_{n+1}(s)((I_1 \times B_1) \times \cdots \times (I_{n+1} \times B_{n+1})) =$$
$$\int_{I_1 \times B_1} M_n(y)((I_2 \times B_2) \times \cdots \times (I_{n+1} \times B_{n+1})) \, M_1(s)(d\langle t, y \rangle) =$$
$$\int_{\mathbb{R}_+ \times S} \chi_{I_1 \times B_1}(t, y) \cdot M_n(y)((I_2 \times B_2) \times \cdots \times (I_{n+1} \times B_{n+1})) \, M_1(s)(d\langle t, y \rangle)$$

(recall that χ indicates the indicator function of a set). We claim that

$$\langle t, y \rangle \mapsto \chi_{I_1 \times B_1}(t, y) \cdot M_n(y)((I_2 \times B_2) \times \cdots \times (I_{n+1} \times B_{n+1}))$$

is a $\mathcal{INV}\left(\mathcal{B}(\mathbb{R}_+ \times S), \Delta_{\mathbb{R}_+} \times \rho_F\right)$-$\mathcal{B}(\mathbb{R}_+)$-measurable function. This inferred from the fact that $I_1 \times B_1$ is $\Delta_{\mathbb{R}_+} \times \rho_F$-invariant, and from the observation that $(\rho_F, \rho_F^{(n)})$ is a congruence for M_n, using Lemma 5.30. Consequently, we may infer from $s \, \rho_F \, s'$ in conjunction with Lemma 5.30 that

$$\int_{\mathbb{R}_+ \times S} \chi_{I_1 \times B_1}(t, y) \cdot M_n(y)((I_2 \times B_2) \times \cdots \times (I_{n+1} \times B_{n+1})) \, M_1(s)(d\langle t, y \rangle) =$$
$$\int_{\mathbb{R}_+ \times S} \chi_{I_1 \times B_1}(t, y) \cdot M_n(y)((I_2 \times B_2) \times \cdots \times (I_{n+1} \times B_{n+1})) \, M_1(s')(d\langle t, y \rangle),$$

which implies equation (‡) also for $n + 1$. □

Reflecting on the proof, we see that the DP-condition on F is needed to establish the initial step in this induction. It is also responsible for maintaining invariance in the induction step through the integral representation rendering the Markov property.

The intermediate goal is to prove that $\rho_F = \rho_{\mathsf{wrap}(F)}$ holds. Because by construction $F \subseteq \mathsf{wrap}(F)$, and because $F \mapsto \rho_F$ is anti-monotonic, we need to show that $\rho_F \subseteq \rho_{\mathsf{wrap}(F)}$ is true for establishing the equality above. We will first investigate ρ_F-invariant Borel sets with respect to a smooth equivalence relation on $(S \times \mathbb{R}_+)^\infty$ related to ρ_F and $\Delta_{\mathbb{R}_+}$.

Some auxiliary operators are introduced: let A, A_1, A_2 be subsets of S, B be a subset of $(S \times \mathbb{R}_+)^\infty$, and $I \subseteq \mathbb{R}_+$ an interval with rational bounds, then

$$P_{\ltimes p}(B) := \{s \in S \mid M_\infty(s)(\{\tau \mid \langle s, \tau \rangle \in B\}) \ltimes p\}$$
$$Q_A := \{s \in S \mid \lim_{t \to \infty} M_\infty(s)(\{\tau \mid \langle s, \tau \rangle @ t \in A\}) \text{ exists}\}$$
$$f_A(s) := \lim_{t \to \infty} M_\infty(s)(\{\tau \mid \langle s, \tau \rangle \in A\}), \text{ if } s \in Q_A$$
$$S_{\ltimes p}(A) := \{s \in Q_A \mid f_A(s) \ltimes p\}$$
$$X(I, A) := \{\sigma \mid \sigma[1] \in A \wedge \delta(\sigma, 0) \in I\}.$$

We observe the following properties:

LEMMA 6.36

Let F be a set of formulas, and recall that $\rho_F \times \Delta_{\mathbb{R}_+}$ denotes the smooth equivalence relation

$$\langle s, t \rangle \, (\rho_F \times \Delta_{\mathbb{R}_+}) \, \langle s', t' \rangle \Longleftrightarrow s \, \rho_F \, s' \wedge t = t',$$

on $S \times \mathbb{R}_+$. Assume that F satisfies the DP-condition.

a. *If $B \in \mathcal{INV}\left(\mathcal{B}((S \times \mathbb{R}_+)^\infty), (\rho_F \times \Delta_{\mathbb{R}_+})^\infty\right)$, then $P_{\ltimes p}(B)$ is a member of the σ-algebra $\mathcal{INV}\left(\mathcal{B}(S), \rho_F\right)$.*

b. *If $A \in \mathcal{INV}\left(\mathcal{B}(S), \rho_F\right)$, then*

    i. *$Q_A \in \mathcal{INV}\left(\mathcal{B}(S), \rho_F\right)$,*

    ii. *$S_{\ltimes p}(A) \in \mathcal{INV}\left(\mathcal{B}(S), \rho_F\right)$,*

    iii. *$X(I, A) \in \mathcal{INV}\left(\mathcal{B}((S \times \mathbb{R}_+)^\infty), (\rho_F \times \Delta_{\mathbb{R}_+})^\infty\right)$.*

c. *If $A_1, A_2 \in \mathcal{INV}\left(\mathcal{B}(S), \rho_F\right)$, then*

$$U(I, A_1, A_2) \in \mathcal{INV}\left(\mathcal{B}((S \times \mathbb{R}_+)^\infty), (\rho_F \times \Delta_{\mathbb{R}_+})^\infty\right).$$

PROOF   1. Since F satisfies the DP-condition, we know from Proposition 6.35 that c_F is a congruence for $M_\infty : S \rightsquigarrow (\mathbb{R}_+ \times S)^\infty$. From Lemma 5.30 and Corollary 2.10 we infer that $s \mapsto M_\infty(s)(B_s)$ is a $\mathcal{INV}\left(\mathcal{B}(S), \rho_F\right)$-$\mathcal{B}(\mathbb{R}_+)$-measurable function, where B_s is as usual the cut $\{\tau \mid \langle s, \tau \rangle \in B\}$ at s. This implies the assertion in part a.

2. Define for $t \in \mathbb{R}_+$ the set $J_A := \{\sigma \mid \sigma @ t \in A\}$, then J_A will be shown to be a member of $\mathcal{INV}\left(\mathcal{B}((S \times \mathbb{R}_+)^\infty), (\rho_F \times \Delta_{\mathbb{R}_+})^\infty\right)$. In fact, suppose

σ $(\rho_F \times \Delta_{\mathbb{R}_+})^\infty$ σ', then $\delta(\sigma, i) = \delta(\sigma', i)$ holds for all i (this is so since the equivalence does not affect the timing information), thus $\mathsf{Stop}(\sigma, r) = \mathsf{Stop}(\sigma', r)$ for all $r \geq 0$. Consequently, we obtain (cf. the proof for Lemma 6.21)

$$\sigma \in J_A \Leftrightarrow \sigma @ t \in A$$
$$\Leftrightarrow \exists k : \mathsf{Stop}(\sigma, t) = k, \sigma[k] \in A$$
$$\Leftrightarrow \exists k : \mathsf{Stop}(\sigma', t) = k, \sigma[k] \in A$$
$$\Leftrightarrow \sigma' \in J_A,$$

establishing the invariance of J_A. Clearly, J_A is a Borel set by Lemma 6.21.

Again, we infer that $s \mapsto M_\infty(s)(\{\tau \mid \langle s, \tau \rangle \in J_A\})$ is $\mathcal{INV}(\mathcal{B}(S), \rho_F)$-$\mathcal{B}(\mathbb{R}_+)$-measurable, hence

$$A_{t,s} := \{s' \mid M_\infty(s')(\{\tau \mid \langle s', \tau \rangle @ t \in A\}) \leq s\}$$

defines an ρ_F-invariant Borel set. We know from the proof of Proposition 6.23 that

$$\liminf_{t \to \infty} M_\infty(s)(\{\tau \mid \langle s, \tau \rangle @ t \in A\}) \leq x \Leftrightarrow s \in \bigcap_{\mathbb{Q} \ni t \geq 0} \bigcup_{\mathbb{Q} \ni r \geq t} A_{r,x},$$

thus

$$s \mapsto \liminf_{t \to \infty} M_\infty(s)(\{\tau \mid \langle s, \tau \rangle @ t \in A\})$$

defines a $\mathcal{INV}(\mathcal{B}(S), \rho_F)$-$\mathcal{B}(\mathbb{R}_+)$-measurable map, so does

$$s \mapsto \limsup_{t \to \infty} M_\infty(s)(\{\tau \mid \langle s, \tau \rangle @ t \in A\}).$$

Since these maps coincide on Q_A, this establishes the first part of b.

3. Represent $X(I, A)$ for the ρ_F-invariant Borel set $A \subseteq S$ and the interval $I \subseteq \mathbb{R}_+$ with rational endpoints as $X(I, A) = (S \times I) \times (A \times \mathbb{R}_+) \times (S \times \mathbb{R}_+)^\infty$, then it is clear that this is a cylinder set which is $(\rho_F \times \Delta_{\mathbb{R}_+})^\infty$-invariant. This establishes the second part of b.

4. Represent for the ρ_F-invariant Borel sets A_1, A_2 and the interval $I \subseteq \mathbb{R}_+$ with rational endpoints the set $U(I, A_1, A_2)$ as in the proof of Lemma 6.25 as

$$U(I, A_1, A_2) = \bigcup_{t \in \mathbb{Q} \cap I} \left(\{\sigma \mid \sigma @ t \in A_1\} \cap \bigcap_{t' \in \mathbb{Q} \cap [0, t]} \{\sigma \mid \sigma @ t' \in A_2\} \right),$$

and observe that the sets involved are all invariant Borel sets, as shown in part 2 of the present proof, then part c follows readily. $\Box$

This Lemma will be instrumental in establishing our main result on bisimulations. Its proof is somewhat awkward due to the necessity of keeping track of many smooth relations at once. It indicates on the other hand that smooth equivalence relations are a versatile tool for these investigations.

6.4.3   Closure Operations

We show that ρ_F coincides with a finer equivalence relation that is generated by F's closure under the operations offered by the logic. This closure is not the only one of interest: we will close F also towards the future; thus, when we know that $\varphi \in F$, then we also know that $\mathcal{X}^I \varphi$ will be a member of F. The reason for this closure under the DP-condition will become apparent soon.

PROPOSITION 6.37

Let $F \neq \emptyset$ be a set of formulas, denote by ρ_F the equivalence relation on the set of states imposed by F, and let $\mathsf{wrap}\,(F)$ be the closure of F under the logic's operators. Then $\rho_F = \rho_{\mathsf{wrap}(F)}$ holds, provided F satisfies the DP-condition.

PROOF   1. Because $\rho_{\mathsf{wrap}(F)} \subseteq \rho_F$ is trivial, we need to establish the other inclusion, and since $\rho_{\mathsf{wrap}(F)}$ is determined by the countable set $\{[\![\varphi]\!] \mid \varphi \in \mathsf{wrap}\,(F)\}$ of Borel sets, it is sufficient to show that $[\![\varphi]\!] \in \mathcal{INV}\,(\mathcal{B}(S), \rho_F)$ for each $\varphi \in \mathsf{wrap}\,(F)$.

2. Since for each $\varphi \in F$ we have trivially $[\![\varphi]\!] \in \mathcal{INV}\,(\mathcal{B}(S), \rho_F)$, an inductive reasoning with Lemma 6.36 on the structure of F-state formulas and of F-path formulas establishes the assertion.   ⬚

As an interesting direct and first consequence of Proposition 6.37 we obtain that the equivalence of states on the atomic propositions determines their equivalence of all formulas, provided the DP-condition is satisfied. If it is not, we force it: Define for a set F of formulas

$$\mathsf{dp}(F) := \bigcap \{G \subseteq \mathfrak{L}_P \mid F \subseteq G, G \text{ has the DP-condition}\}$$

as the smallest set of formulas that satisfy the DP-condition (this construction is sensible because the set $\mathfrak{L}_P$ of all formulas satisfies the condition under consideration).

We obtain from Proposition 6.37 right away:

COROLLARY 6.38

$\rho_{\mathsf{dp}(P)} = \rho_{\mathfrak{L}_P}$.

This result is not yet fully satisfying; in practice it means that one has to have a look at the formulas in DP-closure for concluding whether or not a given property holds for all formulas. It is, however, desirable to restrict oneself to observing properties on the atomic propositions alone, and then to say that this property holds for the entirety of formulas. This is what we investigate now. The basic idea is to find a suitable representation for $\mathsf{dp}(F)$

and then to capitalize on Corollary 5.6 for identifying the equivalence relation as $\rho_{\mathsf{dp}(F)}$.

Let F be a nonempty set of formulas. Define for $\Psi \subseteq \mathcal{L}_P$ the set valued map

$$H(\Psi) := F \cup \{ \bigwedge_{1 \leq i \leq n} \varphi_i \mid n \in \mathbb{N}, \varphi_1, \ldots, \varphi_n \in \Psi \} \cup$$

$$\{ \mathcal{P}_{\bowtie p}(\mathcal{X}^{[a,b]} \varphi) \mid \varphi \in \Psi, a, b, p \text{ rational} \},$$

then the least fixed point

$$H_* := \mu \Psi . H(\Psi)$$

exists by the Kleene-Knaster-Tarski Fixed Point Theorem, and

$$H_* = \bigcup_{n \in \mathbb{N}} H^{(n)}(\emptyset)$$

holds, with $H^{(n)}$ as the nth iterate of H. Similarly, define for a family $\mathcal{A}$ of Borel sets in S

$$h(\mathcal{A}) := \{ [\![\varphi]\!] \mid \varphi \in F \} \cup \mathcal{A} \cup \{ P_{\bowtie p}(X([a,b], A)) \mid A \in \mathcal{A}, a, b, p \text{ rational} \}.$$

Again invoking the Kleene-Knaster-Tarski Theorem, we know that the smallest fixed point

$$\mathcal{C}_* := \mu \mathcal{A} . h(\mathcal{A})$$

exists, and can be computed through

$$\mathcal{C}_* = \bigcup_{n \in \mathbb{N}} h^{(n)}(\emptyset).$$

Here $h^{(n)}$ is of course the nth iterate of h.

As witnessed by the use of the path quantifier, both constructs are closely related:

LEMMA 6.39
Construct the set H_* of formulas and the family $\mathcal{C}_*$ of Borel sets as above. Then

a. $H_* = \mathsf{dp}(F)$,

b. *the $\rho_{\mathsf{dp}(F)}$-invariant sets are generated from H_*, thus*

$$\sigma(\mathcal{C}_*) = \mathcal{INV}\left(\mathcal{B}(S), \rho_{\mathsf{dp}(F)}\right)$$

holds.

PROOF   1. It is clear from the definition of the map H that $\mu\Psi.H(\Psi)$ satisfies the DP-condition, and it is equally clear that each set of formulas that satisfies this condition and contains F contains also $H^{(n)}(\emptyset)$ for each $n \in \mathbb{N}$.

2. If $\varphi \in H_*$, then the representation of $\mu\Psi.H(\Psi)$ shows that $[\![\varphi]\!] \in \mathcal{C}_*$. On the other hand, it is not difficult to see that

$$h(\{[\![\varphi]\!] \mid \varphi \in \Psi\}) \subseteq \{[\![\varphi]\!] \mid \varphi \in H(\Psi)\} \subseteq \sigma(\{[\![\varphi]\!] \mid \varphi \in H_*\}).$$

This shows that

$$\sigma(\{[\![\varphi]\!] \mid \varphi \in H_*\}) = \sigma(\mathcal{C}_*),$$

establishing the second claim.   ▯

Looking at the maps — which yield equivalent representations for the dp()-closure — it is noticeable that H as the version catering for formulas takes the conjunction into account, while its set-theoretic cousin h does not. This is due to the observation that the σ-algebra of invariant Borel sets uniquely determines the equivalence relation by Corollary 5.6, but that this σ-algebra can have many different generators which may or may not be closed with respect to finite intersection.

6.4.4   *F*-Bisimulations

Let us define F-bisimulations in order to put these results into the proper context. Define for $F \subseteq \mathcal{L}_P$ and for each state $s \in S$ the set

$$\mathbf{L}_F(s) := \{\varphi \in F \mid s \models \varphi\}$$

as the set of all formulas in F that are satisfied by s.

DEFINITION 6.40   *Let F be a set of formulas, then a smooth equivalence relation $\equiv_F$ is called an F-bisimulation iff*

a. $\mathbf{L}_F(s) = \mathbf{L}_F(s')$, *whenever* $s \equiv_F s'$.

b. $K(s)(D) = K(s')(D)$, *whenever* $s \equiv_F s'$ *and* $D \in \mathcal{INV}(\mathcal{B}(S), \equiv_F)$.

An F-bisimulation is focussed on the behavior that manifests itself on the states, rather than on paths. Hence we use for its formulation the relation K rather than M. If $\equiv_F$ is an F-bisimulation, condition $b.$ tells us that this relation is in particular a congruence for K (see Definition 5.26), so we may define the factor relation

$$K_{\equiv_F}([s]_{\equiv_F})(D) := K(s)((\eta_{\equiv_F}^{-1}[D]))$$

whenever $D \in \mathcal{B}(S/\equiv_F)$ in a Borel set in the factor space. It has the additional property that the map $\mathbf{L}_F : S \to F$ is constant on the equivalence classes. This observation yields a characterization of F-bisimulations in terms of congruences:

PROPOSITION 6.41

The following statements are equivalent for a smooth equivalence relation ρ on S

a. ρ is an F-bisimulation.

b. ρ is a congruence for K with $s \rho s' \Rightarrow \mathbf{L}_F(s) = \mathbf{L}_F(s')$.

Consequently, F-bisimilar states accept exactly the same formulas in F, and they behave in exactly the same way on the $\equiv_F$-invariant Borel sets. As a first result towards relating the results obtained so far to F-bisimulations, we see that under the mild condition of F being closed under conjunctions, ρ_F is actually one:

PROPOSITION 6.42

The relation ρ_F is an F-bisimulation for each $F \subseteq \mathfrak{L}_P$, provided F is closed under conjunctions.

PROOF The definition of ρ_F guarantees that $\mathbf{L}_F(x) = \mathbf{L}_F(x')$ is true whenever $s \rho_F s'$. Thus we need to show that $K(s)(D) = K(s')(D)$ for $s \rho_F s'$ and for each $D \in \mathcal{INV}(\mathcal{B}(S), \rho_F)$ holds. Define for fixed states s, s' that are ρ_F-related

$$\mathcal{D} := \{D \in \mathcal{INV}(\mathcal{B}(S), \rho_F) \mid K(s)(D) = K(s')(D)\}.$$

Then $\mathcal{D}$ is a σ-algebra, and it will be enough to show that a generator of $\mathcal{INV}(\mathcal{B}(S), \rho_F)$ that is closed under finite intersection is contained in $\mathcal{D}$. We know from Lemma 6.34 that $K(s)(\llbracket\varphi\rrbracket) = K(s')(\llbracket\varphi\rrbracket)$ holds for each $\varphi \in F$. But this implies with the π-λ-Theorem 1.1 the following chain:

$$\mathcal{INV}(\mathcal{B}(S), \rho_F) = \sigma(\{\llbracket\varphi\rrbracket \mid \varphi \in F\}) \subseteq \sigma(\mathcal{D}) \subseteq \mathcal{D} \subseteq \mathcal{INV}(\mathcal{B}(S), \rho_F),$$

establishing the assertion. ▯

The relation ρ_F is provided naturally with F, so it plays a prominent role among all the F-bisimulations (there are other F-bisimulations, e.g., the identity is one, but probably not the most interesting among all the candidates):

DEFINITION 6.43 *The states $s, s' \in S$ are called F-bisimilar iff $s \rho_F s'$ holds.*

This is a characterization of F-bisimilarity:

THEOREM 6.44

Let $\emptyset \neq F \subseteq \mathfrak{L}_P$ be a set of formulas which satisfy the DP-condition, then two states are F-bisimilar iff they satisfy exactly the same formulas in wrap (F).

PROOF This follows immediately from Proposition 6.37 in conjunction with Proposition 6.42. $\Box$

Specializing to the set of atomic formulas, we obtain at once:

COROLLARY 6.45

Two states are dp(P)*-bisimilar iff they satisfy exactly the same formulas in* $\mathfrak{L}_P$.

This is not yet completely satisfying for practical purposes, because one has to construct the closure dp(P) of the set of all atomic propositions, which may be done iteratively through the computation of a fixed point, as the discussion leading to Lemma 6.39 shows. Nevertheless it leads to an infinite process, handling a countable set of objects. But suppose we are in the situation in which both the state transitions K and the jump times L are determined through a rate function R (cp. Observation 1). Now

$$s \models \mathcal{P}_{\bowtie p}(\mathcal{X}^I \varphi) \Leftrightarrow (L(s)(I) \cdot K(s)(\llbracket \varphi \rrbracket)) \bowtie p$$

as an easy computation reveals. Thus the σ-algebra of $\rho_{\mathsf{dp}(P)}$-invariant Borel sets is determined by the ρ_P-invariant Borel sets and by the smallest σ-algebra $\mathcal{T}_R$ on S that renders the map $s \mapsto R(s)(A)$ measurable for each $A \in \mathcal{INV}\,(\mathcal{B}(S), \rho_P)$. This is so by Lemma 6.39. This observation yields

COROLLARY 6.46

If $s \mapsto R(s)(A)$ is a $\mathcal{INV}\,(\mathcal{B}(S), \rho_P)$-$\mathcal{B}(\mathbb{R}_+)$ -measurable map for each ρ_P-invariant Borel set A, then the following conditions are equivalent for any two states $s, s' \in S$:

a. s and s' are P-bisimilar.

b. s and s' satisfy exactly the same formulas in $\mathfrak{L}_P$.

PROOF The condition implies that

$$\mathcal{INV}\,(\mathcal{B}(S), \rho_P) = \mathcal{INV}\,(\mathcal{B}(S), \rho_{\mathsf{dp}(P)}) = \mathcal{INV}\,(\mathcal{B}(S), \rho_{\mathfrak{L}_P}),$$

because all sets that are added when constructing $\mathcal{INV}\,(\mathcal{B}(S), \rho_{\mathsf{dp}(P)})$ through the process described in Lemma 6.39 are $\mathcal{INV}\,(\mathcal{B}(S), \rho_P)$ - measurable. Thus we infer from Corollary 5.6 the first equality. The second equality comes from Corollary 6.45. Given this equality, the assertion follows from Theorem 6.44.

□

The proof capitalizes on the uniqueness of the invariant sets for a smooth equivalence relation: since we are able to identify these sets, we may conclude what shape the relation has. This shows that a closer inspection of the invariant Borel sets bears some — probably unexpected — fruits. The condition imposed in Corollary 6.46 above is satisfied in the finite case whenever the rate function is constant on the equivalence classes for $\equiv_P$. This can be checked quite efficiently once the classes are computed.

Bisimilarity for stochastic relations was discussed extensively in particular in Chapter 5, so the question of relating F-bisimilarity to that more general notion arises. A first step towards a characterization is

PROPOSITION 6.47

Let $\emptyset \neq F \subseteq \mathfrak{L}_P$ be a set of formulas which satisfy the DP-condition, then there exists a smooth 2-bisimulation $N : \rho_F \rightsquigarrow (\Delta_{\mathbb{R}_+} \times \rho_F)^\infty$ for $M : S \rightsquigarrow (\mathbb{R}_+ \times S)^\infty$.

PROOF From Proposition 6.35 we know that $\left(\rho_F, (\Delta_{\mathbb{R}_+} \times \rho_F)^\infty\right)$ is a congruence for M, because F satisfies the DP-condition. Thus the assertion follows from Proposition 5.51. □

The DP-condition turns out to be crucial as a necessary condition for the two notions of bisimilarity to be related. It can be said actually a bit more. We introduce for this the *extension* of F,

$$\mathsf{ext}\,(F) := \{\varphi \mid [\![\varphi]\!] \in \mathcal{INV}\,(\mathcal{B}(S), \rho_F)\}.$$

Thus $\varphi \in \mathsf{ext}\,(F)$ iff $[\![\varphi]\!]$ is ρ_F-invariant, so it is immediate that $F \subseteq \mathsf{ext}\,(F)$, and that $\mathcal{INV}\,(\mathcal{B}(S), \rho_F) = \mathcal{INV}\,(\mathcal{B}(S), \rho_{\mathsf{ext}(F)})$. The reason for introducing the extension is quite obviously of a strategic nature: we cannot lay our hands on F directly, but we can determine whether or not a formula is in $\mathsf{ext}\,(F)$ by having a look at the invariant Borel sets.

PROPOSITION 6.48

The following conditions are equivalent for a set $\emptyset \neq F \subseteq \mathfrak{L}_P$ of formulas:

a. $\mathsf{ext}\,(F)$ satisfies the DP-condition.

b. There exists a smooth 2-bisimulation $N : \rho_F \rightsquigarrow (\Delta_{\mathbb{R}_+} \times \rho_F)^\infty$ for $M_\infty : S \rightsquigarrow (\mathbb{R}_+ \times S)^\infty$.

c. $\left(\rho_F, (\Delta_{\mathbb{R}_+} \times \rho_F)^\infty\right)$ is a congruence for M.

PROOF   1. We know from Proposition 5.51 that the conditions b and c are equivalent, and we know from $\mathcal{INV}\left(\mathcal{B}(S), \rho_F\right) = \mathcal{INV}\left(\mathcal{B}(S), \rho_{\text{ext}(F)}\right)$ that $\rho_F = \rho_{\text{ext}(F)}$ by Corollary 5.6. Thus $a \Rightarrow b$ is just Proposition 6.47.

2. $b \Rightarrow a$ Abbreviate as above $\rho_F^{(\infty)} := (\Delta_{\mathbb{R}_+} \times \rho_F)^\infty$. We claim that the set

$$A_s := \{\tau \mid \langle s, \tau \rangle \in \mathcal{X}^I \, \varphi\}$$

is $\rho_F^{(\infty)}$-invariant, whenever $[\![\varphi]\!] \in \mathcal{INV}\left(\mathcal{B}(S), \rho_F\right)$ and $s \in S$. In fact, let $\langle s, \tau \rangle \in [\![\mathcal{X}^I \, \varphi]\!]$ and assume that $\tau \, \rho_F^{(\infty)} \, \tau'$. By definition, $\langle s, \tau \rangle [1] \models \varphi$ and $\delta(\langle s, \tau \rangle, 0) \in I$. But $\langle s, \tau \rangle [1] \, \rho_F \, \langle s, \tau' \rangle [1]$ and $\delta(\langle s, \tau \rangle, 0) = \delta(\langle s, \tau' \rangle, 0)$, so that we may conclude $\langle s, \tau' \rangle [1] \models \varphi$ and $\delta(\langle s, \tau' \rangle, 0) \in I$. Consequently, $\langle s, \tau' \rangle \in [\![\mathcal{X}^I \, \varphi]\!]$, so that $\tau' \in A_s$. Note that A_s does not really depend on s by the definition of $[\![\mathcal{X}^I \, \varphi]\!]$. Because A_s is a member of $\mathcal{INV}\left(\mathcal{B}((\mathbb{R}_+ \times S)^\infty), \rho_F^{(\infty)}\right)$, we infer from Lemma 5.9, part a, that for $\langle s, s' \rangle \in \rho_F$ the following holds (here $\pi_i : \rho_F^{(\infty)} \to (\mathbb{R}_+ \times S)^\infty$ are the corresponding projections, $i = 1, 2$):

$$
\begin{aligned}
M_\infty(s)(\{\tau \mid \langle s, \tau \rangle \in \mathcal{X}^I \, \varphi\}) &= N(\langle s, s' \rangle)(\pi_1^{-1}[A_s]) \\
&= N(\langle s, s' \rangle)(\pi_1^{-1}[A_s] \cap \rho_F^{(\infty)}) \\
&= N(\langle s, s' \rangle)(\pi_2^{-1}[A_{s'}] \cap \rho_F^{(\infty)}) \\
&= N(\langle s, s' \rangle)(\pi_2^{-1}[A_s']) \\
&= M_\infty(s')(\{\tau \mid \langle s', \tau \rangle \in \mathcal{X}^I \, \varphi\}).
\end{aligned}
$$

But this means that $s \models \mathcal{P}_{\bowtie p}(\mathcal{X}^I \, \varphi)$ iff $s' \models \mathcal{P}_{\bowtie p}(\mathcal{X}^I \, \varphi)$ whenever $s \, \rho_F \, s'$ and $[\![\varphi]\!] \in \mathcal{INV}\left(\mathcal{B}(S), \rho_F\right)$. Consequently, $[\![\mathcal{P}_{\bowtie p}(\mathcal{X}^I \, \varphi)]\!]$ is an ρ_F-invariant Borel set, thus $\mathcal{P}_{\bowtie p}(\mathcal{X}^I \, \varphi) \in \text{ext}(F)$.   □

6.5   Logical Equivalence for μCSL

The logic μ**CSL** to be investigated now will be quite similar to **CSL** (in fact, the latter superficially looks like a fragment of the former). It will contain the μ-operator, and we will have a look at the interplay of bisimilarity, behavioral and logical equivalence. While for **CSL** the question was investigated how far a set of formulas may be pushed for answering questions pertaining to all (state) formulas, we will deal here with the interplay of state and path formulas when it comes to understand what it means that two models have the same theories for states and paths. Thus we have to be careful about the interplay between state and path formulas, for example, we close path formulas under the next operator rather than making the operator a bridge between state and path formulas. We also close path formulas under conjunction. Introducing the μ-operator means that we have to care about variables, rendering models a bit more involved.

The other interesting formulas from **CSL** like path quantification or the steady state operator will be defined for μ**CSL** as well, so that some properties from **CSL** are carried over. The fixed point operator, however, requires special attention when, e.g., measurability is to be established, but somewhat surprisingly it does not substantially enter the discussion on bisimilarity or logical and behavioral equivalence.

We will first define the logic formally, then define models and their morphisms and the interpretation of μCSL. Some standard properties like Borel measurability are established, and logical equivalence is defined. We deal with properties on states and on paths; the corresponding equivalence relations are investigated and related to each other. This happens on the basis of the underlying stochastic relations. The machinery from Chapters 4 and 5 is put into action, and the relations obtained from these constructions are modified so that they fit into the mold of the models for the logic. The main result is that logical equivalence and bisimilarity are equivalent, and that this holds also for behavioral equivalence, provided the factor space induced by the theory of states is a Polish space again (this is so since the projective limit construction, on which interpretations are based, does not seem to work for general analytic spaces, but only for their Polish brethren).

6.5.1 The Logic μCSL

State formulas and path formulas for μCSL are again defined recursively:

State formulas are defined through the syntax

$$\varphi ::= \top \mid a \mid Z \mid \neg \varphi \mid \varphi \wedge \varphi' \mid \mathcal{S}_{\bowtie p}(\varphi) \mid \mathcal{P}_{\bowtie p}(\psi).$$

Here $a \in \mathsf{AP}$ is an atomic proposition, $Z \in \mathsf{SV}$ is a state variable, ψ is a path formula, $\bowtie$ is one of the relational operators $<, \leq, >, \geq$, and $p \in [0, 1]$ is a rational number.

Path formulas are defined through

$$\psi ::= P \mid \neg \psi \mid \psi \wedge \psi' \mid \mathcal{X}^I \psi \mid \varphi \, \mathcal{U}^I \, \varphi' \mid \mu P.\psi$$

with $P \in \mathsf{PV}$ as a path variable, φ, φ' as state formulas, $I \subseteq \mathbb{R}_+$ a closed interval of the real numbers with rational bounds (including $I = \mathbb{R}_+$); these intervals will be called *rational intervals*. The operator μ describes the smallest fixed point; it binds variables in the usual sense. We assume that the variable bound by it is in the range of an even number of negations.

The sets AP, SV and PV are assumed to be mutually disjoint and countable.

6.5.2   Models for μCSL and Their Morphisms

We are ready for the definition of models for μ**CSL** and their morphisms. Since we will work again with projective limits for interpreting path formulas, models will be based on Polish spaces rather more generally on analytic spaces.

DEFINITION 6.49 $\mathcal{M} = (S, M, \mathcal{I}, V)$ *is called a* model for μ**CSL** *iff*

a. S *is a Polish space, the* state space *of* $\mathcal{M}$,

b. $M : S \rightsquigarrow \mathbb{R}_+ \times S$ *is a stochastic relation, the* law of change *of* $\mathcal{M}$,

c. $\mathcal{I} = (\Sigma, \Pi)$ *interprets the variables,*

   (a) $\Sigma : \mathsf{SV} \to \mathcal{B}(S)$ *assigns each state variable a Borel set in* S,

   (b) $\Pi : \mathsf{PV} \to \mathcal{B}(Paths(S))$ *assigns each path variable a Borel set of paths,*

d. $V : \mathsf{AP} \to \mathcal{B}(S)$ *maps each atomic proposition to a Borel set of states.*

Thus a model says how residence times and state changes are to be handled: if $s \in S$ is the present state, then $M(s)(I \times B)$ gives the probability that after $t \in I$ time units a state change will happen, and that the new state will be a member of Borel set $B \subseteq S$. Each model says how the variables are to be interpreted; this is written down through the maps Σ and Π, and we say what sets the atomic propositions are taken from. Note that we assume in each case that the sets under consideration are Borel. Otherwise we could not assign them any probability directly or indirectly; hence this assumption is made for keeping the model within the realm of probabilistic reasoning.

Let $\mathcal{M} = (S, M, (\Sigma, \Pi), V)$ be a model. Given a state variable Z and a Borel set $Q \in \mathcal{B}(S)$, denote by $\mathcal{M}[Z\backslash Q]$ the model $(S, M, (\Sigma', \Pi), V)$ with $\Sigma'(Z) := Q$, otherwise Σ' coincides with Σ. Similarly, the model $\mathcal{M}[P\backslash U]$ is defined for the path variable P and the Borel set $U \in \mathcal{B}((S \times \mathbb{R}_+)^\infty)$. Substituting values in this way may be iterated.

We define a morphism $\Phi : \mathcal{M} \to \mathcal{N}$ for the models $\mathcal{M}$ and $\mathcal{N}$. It is based on a map $\Phi : S \to S'$ between state spaces, which is extended to a map $\Phi_\infty : (S \times \mathbb{R}_+)^\infty \to (S' \times \mathbb{R}_+)^\infty$ upon setting

$$\Phi_\infty(\langle s_0, t_0, s_1, t_1, \dots \rangle) := \langle \Phi(s_0), t_0, \Phi(s_1), t_1, \dots \rangle,$$

thus we transform the states according to Φ but leave the residence times alone; define additionally $id_{\mathbb{R}_+} \times \Phi : \langle t, s \rangle \mapsto \langle t, \Phi(s) \rangle$, and similarly, $\Phi \times id_{\mathbb{R}_+}$.

DEFINITION 6.50   *Let* $\mathcal{M} = (S, M, \mathcal{I}, V)$ *and* $\mathcal{N} = (S', N, \mathcal{I}', V')$ *be models for* μ**CSL**. *Then* $\Phi : \mathcal{M} \to \mathcal{N}$ *is called a* morphism *from* $\mathcal{M}$ *to* $\mathcal{N}$ *iff*

a. $\Phi : S \to S'$ *is a surjective and Borel measurable map between the state spaces,*

b. $(\Phi, id_{\mathbb{R}_+} \times \Phi) : M \to N$ *is a morphism for the associated stochastic relations* M *and* N,

c. $\Phi^{-1}[\Sigma'(Z)] = \Sigma(Z)$ *for each state variable* Z,

d. $\Phi_\infty^{-1}[\Pi'(P)] = \Pi(P)$ *for each path variable* P,

e. $\Phi^{-1}[V'(a)] = V(a)$ *for each atomic proposition* a.

We require the map underlying a morphism to be onto since we want to be able to trace each state in S' back to a state in S, inheriting the corresponding property from the basic stochastic relations. Condition b says that this diagram is commutative:

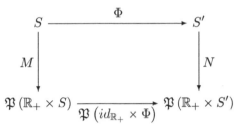

Thus we have in particular

$$N(\Phi(s))(I \times B) = M(s)\left(I \times \Phi^{-1}[B]\right)$$

for every state $s \in S$, every rational interval I and every Borel set $B \in \mathcal{B}(S)$. Conditions c to e relate the interpretations of variables and atomic propositions. For example, condition c says that for a state s and a state variable Z we have $s \in \Sigma(Z)$ iff $\Phi(s) \in \Sigma'(Z)$.

Morphisms are compatible with substitutions.

LEMMA 6.51

Let $\Phi : \mathcal{M} \to \mathcal{N}$ *be a morphism for the models* $\mathcal{M}$ *and* $\mathcal{N}$, *assume that* S' *is the state space for* $\mathcal{N}$. *Then*

$$\Phi : \mathcal{M}[Z\backslash\Phi^{-1}[Q], P\backslash\Phi_\infty^{-1}[R]] \to \mathcal{N}[Z\backslash Q, P\backslash R]$$

is a morphism whenever Z *is a state variable,* P *is a path variable,* $Q \in \mathcal{B}(S')$ *and* $R \in \mathcal{B}((S' \times \mathbb{R}_+)^\infty)$ *are Borel sets.*

We will show now that model morphisms may be interpreted as morphisms between these projective limits. To be specific:

PROPOSITION 6.52

Let $\mathcal{M}$ *and* $\mathcal{N}$ *be models,* $\Phi : \mathcal{M} \to \mathcal{N}$ *be a morphism from* $\mathcal{M}$ *to* $\mathcal{N}$, *then* $(\Phi, \Phi_\infty) : \mathcal{M}_\infty \to \mathcal{N}_\infty$ *is a morphism between the stochastic relations* $\mathcal{M}_\infty$ *and* $\mathcal{N}_\infty$.

PROOF   0. We have to show that $N_\infty \circ \Phi = \mathfrak{P}(\Phi_\infty) \circ M_\infty$, equivalently, that for each $s \in S$ and each Borel set $F \subseteq \mathbb{R}_+ \times (S' \times \mathbb{R}_+)^\infty$ the equation

$$N_\infty(\Phi(s))(F) = M_\infty(s)(\Phi_\infty^{-1}[F])$$

holds. We consider first the case that

$$F = \prod_{j=1}^{n} (I_j \times B'_j) \times \prod_{j>n} (\mathbb{R}_+ \times S)$$

holds, where $I_1, \ldots, I_n$ are rational intervals, and $B'_1, \ldots, B'_n$ are Borel sets in S'. Since we are dealing with projective limits, and since in this case F is a cylinder set, we have

$$N_\infty(\Phi(s))(F) = N_n(\Phi(s)) \left(\prod_{j=1}^{n} (I_j \times B'_j) \right) ;$$

similarly we see

$$M_\infty(s)(\Phi_\infty^{-1}[F]) = M_n(s) \left(\prod_{j=1}^{n} (I_j \times \Phi^{-1}[B'_j]) \right) .$$

Hence it is enough to show in this case that

$$N_n(\Phi(s)) \left(\prod_{j=1}^{n} (I_j \times B'_j) \right) = M_n(s) \left(\prod_{j=1}^{n} (I_j \times \Phi^{-1}[B'_j]) \right)$$

holds. This is done by induction on n.

1.   The induction's beginning at $n = 1$ is trivial by the definition of a morphism for models. The induction step works as follows:

$$N_{n+1}(\Phi(s)) \left((I_1 \times B'_1) \times \cdots \times (I_{n+1} \times B'_{n+1}) \right) =$$

$$\int_{\prod_{j=1}^{n}(I_j \times B'_j)} N(\hbar_{S'}(\mathbf{w}'))(I_{n+1} \times B'_{n+1}) \, N_n(\Phi(s))(d\mathbf{w}') =$$

$$\int_{(\mathbb{R}_+ \times S')^n} \chi_{\prod_{j=1}^{n}(I_j \times B'_j)}(\mathbf{w}') N(\hbar_{S'}(\mathbf{w}'))(I_{n+1} \times B'_{n+1}) N_n(\Phi(s))(d\mathbf{w}') =$$

$$\int_{(\mathbb{R}_+ \times S)^n} \chi_{\prod_{j=1}^{n}(I_j \times \Phi^{-1}[B'_j])}(\mathbf{w}) N(\hbar_{S'}(\mathbf{w}))(I_{n+1} \times \Phi^{-1}[B'_{n+1}]) M_n(s)(d\mathbf{w}) =$$

$$\int_{\prod_{j=1}^{n}(I_j \times \Phi^{-1}[B'_j])} M(\hbar_S(\mathbf{w}))(I_{n+1} \times \Phi^{-1}[B'_{n+1}]) M_n(s)(d\mathbf{w}) =$$

$$M_{n+1}(s)((I_1 \times \Phi^{-1}[B'_1]) \times \cdots \times (I_{n+1} \times \Phi^{-1}[B'_{n+1}])$$

by the Change of Variables Formula (Proposition 1.95) and by the fact that Φ induces a morphism $(\Phi, id_{\mathbb{R}_+} \times \Phi) : M \to N$.

2. Now define

$$\mathcal{D} := \{F \in \mathcal{B}((S \times \mathbb{R}_+)^\infty) \mid N_\infty(\Phi(s))(F) = M_\infty(s)(\Phi_\infty^{-1}[F])\}$$

then part 1. of the proof shows that $\mathcal{D}$ contains all the cylinder sets, and it is clear that $\mathcal{D}$ is a σ-algebra. Thus $\mathcal{D}$ contains the σ-algebra generated from the cylinder sets, which are just the Borel sets. Hence $\mathcal{D}$ coincides with $\mathcal{B}((S \times \mathbb{R}_+)^\infty)$. ⬜

6.5.3 Interpreting μCSL

We are now ready for an interpretation of μ**CSL**. Fix a model $\mathcal{M} = (S, M, \mathcal{I}, V)$ over the Polish space S and let $M_\infty : S \rightsquigarrow \mathbb{R}_+ \times (S \times \mathbb{R}_+)^\infty$ be the associated stochastic relation that relates (initial) states to paths.

6.5.3.1   The semantics

The semantics of μ**CSL** is then described recursively through relation $\models$ between states resp. paths, and formulas as described below. We again denote by

$$\llbracket \varphi \rrbracket_\mathcal{M} := \{s \in S \mid \mathcal{M}, s \models \varphi\}$$

and

$$\llbracket \psi \rrbracket_\mathcal{M} := \{\sigma \in (S \times \mathbb{R}_+)^\infty \mid \mathcal{M}, \sigma \models \psi\}$$

the set of all states resp. paths for which the respective formula holds. The semantics is very similar to that for **CSL** given in Section 6.4.1; it is nevertheless given in full for the sake of completeness.

a. $\mathcal{M}, s \models \top$ is true for all $s \in S$.

b. $\mathcal{M}, s \models a$ iff $s \in V(a)$.

c. $\mathcal{M}, s \models Z$ iff $s \in \Sigma(Z)$ for $Z \in$ SV.

d. $\mathcal{M}, s \models \varphi_1 \wedge \varphi_2$ iff $\mathcal{M}, s \models \varphi_1$ and $\mathcal{M}, s \models \varphi_2$.

e. $\mathcal{M}, s \models \neg\varphi$ iff $\mathcal{M}, s \models \varphi$ is false.

f. $\mathcal{M}, s \models \mathcal{S}_{\bowtie p}(\varphi)$ iff $\lim_{t\to\infty} M_\infty(s)(\{\tau \mid \langle s, \tau\rangle @ t \models \varphi\})$ exists and is $\bowtie p$.

g. $\mathcal{M}, s \models \mathcal{P}_{\bowtie p}(\psi)$ iff $M_\infty(s)(\{\tau \mid \langle s, \tau\rangle \models \psi\}) \bowtie p$.

h. $\mathcal{M}, \sigma \models P$ iff $\sigma \in \Pi(P)$ for $P \in$ PV.

i. $\mathcal{M}, \sigma \models \psi_1 \wedge \psi_2$ iff $\mathcal{M}, \sigma \models \psi_1$ and $\mathcal{M}, \sigma \models \psi_2$.

j. $\mathcal{M}, \sigma \models \neg\psi$ iff $\mathcal{M}, \sigma \models \psi$ is false.

k. $\mathcal{M}, \sigma \models \mathcal{X}^I \psi$ iff $\mathcal{M}, \sigma[1 \ldots] \models \psi$ and $\delta(\sigma, 0) \in I$.

l. $\mathcal{M}, \sigma \models \varphi_1 \, \mathcal{U}^I \, \varphi_2$ iff $\exists t \in I : \mathcal{M}, \sigma@t \models \varphi_2$ and $\forall t' \in [0, t[: \mathcal{M}, \sigma@t' \models \varphi_1$.

m. $\mathcal{M}, \sigma \models \mu P.\psi$ iff $\sigma \in \bigcup_{i \geq 0} R_i$, where R_i is recursively determined through

$$R_0 := [\![\psi]\!]_{\mathcal{M}[P\backslash\emptyset]}, R_{i+1} := [\![\psi]\!]_{\mathcal{M}[P\backslash R_i]}.$$

Define the *theory* $Th_{\mathcal{M}}(s)$ *of a state* s as above as the formulas which hold in s,

$$Th_{\mathcal{M}}(s) := \{\varphi \mid \varphi \text{ is a state formula}, \mathcal{M}, s \models \varphi\}.$$

Similarly, the *theory* $Th_{\mathcal{M}}(\sigma)$ *of a path* σ is defined:

$$Th_{\mathcal{M}}(\sigma) := \{\psi \mid \psi \text{ is a path formula}, \mathcal{M}, \sigma \models \psi\}.$$

We need to show that the sets of states and paths, resp., in which formulas hold are Borel measurable. This is done exactly as in Section 6.4.1, in particular in Proposition 6.31, with the exception of the μ-operator, which needs to be treated separately.

For this, define a model $\mathcal{M}'$ being *based on model* $\mathcal{M}$ iff either $\mathcal{M}'$ coincides with $\mathcal{M}$, or if $\mathcal{M}'$ is of the form $\mathcal{M}_0[P\backslash R]$ with a Borel set $R \subseteq (S \times \mathbb{R}_+)^\infty$, P a path variable, and model $\mathcal{M}_0$ based on $\mathcal{M}$.

LEMMA 6.53

Assume that $[\![\psi]\!]_{\mathcal{M}'} \in \mathcal{B}((S \times \mathbb{R}_+)^\infty)$ holds for every model $\mathcal{M}'$ that is based on $\mathcal{M}$. Then $[\![\mu P.\psi]\!]_{\mathcal{M}} \in \mathcal{B}((S \times \mathbb{R}_+)^\infty)$.

PROOF   Define inductively, as above $R_0 := [\![\psi]\!]_{\mathcal{M}[P\backslash\emptyset]}$, and $R_{i+1} := [\![\psi]\!]_{\mathcal{M}[P\backslash R_i]}$. An easy induction on i shows that $\mathcal{M}[P\backslash R_i]$ is based on $\mathcal{M}$, so $[\![\mu P.\psi]\!]_{\mathcal{M}}$ is the countable union of Borel sets, which is a Borel set again.   ∎

PROPOSITION 6.54

Let $\mathcal{M}'$ be a model based on $\mathcal{M}$. Then

a. $[\![\varphi]\!]_{\mathcal{M}'} \in \mathcal{B}(S)$ *for all state formulas* φ,

b. $[\![\psi]\!]_{\mathcal{M}'} \in \mathcal{B}((S \times \mathbb{R}_+)^\infty)$ *for all path formulas* ψ.

PROOF   Using structural induction on path formulas and Lemma 6.53.   ∎

Consequently, we get for the given model:

COROLLARY 6.55
$[\![\varphi]\!]_{\mathcal{M}} \in \mathcal{B}(S)$ *for all state formulas* φ, *and* $[\![\psi]\!]_{\mathcal{M}} \in \mathcal{B}((S \times \mathbb{R}_+)^\infty)$ *for all path formulas* ψ.

Of course it is important to know that the sets under consideration are Borel, for otherwise the corresponding sets are not in the range of the corresponding probability, and one cannot compute probabilities like $M_\infty(s)(\{\tau \mid \langle s, \tau \rangle \models \psi\})$.

6.5.3.2 Fixed point properties

The μ-operator plays a special rôle: intuitively, it models the smallest fixed point. This is true here as well:

PROPOSITION 6.56
$[\![\mu P.\psi]\!]_{\mathcal{M}}$ *is the smallest fixed point of* $R \mapsto [\![\psi]\!]_{\mathcal{M}[P \backslash R]}$.

For establishing Proposition 6.56 we need some auxiliary considerations.

DEFINITION 6.57 *Let P be a path variable, ψ be a path formula. Then*

a. ψ *is said to be P-even iff each free occurrence of P lies within an even number of negations.*

b. ψ *is said to be P-odd iff each free occurrence of P lies within an odd number of negations.*

Recall that the variable P bound by the fixed point operator $\mu P.\psi$ is required to be in the range of an even number of negations, so that ψ then should be P-even.

In a similar way, we describe the effect of a substitution, looking for growing or shrinking sets.

DEFINITION 6.58 *Let P be a path variable, ψ be a path formula. Then*

a. ψ *is called P-monotone iff for $R \mapsto [\![\psi]\!]_{\mathcal{M}_0[P \backslash R]}$ is a monotone map for each model $\mathcal{M}_0$ based on $\mathcal{M}$.*

b. ψ *is called P-antitone iff for $R \mapsto [\![\psi]\!]_{\mathcal{M}_0[P \backslash R]}$ is an antitone map for each model $\mathcal{M}_0$ based on $\mathcal{M}$.*

The map $R \mapsto [\![\psi]\!]_{\mathcal{M}_0[P \backslash R]}$ is defined from $\mathcal{B}((S \times \mathbb{R}_+)^\infty)$ into itself.

Now define $\mathfrak{PF}_0$ as the set of all path formulas ψ, such that for all path variables P the following holds:

1. if ψ is P-even, then ψ is P-monotone,

2. if ψ is P-odd, then ψ is P-antitone.

LEMMA 6.59

The set $\mathfrak{PF}_0$ has the following properties:

a. If P is a path variable, then $P \in \mathfrak{PF}_0$.

b. $\mathfrak{PF}_0$ is closed under negation, conjunction and under the next operator.

c. $\mathfrak{PF}_0$ is closed under the until operator for state variables: if φ_1, φ_2 are state variables, and I is a rational interval, then $\varphi_1 \, \mathcal{U}^I \, \varphi_2 \in \mathfrak{PF}_0$.

PROOF   1. The assertion is evident for path variables, and for the negation.

2. Assume that $\psi_1, \psi_2 \in \mathfrak{PF}_0$ such that $\psi := \psi_1 \wedge \psi_2$ is P-even, then both ψ_1 and ψ_2 are P-even (the conjunction serving as a demarcation line for negations), so $\psi \in \mathfrak{PF}_0$; similarly we show that $\psi \in \mathfrak{PF}_0$, provided ψ is P-odd. Thus $\mathfrak{PF}_0$ is closed under conjunction. The argumentation both for the next operator and for the until operator is the same.   ▯

The discussion for the μ-operator needs a simple case analysis, so it is separated.

LEMMA 6.60

Let $\psi' \in \mathfrak{PF}_0$, and assume that Q is a path variable. Then $\mu Q.\psi' \in \mathfrak{PF}_0$.

PROOF   Put $\psi := \mu Q.\psi'$. Assume that the path variable P is different from Q, then ψ and ψ' share the characteristic of being P-even or P-odd, so the assertion follows directly.

Now consider the case that P equals Q, then we see that

$$[\![\psi']\!]_{(\mathcal{M}_0[Q \backslash R_1])[Q \backslash R_2]} = [\![\psi']\!]_{\mathcal{M}_0[Q \backslash R_2]}.$$

But this means that $R \mapsto [\![\psi]\!]_{\mathcal{M}_0[Q \backslash R]}$ is independent of R, hence is monotone as well as antitone.   ▯

We are poised for giving a proof to Proposition 6.56.

PROOF   (of Proposition 6.56) From the previous lemmata we infer that $\mathfrak{PF}_0$ equals the set of all path formulas. Since the variable P bound by the μ-operator is in the range of an even number of negations, we know that ψ is P-monotone. Since all operations in the logic are finitary, we infer that

$R \mapsto [\![\psi]\!]_{\mathcal{M}[P \setminus R]}$ is $\cup$-continuous. Thus the assertion follows from the classical Knaster-Tarski Fixed Point Theorem. $\quad\square$

A little speculation. This fixed point property will play no rôle in the discussions and constructions to follow. It has been discussed to provide a (traditionally oriented) motivation for the construction of the semantics for the μ-operator. Other, similar definitions for the semantics are conceivable without altering the probabilistic properties of the semantics with the provision that it operates within the realm of measurable operations. This is a brief list:

Intersection $R_0 := [\![\psi]\!]_{\mathcal{M}[P \setminus S]}$, and continuing inductively, define R_{i+1} as $[\![\psi]\!]_{\mathcal{M}[P \setminus R_i]}$. The sequence $(R_n)_{n \in \mathbb{N}}$ may or may not be decreasing. Put $[\![\mu P.\psi]\!]_{\mathcal{M}} := \bigcap_{j \geq 0} R_j$.

Lower limit Define the sequence $(R_n)_{n \in \mathbb{N}}$ as above, and put $[\![\mu P.\psi]\!]_{\mathcal{M}} := \bigcup_{i \geq 0} \bigcap_{j \geq i} R_j$.

Upper limit Put $[\![\mu P.\psi]\!]_{\mathcal{M}} := \bigcap_{i \geq 0} \bigcup_{j \geq i} R_j$.

Note that both the lower limit and the upper limit could be defined with an arbitrary Borel set S^* as the starting point for defining R_0. Since we will be working with sub-σ-algebras, S^* could be taken from one of them as well and deliver a member of the same σ-algebra as the set of states in which the formula is valid, as the proofs below indicate. But we do not want to dwell upon this marginal point.

Interpreting R_i as the set of states for which ψ is valid after i process steps, the *intersection semantics* would of course give the largest fixed point. The *lower limit semantics* would describe those states in which ψ eventually holds, whereas the *upper limit semantics* would describe those states in which ψ holds infinitely often. Given a satisfactory interpretation of these semantics, it is only important from a probabilistic point of view that the operations performed are those of a σ-algebra.

Returning. But let us return to the interpretation of the μ-operator as the smallest fixed point given in Section 6.5.3.1, which will be the one assumed for the rest of the paper.

We will show now that theories are invariant under model morphisms, specifically we will prove:

PROPOSITION 6.61
*Let $\mathcal{M}$ and $\mathcal{M}'$ be models for μ**CSL**, and assume that $\Phi : \mathcal{M} \to \mathcal{M}'$ is a morphism. Then we have*

a. $[\![\varphi]\!]_{\mathcal{M}} = \Phi^{-1}[[\![\varphi]\!]_{\mathcal{M}'}]$ *for all state formulas φ.*

b. $[\![\psi]\!]_{\mathcal{M}} = \Phi_\infty^{-1} [[\![\psi]\!]_{\mathcal{M}'}]$ *for all state formulas* ψ.

This will be established through a series of auxiliary lemmata. To make the notation a bit less heavy, denote by τ and τ' generic elements of $\mathbb{R}_+ \times (S \times \mathbb{R}_+)^\infty$ resp. $\mathbb{R}_+ \times (S' \times \mathbb{R}_+)^\infty$, s and s' are typical states in S and S', respectively. We assume unless further notice that $\mathcal{M}$ and $\mathcal{M}'$ are models, and that $\Phi : \mathcal{M} \to \mathcal{M}'$ is a morphism.

We start with an auxiliary statement that will come in handy when manipulating probabilities involving Borel subsets of paths.

LEMMA 6.62
Let $B' \in \mathcal{B}((S' \times \mathbb{R}_+)^\infty)$, then

$$M_\infty(s)(\{\tau \mid \langle s, \tau \rangle \in \Phi_\infty^{-1} [B']\}) = M_\infty'(\Phi(s))(\{\tau' \mid \langle \Phi(s), \tau' \rangle \in B'\}).$$

PROOF   An easy calculation yields

$$\langle s, \tau \rangle \in \Phi_\infty^{-1} [B'] \Leftrightarrow \tau \in (id_{\mathbb{R}_+} \times \Phi_\infty)^{-1} \left[B'_{\Phi(s)}\right].$$

Consequently, because $(\Phi, \Phi_\infty) : M_\infty \to M_\infty'$ is a morphism between the stochastic relations,

$$\begin{aligned}
M_\infty(s) \left(\{\tau \mid \langle s, \tau \rangle \in \Phi_\infty^{-1} [B']\}\right) &= M_\infty(s)((id_{\mathbb{R}_+} \times \Phi_\infty)^{-1} \left[B'_{\Phi(s)}\right]) \\
&= M_\infty'(\Phi(s))(B'_{\Phi(s)}) \\
&= M_\infty'(\Phi(s))(\{\tau' \mid \langle \Phi(s), \tau' \rangle \in B'\}).
\end{aligned}$$

This establishes the desired equality. $\square$

LEMMA 6.63
Assume that ψ is a path formula with $[\![\psi]\!]_{\mathcal{M}} = \Phi_\infty^{-1} [[\![\psi]\!]_{\mathcal{M}'}]$. Then

a. $[\![\mathcal{P}_{\bowtie p}(\psi)]\!]_{\mathcal{M}} = \Phi^{-1} [[\![\mathcal{P}_{\bowtie p}(\psi)]\!]_{\mathcal{M}'}]$,

b. $[\![\mathcal{X}^I \psi]\!]_{\mathcal{M}} = \Phi^{-1} [[\![\mathcal{X}^I \psi]\!]_{\mathcal{M}'}]$.

PROOF   1. Applying Lemma 6.62 to the definition of the semantics of the path quantifier, we obtain

$$\begin{aligned}
[\![\mathcal{P}_{\bowtie p}(\psi)]\!]_{\mathcal{M}} &= \{s \mid M_\infty(s)(\{\tau \mid \langle s, \tau \rangle \in [\![\psi]\!]_{\mathcal{M}}\}) \bowtie p\} \\
&= \{s \mid M_\infty(s)(\{\tau \mid \langle s, \tau \rangle \in \Phi_\infty^{-1} [[\![\psi]\!]_{\mathcal{M}'}]\}) \bowtie p\} \\
&= \{s \mid M_\infty'(\Phi(s))(([\![\psi]\!]_{\mathcal{M}'})_{\Phi(s)}) \bowtie p\} \\
&= \Phi^{-1} [\{s' \mid M_\infty'(s')(\{\tau' \mid \langle s', \tau' \rangle \in [\![\psi]\!]_{\mathcal{M}'}\}) \bowtie p\}] \\
&= \Phi^{-1} [[\![\mathcal{P}_{\bowtie p}(\psi)]\!]_{\mathcal{M}'}].
\end{aligned}$$

2. The assertion is obvious for the next operator. ⊔

The first part of this proof is like a pattern for the proofs to the following statements.

LEMMA 6.64
  Assume that $[\![\varphi_i]\!]_{\mathcal{M}} = \Phi^{-1}[[\varphi_i]\!]_{\mathcal{M}'}]$ ($i = 1, 2$) for the state formulas φ_1, φ_2. Then

a. $[\![\mathcal{S}_{\ltimes p}(\varphi_1)]\!]_{\mathcal{M}} = \Phi^{-1}[[\mathcal{S}_{\ltimes p}(\varphi_1)]\!]_{\mathcal{M}'}]$,

b. $[\![\varphi_1 \mathcal{U}^I \varphi_2]\!]_{\mathcal{M}} = \Phi^{-1}[[\varphi_1 \mathcal{U}^I \varphi_2]\!]_{\mathcal{M}'}]$.

PROOF The assertion for the steady state operator follows from the observation

$$M_{\infty}(s)(\{\tau \mid \langle s, \tau\rangle @t \in [\![\varphi]\!]_{\mathcal{M}}\}) = M_{\infty}'(\Phi(s))(\{\tau' \mid \langle\Phi(s), \tau'\rangle @t \in [\![\varphi]\!]_{\mathcal{M}'}\})$$

which is established using Lemma 6.62. The assertion for the until operator follows from the observation that $\sigma @t \in [\![\varphi_i]\!]_{\mathcal{M}}$ iff $\Phi_{\infty}(\sigma)@t \in [\![\varphi_i]\!]_{\mathcal{M}'}$. ⊔

LEMMA 6.65
  Assume that ψ is a path formula with $[\![\psi]\!]_{\mathcal{M}} = \Phi_{\infty}^{-1}[[\psi]\!]_{\mathcal{M}'}]$. Then

$$[\![\mu P.\psi]\!]_{\mathcal{M}} = \Phi^{-1}[[\mu P.\psi]\!]_{\mathcal{M}'}].$$

PROOF 0. We observe first that $[\![\psi]\!]_{\mathcal{M}} = \Phi_{\infty}^{-1}[[\psi]\!]_{\mathcal{M}'}]$ entails the equality $[\![\psi]\!]_{\mathcal{M}[P \setminus \Phi_{\infty}^{-1}[R']]} = \Phi_{\infty}^{-1}[[\psi]\!]_{\mathcal{M}'[P \setminus R']}]$ for the Borel set $R' \subseteq (S' \times \mathbb{R}_+)^{\infty}$.
  1. Put $R_0 := [\![\psi]\!]_{\mathcal{M}[P \setminus \emptyset]}$ and $R_0' := [\![\psi]\!]_{\mathcal{M}'[P \setminus \emptyset]}$, then the assumption and part 0. together imply $R_0 = \Phi_{\infty}^{-1}[R_0']$. Arguing inductively and assuming that we have shown $R_i = \Phi_{\infty}^{-1}[R_i']$, we see

$$R_{i+1} := [\![\psi]\!]_{\mathcal{M}[P \setminus R_i]} = [\![\psi]\!]_{\mathcal{M}[P \setminus \Phi_{\infty}^{-1}[R_i']]} = \Phi_{\infty}^{-1}[[\psi]\!]_{\mathcal{M}'[P \setminus R_i']}] = \Phi_{\infty}^{-1}[R_{i+1}'].$$

This establishes the claim, then:

$$[\![\mu P.\psi]\!]_{\mathcal{M}} = \bigcup_{i \geq 0} R_i = \bigcup_{i \geq 0} \Phi_{\infty}^{-1}[R_i'] = \Phi_{\infty}^{-1}\left[\bigcup_{i \geq 0} R_i'\right] = \Phi_{\infty}^{-1}[[\mu P.\psi]\!]_{\mathcal{M}'}].$$

⊔

We are now in a position to prove that a morphism preserves the sets of validity.

PROOF   (of Proposition 6.61) The proof is done by induction on the structure of the formulas. The cases of atomic propositions, of state and of path variables are covered by the properties of morphisms; the case of Boolean connectives are obvious. Structured formulas are dealt with by Lemmas 6.63 - 6.65. $\square$

6.5.4   Congruences

We will define two equivalence relations on states respectively on paths. These relations will be studied carefully, since they will be fundamental for discussing bisimilarity and behavioral as well as logical equivalence later on.

Fix the model $\mathcal{M} = (S, M, \mathcal{I}, V)$ and define

$$s \; \zeta_{\mathcal{M}} \; s' \Leftrightarrow Th_{\mathcal{M}}(s) = Th_{\mathcal{M}}(s'),$$
$$\sigma \; \omega_{\mathcal{M}} \; \sigma' \Leftrightarrow Th_{\mathcal{M}}(\sigma) = Th_{\mathcal{M}}(\sigma').$$

Then both $\zeta_{\mathcal{M}}$ and $\omega_{\mathcal{M}}$ are smooth equivalence relations on S resp. $(S \times \mathbb{R}_+)^{\infty}$. This is so since there are only countably many formulas, and because we have

$$s \; \zeta_{\mathcal{M}} \; s' \Leftrightarrow [\mathcal{M}, s \models \varphi \Leftrightarrow \mathcal{M}, s' \models \varphi] \text{ for all state formulas } \varphi,$$
$$\Leftrightarrow [s \in [\![\varphi]\!]_{\mathcal{M}} \Leftrightarrow s' \in [\![\varphi]\!]_{\mathcal{M}}] \text{ for all state formulas } \varphi.$$

From this it is clear that the countable set $\{[\![\varphi]\!]_{\mathcal{M}} \mid \varphi \text{ is a state formula}\}$ determines the relation $\zeta_{\mathcal{M}}$, and that

$$\mathcal{INV}\,(\mathcal{B}(S), \zeta_{\mathcal{M}}) = \sigma(\{[\![\varphi]\!]_{\mathcal{M}} \mid \varphi \text{ is a state formula}\}).$$

In a similar way we see that $\omega_{\mathcal{M}}$ is smooth, and that

$$\mathcal{INV}\,(\mathcal{B}((S \times \mathbb{R}_+)^{\infty}), \omega_{\mathcal{M}}) = \sigma(\{[\![\psi]\!]_{\mathcal{M}} \mid \psi \text{ is a path formula}\})$$

holds as well. These two relations will be studied now in some detail. It will turn out that the relationship of $\zeta_{\mathcal{M}}$ and $\omega_{\mathcal{M}}$ is closer than meets the eye.

First we show that they form essentially a congruence for M_{∞}.

PROPOSITION 6.66
   The pair $(\zeta_{\mathcal{M}}, \Delta_{\mathbb{R}_+} \times \omega_{\mathcal{M}})$ of smooth equivalence relations is a congruence for $M_{\infty} : S \rightsquigarrow \mathbb{R}_+ \times (S \times \mathbb{R}_+)^{\infty}$.

PROOF   0. We show first that

$$M_{\infty}(s_1)(I \times [\![\psi]\!]_{\mathcal{M}}) = M_{\infty}(s_2)(I \times [\![\psi]\!]_{\mathcal{M}})$$

for rational intervals I, and for path formulas ψ, whenever $s_1 \; \zeta_{\mathcal{M}} \; s_2$. Based on this, we show that the equality in question holds for all invariant sets.

1. Fix a rational interval I and a path formula ψ, and assume $s_1 \zeta_{\mathcal{M}} s_2$. Thus we have for each rational number p: $s_1 \in [\![\mathcal{P}_{\leq p}(\psi)]\!]_{\mathcal{M}} \Leftrightarrow s_2 \in [\![\mathcal{P}_{\leq p}(\psi)]\!]_{\mathcal{M}}$, equivalently, $M_\infty(s_1)(I \times [\![\psi]\!]_{\mathcal{M}}) \leq p \Leftrightarrow M_\infty(s_2)(I \times [\![\psi]\!]_{\mathcal{M}}) \leq p$. This means that these probabilities are equal.

2. Now consider as in the proof of Proposition 6.52 the set

$$\mathcal{D} := \{D \in \mathcal{B}(\mathbb{R}_+) \otimes \mathcal{INV}\,(\mathcal{B}((S \times \mathbb{R}_+)^\infty), \omega_{\mathcal{M}}) \mid M_\infty(s_1)(D) = M_\infty(s_2)(D)\}.$$

Here still $s_1 \zeta_{\mathcal{M}} s_2$ is assumed. The first part of this proof shows that $\mathcal{D}$ contains the set

$$\{I \times [\![\psi]\!]_{\mathcal{M}} \mid I \text{ is a rational interval}, \psi \text{ is a path formula}\},$$

which is a generator for $\mathcal{B}(\mathbb{R}_+) \otimes \mathcal{INV}\,(\mathcal{B}((S \times \mathbb{R}_+)^\infty), \omega_{\mathcal{M}})$. By the π-λ-Theorem 1.1, $\mathcal{D}$ equals the latter σ-algebra. This implies the assertion, since we see from Lemma 5.14 that

$$\mathcal{INV}\,(\mathcal{B}(\mathbb{R}_+ \times (S \times \mathbb{R}_+)^\infty), \Delta_{\mathbb{R}_+} \times \omega_{\mathcal{M}}) =$$
$$\mathcal{INV}\,(\mathcal{B}(\mathbb{R}_+), \Delta_{\mathbb{R}_+}) \otimes \mathcal{INV}\,(\mathcal{B}((S \times \mathbb{R}_+)^\infty), \omega_{\mathcal{M}}) =$$
$$\mathcal{B}(\mathbb{R}_+) \otimes \mathcal{INV}\,(\mathcal{B}((S \times \mathbb{R}_+)^\infty), \omega_{\mathcal{M}}).$$

$\square$

6.5.4.1 Relating the relations

We will show that two infinite paths are $\omega_{\mathcal{M}}$-equivalent iff their state components are $\zeta_{\mathcal{M}}$-equivalent (and the timing information is identical). This will support the investigation of logical equivalence later on, mainly since the information available for states is easier to handle than the one for infinite paths. It turns out, however, that this equality is not easily obtained and requires a careful look at the invariant Borel sets.

One inclusion is rather immediate.

LEMMA 6.67
$\omega_{\mathcal{M}} \subseteq (\zeta_{\mathcal{M}} \times \Delta_{\mathbb{R}_+})^\infty$.

PROOF 0. Fix infinite paths σ and σ' with $\sigma \, \omega_{\mathcal{M}} \, \sigma'$, then we have to establish that both $\delta(\sigma, i) = \delta(\sigma', i)$ and $\sigma[i] \zeta_{\mathcal{M}} \sigma'[i]$ hold for each $i \geq 0$. It is obvious that the timing information for σ and σ' coincides, so we have to take care of the state components. Define inductively for rational intervals $I_1, I_2, \ldots$, and for the path formula ψ the path formula

$$X_1^{I_1}\psi := \mathcal{X}^{I_1}\,\psi,$$
$$X_{n+1}^{I_1,\ldots I_{n+1}}\psi := \mathcal{X}^{I_{n+1}}\left(X_n^{I_1,\ldots,I_n}\psi\right).$$

1. Since $\sigma \, \omega_{\mathcal{M}} \, \sigma'$ we know that for an arbitrary state formula φ and for an arbitrary rational time t

$$\mathcal{M}, \sigma \models \varphi \, \mathcal{U}^{[0,t]} \, \top \Leftrightarrow \mathcal{M}, \sigma' \models \varphi \, \mathcal{U}^{[0,t]} \, \top$$

thus $\sigma[0] \in [\![\varphi]\!]_{\mathcal{M}}$ iff $\sigma'[0] \in [\![\varphi]\!]_{\mathcal{M}}$, hence $\sigma[0] \, \zeta_{\mathcal{M}} \, \sigma'[0]$.

2. Now let $i > 0$, then we have for arbitrary rational times $t > 0$ and for an arbitrary state formula φ the following equivalences

$$\mathcal{M}, \sigma[i] \models \varphi \Leftrightarrow \mathcal{M}, \sigma \models X_i^{\mathbb{R}_+^i}(\varphi \, \mathcal{U}^{[0,t]} \, \top)$$
$$\Leftrightarrow \mathcal{M}, \sigma' \models X_i^{\mathbb{R}_+^i}(\varphi \, \mathcal{U}^{[0,t]} \, \top)$$
$$\Leftrightarrow \mathcal{M}, \sigma'[i] \models \varphi.$$

Consequently, $\sigma[i] \, \zeta_{\mathcal{M}} \, \sigma'[i]$. This establishes the claim.   ⬚

This has two interesting consequences:

COROLLARY 6.68
We have

$$\mathcal{INV}\left(\mathcal{B}((S \times \mathbb{R}_+)^\infty), (\zeta_{\mathcal{M}} \times \Delta_{\mathbb{R}_+})^\infty\right) \subseteq \mathcal{INV}\left(\mathcal{B}((S \times \mathbb{R}_+)^\infty), \omega_{\mathcal{M}}\right)$$

and

$$M_\infty : (S, \mathcal{INV}(\mathcal{B}(S), \zeta_{\mathcal{M}})) \rightsquigarrow$$
$$(\mathbb{R}_+ \times (S \times \mathbb{R}_+)^\infty, \mathcal{INV}\left(\mathcal{B}(\mathbb{R}_+ \times (S \times \mathbb{R}_+)^\infty), (\Delta_{\mathbb{R}_+} \times \zeta_{\mathcal{M}})^\infty\right))$$

is a stochastic relation.

PROOF   The first assertion follows from the observation that

$$\rho \mapsto \mathcal{INV}\left(\mathcal{B}((S \times \mathbb{R}_+)^\infty), \rho\right)$$

is antitone. The second one follows from the first: since $(\zeta_{\mathcal{M}}, \Delta_{\mathbb{R}_+} \times \omega_{\mathcal{M}})$ is a congruence for M_∞ by Proposition 6.66, we know from Lemma 5.30 that

$$M_\infty : (S, \mathcal{INV}(\mathcal{B}(S), \zeta_{\mathcal{M}})) \rightsquigarrow$$
$$(\mathbb{R}_+ \times (S \times \mathbb{R}_+)^\infty, \mathcal{INV}\left(\mathcal{B}(\mathbb{R}_+ \times (S \times \mathbb{R}_+)^\infty), \Delta_{\mathbb{R}_+} \times \omega_{\mathcal{M}}\right))$$

is a stochastic relation. Consequently it is also a stochastic relation when we choose a smaller σ-algebra on the target space.   ⬚

PROPOSITION 6.69
$\omega_{\mathcal{M}} = (\zeta_{\mathcal{M}} \times \Delta_{\mathbb{R}_+})^\infty.$

The proof of this statement is again preceded by a series of lemmata.

DEFINITION 6.70   *A model $\mathcal{M}[P_1 \backslash R_1, \ldots, P_n \backslash R_n]$ is called $\mathcal{M}$-invariant iff*

a. *$P_1, \ldots, P_n$ are mutually distinct path variables,*

b. *$R_i \in \mathcal{INV}\left(\mathcal{B}((S \times \mathbb{R}_+)^\infty), (\zeta_\mathcal{M} \times \Delta_{\mathbb{R}_+})^\infty\right)$ for $1 \le i \le n$.*

Fix for the auxiliary statements to come $\mathcal{K}$ as an $\mathcal{M}$-invariant model.

LEMMA 6.71
   $[\![\mathcal{P}_{\ltimes p}(\psi)]\!]_\mathcal{K} \in \mathcal{INV}\left(\mathcal{B}(S), \zeta_\mathcal{M}\right)$ *provided* $[\![\psi]\!]_\mathcal{K} \in \mathcal{INV}\left(\mathcal{B}((S \times \mathbb{R}_+)^\infty), \omega_\mathcal{M}\right)$
holds for the path formula ψ.

PROOF   By Corollary 2.10, the map $s \mapsto M_\infty(s)(\{\tau \mid \langle s, \tau \rangle \in [\![\psi]\!]_\mathcal{K}\})$ is $\mathcal{INV}\left(\mathcal{B}(S), \zeta_\mathcal{M}\right)$-measurable, because

$$M_\infty : (S, \mathcal{INV}\left(\mathcal{B}(S)\right), \zeta_\mathcal{M}) \rightsquigarrow$$
$$(\mathbb{R}_+ \times (S \times \mathbb{R}_+)^\infty, \mathcal{INV}\left(\mathcal{B}(\mathbb{R}_+ \times (S \times \mathbb{R}_+)^\infty)\right), \Delta_{\mathbb{R}_+} \times \omega_\mathcal{M})$$

is a stochastic relation by Lemma 5.30. This implies the assertion, since

$$[\![\mathcal{P}_{\ltimes p}(\psi)]\!]_\mathcal{K} = \{s \mid M_\infty(s)(\{\tau \mid \langle s, \tau \rangle \in [\![\psi]\!]_\mathcal{K}\}) \ltimes p\}.$$

$\square$

LEMMA 6.72
   $[\![\mathcal{S}_{\ltimes p}(\varphi)]\!]_\mathcal{K} \in \mathcal{INV}\left(\mathcal{B}(S), \zeta_\mathcal{M}\right)$, *provided* $[\![\varphi]\!]_\mathcal{K} \in \mathcal{INV}\left(\mathcal{B}(S), \zeta_\mathcal{M}\right)$ *holds for the state formula φ.*

PROOF   0. We show first that

$$B_A := \{\tau \mid \langle s', \tau \rangle @t \in A\}$$

defines a member of $\mathcal{INV}\left(\mathcal{B}(\mathbb{R}_+ \times (S \times \mathbb{R}_+)^\infty), (\Delta_{\mathbb{R}_+} \times \zeta_\mathcal{M})^\infty\right)$ for any $\zeta_\mathcal{M}$-invariant Borel set $A \in \mathcal{INV}\left(\mathcal{B}(S), \zeta_\mathcal{M}\right)$, for every time t, and for any $s' \in S$. In fact, let $\langle s', \tau \rangle @t = \tau[k] \in A$, and assume that $\tau \ (\Delta_{\mathbb{R}_+} \times \zeta_\mathcal{M})^\infty \ \tau'$, then $\langle s', \tau' \rangle @t = \tau'[k]$. Consequently, B_A is $(\Delta_{\mathbb{R}_+} \times \zeta_\mathcal{M})^\infty$-invariant; it is a Borel set by Lemma 6.21.

1. We infer from Corollary 6.68 that for every $A \in \mathcal{INV}\left(\mathcal{B}(S), \zeta_{\mathcal{M}}\right)$ and for every time t the set $\{s \mid M_\infty(s)(\{\tau \mid \langle s, \tau \rangle @ t \in A\}) \ltimes p\}$ is a member of $\mathcal{INV}\left(\mathcal{B}(S), \zeta_{\mathcal{M}}\right)$. Consequently, the maps

$$f_A : s \mapsto \liminf_{t \to \infty} M_\infty(s)(\{\tau \mid \langle s, \tau \rangle @ t \in A\})$$

$$g_A : s \mapsto \limsup_{t \to \infty} M_\infty(s)(\{\tau \mid \langle s, \tau \rangle @ t \in A\})$$

both define $\mathcal{INV}\left(\mathcal{B}(S), \zeta_{\mathcal{M}}\right)$-measurable maps (see Proposition 6.23). This implies the assertion upon setting $A = [\![\varphi]\!]_\mathcal{K}$, since

$$[\![\mathcal{S}_{\ltimes p}(\varphi)]\!]_\mathcal{K} = \{s \mid f_{[\varphi]_\mathcal{K}}(s) = g_{[\varphi]_\mathcal{K}}(s)\} \cap \{s \mid f_{[\varphi]_\mathcal{K}}(s) \ltimes p\}$$

is the intersection of $\zeta_{\mathcal{M}}$-invariant Borel sets.   ∎

LEMMA 6.73

If $[\![\varphi_1]\!]_\mathcal{K}, [\![\varphi_2]\!]_\mathcal{K} \in \mathcal{INV}\left(\mathcal{B}(S), \zeta_{\mathcal{M}}\right)$ for the state variables φ_1, φ_2, then

$$[\![\varphi_1 \, \mathcal{U}^I \, \varphi_2]\!]_\mathcal{K} \in \mathcal{INV}\left(\mathcal{B}((S \times \mathbb{R}_+)^\infty), (\zeta_{\mathcal{M}} \times \Delta_{\mathbb{R}_+})^\infty\right).$$

PROOF   0. We show exactly as in the proof of Lemma 6.72 that

$$\{\sigma \mid \sigma @ t \in A\} \in \mathcal{INV}\left(\mathcal{B}((S \times \mathbb{R}_+)^\infty), (\zeta_\times \Delta_{\mathbb{R}_+})^\infty\right),$$

provided $A \in \mathcal{INV}\left(\mathcal{B}(S), \zeta_{\mathcal{M}}\right)$.

1. Because

$$[\![\varphi_1 \, \mathcal{U}^I \, \varphi_2]\!]_\mathcal{K} = \bigcup_{t \in \mathbb{Q} \cap I} \left(\{\sigma \mid \sigma @ t \in [\![\varphi_2]\!]_\mathcal{K}\} \cap \bigcap_{t' \in \mathbb{Q} \cap [0,t]} \{\sigma \mid \sigma @ t' \in [\![\varphi_1]\!]_\mathcal{K}\} \right),$$

we infer from Lemma 6.25 the assertion.   ∎

LEMMA 6.74

Assume that $[\![\psi]\!]_\mathcal{L} \in \mathcal{INV}\left(\mathcal{B}((S \times \mathbb{R}_+)^\infty), (\zeta_{\mathcal{M}} \times \Delta_{\mathbb{R}_+})^\infty\right)$ for the path formula ψ and for every $\mathcal{M}$-invariant model $\mathcal{L}$. Then

$$[\![\mu P.\psi]\!]_\mathcal{K} \in \mathcal{INV}\left(\mathcal{B}((S \times \mathbb{R}_+)^\infty), (\zeta_{\mathcal{M}} \times \Delta_{\mathbb{R}_+})^\infty\right).$$

PROOF   Define inductively

$$\mathcal{K}_0 := \mathcal{K}[P \backslash \emptyset], R_0 := [\![\psi]\!]_{\mathcal{K}_0}, \mathcal{K}_{i+1} := \mathcal{K}[P \backslash R_i], R_{i+1} := [\![\psi]\!]_{\mathcal{K}_{i+1}}.$$

Then an easy inductive argument shows that

1. $\mathcal{K}_i$ is an $\mathcal{M}$-invariant model,

2. $R_i \in \mathcal{INV} \left(\mathcal{B}((S \times \mathbb{R}_+)^\infty), (\zeta_\mathcal{M} \times \Delta_{\mathbb{R}_+})^\infty \right)$.

Since $[\![\mu P.\psi]\!]_\mathcal{K} = \bigcup_{i \geq 0} R_i$, the assertion is established.   ⬚

We have now enough details for establishing that the equivalence relation $\omega_\mathcal{M}$ coincides with $(\zeta_\mathcal{M} \times \Delta_{\mathbb{R}_+})^\infty$.

PROOF   (of Proposition 6.69) By Corollary 5.6 it is enough to show that

$$\mathcal{INV} \left(\mathcal{B}((S \times \mathbb{R}_+)^\infty), \omega_\mathcal{M} \right) = \mathcal{INV} \left(\mathcal{B}((S \times \mathbb{R}_+)^\infty), (\zeta_\mathcal{M} \times \Delta_{\mathbb{R}_+})^\infty \right)$$

are identical; in view of Corollary 6.68 it is enough to show that

$$\mathcal{INV} \left(\mathcal{B}((S \times \mathbb{R}_+)^\infty), \omega_\mathcal{M} \right) \subseteq \mathcal{INV} \left(\mathcal{B}((S \times \mathbb{R}_+)^\infty), (\zeta_\mathcal{M} \times \Delta_{\mathbb{R}_+})^\infty \right)$$

holds. Since the σ-algebra $\mathcal{INV} \left(\mathcal{B}((S \times \mathbb{R}_+)^\infty), \omega_\mathcal{M} \right)$ is generated by the sets $[\![\psi]\!]_\mathcal{M}$ for path variables ψ, it is enough to show that these sets are $(\zeta_\mathcal{M} \times \Delta_{\mathbb{R}_+})^\infty$-invariant. But this is now immediate from Lemma 6.71–Lemma 6.74.
⬚

The consequence of this equality is that we may check the equivalence of paths locally, i.e., through the equivalence of states. This represents a considerable reduction in complexity, because the equivalence relation $\omega_\mathcal{M}$ that operates on infinite paths is uniquely determined through the relation $\zeta_\mathcal{M}$ which in turn operates on states. This will be reflected in the representation of the equivalence classes, as we will see in Corollary 6.75. The reduction makes checking some properties of course much easier. It has also technical advantages when it comes to check the semi-pullback of two models, as we will in the next section.

We give a first consequence of this equality in terms of a representation of the equivalence classes.

COROLLARY 6.75
Given $\sigma \in (S \times \mathbb{R}_+)^\infty$, the $\omega_\mathcal{M}$-class of $\sigma = s_0 \xrightarrow{t_0} s_1 \xrightarrow{t_1} \ldots$ can be represented as

$$[\sigma]_{\omega_\mathcal{M}} = \prod_{j \geq 0} \left([s_j]_{\zeta_\mathcal{M}} \times \{t_j\} \right).$$

Moreover, we have Borel isomorphisms between these analytic spaces

$$(S \times \mathbb{R}_+)^\infty / \omega_\mathcal{M} \cong \left((S \times \mathbb{R}_+)/(\zeta_\mathcal{M} \times \Delta_{\mathbb{R}_+}) \right)^\infty \cong \left((S/\zeta_\mathcal{M}) \times \mathbb{R}_+ \right)^\infty.$$

We are now in a position to define the logical equivalence of models, and to relate it to spans of morphisms.

6.5.5   Logical Equivalence and Bisimilarity

Logical equivalence between two models says roughly that, given a state in one model, there exists a state in the other model so that in both exactly the same formulas are valid, similarly for paths. This equivalence is modeled after the corresponding equivalence that has been investigated in modal logics; see Definition 6.14. We have seen that it is closely tied to the notion of bisimulation through the Hennessy-Milner Theorem, both for the classical case of nondeterministic Kripke models (see the discussion in (Blackburn et al., 2001, Chapter 2.2), in particular Theorem 2.24), and for stochastic Kripke models in Theorem 6.19. The relationship of this equivalence to bisimulations will be discussed now.

Let $\mathcal{M} = (S, M, \mathcal{I}, V)$ and $\mathcal{N} = (S', N, \mathcal{J}, W)$ be models for μ**CSL**. We assume that $\mathcal{M}$ is *nondegenerate*, i.e., that there exists a state formula φ with $\emptyset \neq [\![\varphi]\!]_{\mathcal{M}} \neq S$. Being nondegenerate implies that the factor space $S/\zeta_{\mathcal{M}}$ is not trivial. Corollary 6.75 entails that there exists also a path formula ψ such that $\emptyset \neq [\![\psi]\!]_{\mathcal{M}} \neq (S \times \mathbb{R}_+)^{\infty}$.

6.5.5.1   Basic definitions

Define the models $\mathcal{M}$ and $\mathcal{N}$ as logical equivalent iff they accept exactly the same formulas. This is similar to logical equivalence for Kripke models; see the discussion in Section 6.2.3. In addition and contrast, however, it has to take two levels into account, since we are dealing here with state formulas and with path formulas, so that formulas may hold in states or on paths — this situation is familiar from model checking where one has this dichotomy as well.

DEFINITION 6.76   *The models $\mathcal{M}$ and $\mathcal{N}$ are called* logical equivalent *($\mathcal{M} \approx \mathcal{N}$) iff both*

$$\{Th_{\mathcal{M}}(s) \mid s \in S\} = \{Th_{\mathcal{N}}(s') \mid s' \in S'\}$$

and

$$\{Th_{\mathcal{M}}(\sigma) \mid \sigma \in (S \times \mathbb{R}_+)^{\infty}\} = \{Th_{\mathcal{N}}(\sigma') \mid \sigma' \in (S' \times \mathbb{R}_+)^{\infty}\}$$

hold.

Thus the models are logical equivalent iff these conditions are satisfied:

1. Given a state $s \in S$, there exists a state $s' \in S'$ such that $[\mathcal{M}, s \models \varphi \Leftrightarrow \mathcal{N}, s' \models \varphi]$ holds for all state formulas φ, and vice versa,

2. Given a path $\sigma \in (S \times \mathbb{R}_+)^{\infty}$, there exists a path $\sigma' \in (S' \times \mathbb{R}_+)^{\infty}$ such that $[\mathcal{M}, \sigma \models \psi \Leftrightarrow \mathcal{N}, \sigma' \models \psi]$ holds for all path formulas ψ, and vice versa.

If we can find a morphism between $\mathcal{M}$ and $\mathcal{N}$, then these models are equivalent (see Lemma 6.17, Lemma 5.23).

PROPOSITION 6.77
Let $\Phi : \mathcal{M} \to \mathcal{N}$ be a morphism. Then $\mathcal{M}$ and $\mathcal{N}$ are logical equivalent.

PROOF   Proposition 6.61 implies that $\mathcal{M}, s \models \varphi$ iff $\mathcal{N}, \Phi(s) \models \varphi$ for each state formula φ and each state $s \in S$, and that $\mathcal{M}, \sigma \models \psi$ iff $\mathcal{N}, \Phi_\infty(\sigma) \models \psi$ for each path formula ψ and each path $\sigma \in (S \times \mathbb{R}_+)^\infty$. Since both Φ and Φ_∞ are onto, the assertion follows. $\Box$

We will show that logical equivalent models are bisimilar. Bisimilarity is again introduced as a span of morphisms.

DEFINITION 6.78   *Let $\mathcal{M}$ and $\mathcal{N}$ be nondegenerate models for μCSL.*

a. $\mathcal{M}$ *and* $\mathcal{N}$ *are said to be* bisimilar *iff there exists a model $\mathcal{Q}$ for μCSL and morphisms*

$$\mathcal{M} \xleftarrow{\quad \Phi \quad} \mathcal{Q} \xrightarrow{\quad \Psi \quad} \mathcal{N}.$$

b. $\mathcal{M}$ *and* $\mathcal{N}$ *are said to be* behavioral equivalent *iff there exists a model $\mathcal{Q}$ for μCSL and morphisms*

$$\mathcal{M} \xrightarrow{\quad \Gamma \quad} \mathcal{R} \xleftarrow{\quad \Lambda \quad} \mathcal{N}.$$

It is clear that bisimilar models are logical equivalent, because this notion of equivalence is transitive; see Proposition 6.61. Now suppose model $\mathcal{M} = (S, M, \mathcal{I}, V)$ is bisimilar to model $\mathcal{N} = (S', N, \mathcal{J}, W)$ with mediating model $\mathcal{Q}$ over the state space S'' and the morphisms according to Definition 6.78. Then the condition on bisimilarity implies

1. $M(\Phi(s''))(I \times B) = N(\Psi(s''))(I \times B')$ for every state $s'' \in S''$, every rational interval I and all common events $B \in \mathcal{B}(S), B' \in \mathcal{B}(S')$ (thus every pair of events B, B' such that $\Phi^{-1}[B] = \Psi^{-1}[B']$, as the discussion following Definition 5.37 indicates). Consequently, the probability for $\mathcal{M}$ changing the state during interval I and entering a state in B from state $\Phi(s'')$ equals the probability for $\mathcal{N}$ to change the state during time interval I and entering a state in B' from state $\Psi(s'')$. This illustrates again the mediating work done through model $\mathcal{Q}$.

2. for a state $s'' \in S''$, $\Phi(s'')$ is a member of the valuation for a state variable Z in model $\mathcal{M}$ iff $\Psi(s'')$ is a member for this variable in model $\mathcal{N}$, and similar for path variables, and for atomic propositions.

The situation is a bit different with behavioral equivalence. By implication, the model in range of the cospan is based on a Polish space. Otherwise we could not always conclude that behavioral equivalent models are logical equivalent as well. This is so since computing the set of all states in which a given formula is valid requires the knowledge of a projective limit, and we did establish the existence of such a limit only for the case of Polish spaces, not for analytic ones. On the other hand, these cospans are constructed usually through factoring (see, e.g., Proposition 5.39 and Proposition 6.86 below), and the factor space of a Polish space is not always a Polish one. So we need to exercise some care. Specifically, the behavioral equivalence of the models means that, if $\Gamma(s) = \Lambda(s')$ for states $s \in S$, $s' \in S'$

1. then $M(s)(I \times \Gamma^{-1}[B'']) = N(s')(I \times \Lambda^{-1}[B''])$, whenever $B'' \in \mathcal{B}(S'')$ is a Borel set in S'', and I is a rational interval. Consequently, the probability for M changing the state during interval I and entering a state $s^{\bullet} \in S$ with $\Gamma(s^{\bullet}) \in B''$ from state s equals the probability for N to change the state during time interval I and entering a state $s^{\star} \in S'$ with $\Lambda(s^{\star}) \in B''$ from state s'.

2. s is a member for the valuation of a state variable Z in model M iff s' is a member for the valuation of a state variable Z in model N, similar for path variables, and for atomic propositions.

Returning to the general discussion, fix the models $M = (S, M, \mathcal{I}, V)$ and $N = (S', N, \mathcal{J}, W)$ such that $M \approx N$, hence both models are logical equivalent. Each model has the equivalence relations ζ_M and ω_M resp. ζ_N and ω_N associated with it, as defined in Section 6.5.4.

The stochastic relations M_{∞} and N_{∞} will be investigated with respect to bisimilarity first, and it will be shown first that they are bisimilar *as stochastic relations on Polish spaces*.

LEMMA 6.79
ζ_M and ζ_N spawn each other, so do ω_M and ω_N.

PROOF   0. We will show only that ζ_M spawns ζ_N; interchanging the rôles of M and N will show that ζ_N spawns ζ_M. The argumentation for ω_M and ω_N is nearly verbatim the same, so the reader is invited to fill in the details.

1.   Define for the state $s \in S$ the map $\Upsilon([s]_{\zeta_M}) := [s']_{\zeta_N}$, whenever $Th_M(s) = Th_N(s')$. Because $s_1 \, \zeta_M \, s_2$ iff $Th_M(s_1) = Th_M(s_2)$, and similar for N, the map is well defined. For the state formula φ its class $[\![\varphi]\!]_M$ can be represented as

$$\bigcup \{[s]_{\zeta_M} \mid M, s \models \varphi\},$$

thus it is readily verified that $\Upsilon_{[\varphi]_M} = [\![\varphi]\!]_N$. Consequently,

$$\{\Upsilon_{[\varphi]_M} \mid \varphi \text{ is a state formula}\}$$

is a generator of $\mathcal{INV}\,(\mathcal{B}(S'),\zeta_\mathcal{N})$. This generator is closed under intersections, since the conjunction of two state formulas is again one. ▯

We know that both $(\zeta_\mathcal{M},\Delta_{\mathbb{R}_+}\times\omega_\mathcal{M})$ and $(\zeta_\mathcal{N},\Delta_{\mathbb{R}_+}\times\omega_\mathcal{N})$ are congruences for the stochastic relations M_∞ resp. N_∞. We will show in Proposition 6.80 that they are simulation equivalent, so that the situation is here very similar to that prevailing for logical equivalent modal logics in Section 6.2.3.

PROPOSITION 6.80

Let $\mathcal{M}$ and $\mathcal{N}$ be logical equivalent models. Then the congruences $c_\mathcal{M} := (\zeta_\mathcal{M},\Delta_{\mathbb{R}_+}\times\omega_\mathcal{M})$ and $c_\mathcal{N} := (\zeta_\mathcal{N},\Delta_{\mathbb{R}_|}\times\omega_\mathcal{N})$ are simulation equivalent.

PROOF 1. We know that $\zeta_\mathcal{M}$ and $\zeta_\mathcal{N}$ are in a mutually spawning relationship, so are $\omega_\mathcal{M}$ and $\omega_\mathcal{N}$. Consequently, $\Delta_{\mathbb{R}_+}\times\omega_\mathcal{M}$ and $\Delta_{\mathbb{R}_+}\times\omega_\mathcal{N}$ are related through spawning as well, where

$$\{I\times[\![\psi]\!]_\mathcal{M}\mid I\text{ is a rational interval, }\psi\text{ is a path formula}\}$$

and

$$\{I\times[\![\psi]\!]_\mathcal{N}\mid I\text{ is a rational interval, }\psi\text{ is a path formula}\}$$

arc the generators that relate to each other.

2. Using the map $\Upsilon:S/\zeta_\mathcal{M}\to S'/\zeta_\mathcal{N}$ defined in the proof of Lemma 6.79, we show that

$$M_\infty(s)(I\times[\![\psi]\!]_\mathcal{M})=N_\infty(s')(I\times[\![\psi]\!]_\mathcal{N})$$

for each $s\in S$, $s'\in\Upsilon([s]_{\zeta_\mathcal{M}})$, and for each rational interval I and each path formula ψ. Because $s'\in\Upsilon([s]_{\zeta_\mathcal{M}})$ means $Th_\mathcal{M}(s)=Th_\mathcal{N}(s')$, we obtain for an arbitrary rational number p:

$$\begin{aligned}M_\infty(s)(I\times[\![\psi]\!]_\mathcal{M})\le p&\Leftrightarrow\mathcal{M},s\models\mathcal{P}_{\le p}(\mathcal{X}^I\,\psi)\\&\Leftrightarrow\mathcal{N},s'\models\mathcal{P}_{\le p}(\mathcal{X}^I\,\psi)\\&\Leftrightarrow N_\infty(s')(I\times[\![\psi]\!]_\mathcal{M})\le p,\end{aligned}$$

consequently, both probabilities are identical. This implies that $(\zeta_\mathcal{M},\Delta_{\mathbb{R}_+}\times\omega_\mathcal{M})$ simulates $(\zeta_\mathcal{N},\Delta_{\mathbb{R}_+}\times\omega_\mathcal{N})$. Interchanging the rôles of $\mathcal{M}$ and $\mathcal{N}$ gives the result now. ▯

This yields the properties we are interested in for the associated stochastic relations.

PROPOSITION 6.81

Let $\mathcal{M}$ and $\mathcal{N}$ be logical equivalent models. Then the associated stochastic relations $M_\infty:S\rightsquigarrow(\mathbb{R}_+\times S)^\infty$ and $N_\infty:S'\rightsquigarrow(\mathbb{R}_+\times S')^\infty$ are bisimilar and behavioral equivalent.

PROOF   Since we have identified simulation equivalent congruences on the stochastic relations in question, the assertion follows from Proposition 5.39 for bisimilarity and from Proposition 5.49 for behavioral equivalence.   ▢

6.5.5.2   Tuning the mediator

This result is quite gratifying when being looked at from the point of view of stochastic relations: Given two models for μ**CSL** that are logical equivalent, we can show that the associated *stochastic relations* are bisimilar. It does not help us, however, in this present and preliminary form in finding a *model* that mediates between $\mathcal{M}$ and $\mathcal{N}$ (a similar situation has been encountered already with stochastic Kripke models in Section 6.2.3). An analysis of the construction leading to the mediating relation will again provide information for the construction of a model $\mathcal{L}$ and the desired morphisms $\mathcal{L} \to \mathcal{M}$ and $\mathcal{L} \to \mathcal{N}$. The construction leading to Proposition 6.81 is again based on a semi-pullback construction. This together with the proof of Lemma 4.9 yields a finer description of the mediating stochastic relation.

LEMMA 6.82

Let $\mathcal{M}$ and $\mathcal{N}$ be models for μ**CSL** with $\mathcal{M} \approx \mathcal{N}$. Define

$$A := \{ \langle s, s' \rangle \in S \times S' \mid Th_{\mathcal{M}}(s) = Th_{\mathcal{N}}(s') \},$$
$$B := \{ \langle \langle t, \sigma \rangle, \langle t, \sigma' \rangle \rangle \in (\mathbb{R}_+ \times (S \times \mathbb{R}_+)^\infty) \times (\mathbb{R}_+ \times (S' \times \mathbb{R}_+)^\infty) \mid$$
$$Th_{\mathcal{M}}(\sigma) = Th_{\mathcal{N}}(\sigma') \}.$$

Then A and B are Polish spaces, and there exists a stochastic relation L_0 : $A \rightsquigarrow B$ that mediates between M_∞ and N_∞. The morphisms are composed from the corresponding projections.

We know from Proposition 6.69 that $\omega_{\mathcal{M}} = (\zeta_{\mathcal{M}} \times \Delta_{\mathbb{R}_+})^\infty$ similar for $\zeta_{\mathcal{N}}$ and $\omega_{\mathcal{N}}$. Thus B is essentially the set of all paths over A.

COROLLARY 6.83

Define A and B according to Lemma 6.82. There exists a bijection $\Lambda : B \to (A \times \mathbb{R}_+)^\infty$ that is also a Borel isomorphism.

Define $L' := \mathfrak{P}(\Lambda) \circ L_0$; then this is a stochastic relation $L' : A \rightsquigarrow \mathbb{R}_+ \times (A \times \mathbb{R}_+)^\infty$ that mediates between M_∞ and N_∞. But, still, this is not enough, because we cannot ascertain that L' is actually generated from a model, because we do not know whether or not L' is actually a projective limit of some sorts. Alas, the semi-pullback is a rather flexible construction, and we will show now that we may construct from L' a mediator L_0 with the desired shape, viz., $L_0 = L_\infty$ for some stochastic relation $L : A \rightsquigarrow \mathbb{R}_+ \times A$.

In fact, put for $\langle s, s' \rangle \in A$ and for $E \in \mathcal{B}(\mathbb{R}_+ \times A)$

$$L(s, s')(E) := L'(s, s') \left(E \times \prod_{j>1} (\mathbb{R}_+ \times A) \right).$$

Thus the semi-pullback is restricted to its first component, yielding a stochastic relation $L : A \rightsquigarrow \mathbb{R}_+ \times A$, for which the projective limit can be constructed. This is what we will have a closer look at now.

Define for $n \in \mathbb{N}$ the map $\ell_n : (\mathbb{R}_+ \times A)^n \to (\mathbb{R}_+ \times S)^n$ through

$$\ell_n(t_1, s_1, s'_1, \ldots, t_n, s_n, s'_n) := \langle t_1, s_1, \ldots, t_n, s_n \rangle,$$

the map $r_n : (\mathbb{R}_+ \times A)^n \to (\mathbb{R}_+ \times S')^n$ is defined analogously.

LEMMA 6.84

Define $L_n : A \rightsquigarrow (\mathbb{R}_+ \times A)^n$ inductively from L in the same way as M_n is defined from M in Proposition 6.26, and let π_i be the ith projection. Then the diagram

commutes for every $n \in \mathbb{N}$.

PROOF 1. The proof proceeds by induction on n. For $n = 1$ there is not much to show: By construction, L' mediates between M_∞ and N_∞, and the latter relations are projective limits, so that for $\langle s, s' \rangle \in A$ and $E \in \mathcal{B}(\mathbb{R}_+ \times S)$

$$M_1(s)(E) = M_\infty(s) \left(E \times \prod_{j>1} (\mathbb{R}_+ \times S) \right)$$

$$= L'(s, s') \left(\ell_1^{-1}[E] \times \prod_{j>1} (\mathbb{R}_+ \times A) \right)$$

$$= L_1(s, s')(\ell_1^{-1}[E]).$$

Similarly, the right hand side of the diagram above is shown to commute for $n = 1$.

2. Now assume the assertion is established for n, then we get from the induction hypothesis together with the Change of Variables Formula (Proposition 1.95) for $g : (\mathbb{R}_+ \times S)^n \to \mathbb{R}$ measurable and bounded, and for $\langle s, s' \rangle \in A$ the equality

$$\int_{(\mathbb{R}_+ \times S)^n} g(\mathbf{v}) \, M_n(s)(d\mathbf{v}) = \int_{(\mathbb{R}_+ \times A)^n} (g \circ \ell_n)(\mathbf{w}) \, L_n(s, s')(d\mathbf{w}).$$

This is shown first for $g = \chi_A$ for $A \in \mathcal{B}((\mathbb{R}_+ \times S)^n)$, whence it is equivalent to the induction hypothesis, then it is shown for step functions by the linearity of the integral, subsequently for nonnegative measurable and bounded g by the Monotone Convergence Theorem, and finally for general g by decomposing the map into a positive and a negative part.

3. But now we can perform the induction step: Let $\langle s_0, s_0' \rangle \in A$ and $F \in \mathcal{B}((\mathbb{R}_+ \times S)^{n+1})$ be a Borel set, then

$$M_{n+1}(s_0)(F)$$

$$= \int_{(\mathbb{R}_+ \times S)^n} M(\hbar_S(\mathbf{v}))(\{\langle t, s \rangle \mid \langle \mathbf{v}, t, s \rangle \in F\}) \, M_n(s_0)(d\mathbf{v})$$

$$= \int_{(\mathbb{R}_+ \times A)^n} M(\pi_1(\hbar_A(\mathbf{w})))(\{\langle t, s \rangle \mid \langle \mathbf{w}, t, s, s' \rangle \in \ell_{n+1}^{-1}[F]\}) L_n(s_0, s_0')(d\mathbf{w})$$

$$= \int_{(\mathbb{R}_+ \times A)^n} L(\hbar_A(\mathbf{w}))(\{\langle t, s, s' \rangle \mid \langle \mathbf{w}, t, s, s' \rangle \in \ell_{n+1}^{-1}[F]\}) L_n(s_0, s_0')(d\mathbf{w})$$

$$= L_{n+1}(s_0, s_0')(\ell_{n+1}^{-1}[F]).$$

$\square$

Now extend ℓ_n and r_n to the corresponding infinite products, yielding maps ℓ_∞ resp. r_∞.

PROPOSITION 6.85

Assume that $\mathcal{M}$ and $\mathcal{N}$ are Hennessy-Milner equivalent; construct the Polish space A and stochastic relation $L : A \rightsquigarrow \mathbb{R}_+ \times A$ as above. Then

i. $(\pi_1, \ell_\infty) : L_\infty \to M_\infty$ and $(\pi_2, r_\infty) : L_\infty \to N_\infty$ are morphisms.

ii. M_∞ and N_∞ are bisimilar with L_∞ as a mediator.

PROOF   1. We establish first that both $M_\infty \circ \pi_1 = \mathfrak{P}(\ell_\infty) \circ L_\infty$ and $M_\infty \circ \pi_2 = \mathfrak{P}(r_\infty) \circ L_\infty$ hold, and deal only with the first equality (the second one is established in exactly the same way, *mutatis mutandis*). In order to prove the first equality we have to show that

$$M_\infty(s)(F) = L_\infty(s, s')(\ell_\infty^{-1}[F])$$

holds, whenever $F \in \mathcal{B}((S \times \mathbb{R}_+)^\infty)$ and $\langle s, s' \rangle \in A$. By an argument exactly as in the proof of Proposition 6.66 we may capitalize on the fact that we are dealing with projective limits, permitting us to put the focus on cylinder sets. But by showing that the double diagram above is commutative for each $n \in \mathbb{N}$ we have established the claim for these sets.

2. The σ-algebra of common events is nontrivial. This will be established now. Because the models under consideration are nondegenerate, there exists a state formula ψ with $\emptyset \neq [\![\psi]\!]_\mathcal{M} \neq (S \times \mathbb{R}_+)^\infty$. We know that $\Delta_{\mathbb{R}_+} \times \zeta_\mathcal{M}$ spawns $\Delta_{\mathbb{R}_+} \times \zeta_\mathcal{N}$ via $(\Upsilon, \{I \times [\![\psi']\!]_\mathcal{M} \mid I \text{ rational}, \psi' \text{ is a path formula}\})$ for some suitably chosen map Υ. Thus

$$\ell_\infty^{-1} [I \times [\![\psi]\!]_\mathcal{M}] = r_\infty^{-1} [I \times [\![\psi]\!]_\mathcal{N}],$$

because $[\![\psi]\!]_\mathcal{N} = \Upsilon_{[\![\psi]\!]_\mathcal{M}}$; see the proof of Lemma 6.79. It is also immediate that

$$\emptyset \neq \ell_\infty^{-1} [I \times [\![\varphi]\!]_\mathcal{M}] \neq (A \times \mathbb{R}_+)^\infty.$$

Consequently, the σ-algebra

$$\ell_\infty^{-1} [\mathcal{B}(\mathbb{R}_+ \times (S \times \mathbb{R}_+)^\infty)] \cap r_\infty^{-1} [\mathcal{B}(\mathbb{R}_+ \times (S' \times \mathbb{R}_+)^\infty)]$$

is nontrivial. □

We are nearly ready for the main result, but we did not yet deal with behavioral equivalence. This will happen now.

PROPOSITION 6.86
*Let $\mathcal{M}$ and $\mathcal{N}$ be logical equivalent models for μ**CSL***. *If $S/\zeta_\mathcal{M}$ is a Polish space, then $\mathcal{M}$ and $\mathcal{N}$ are behavioral equivalent.*

PROOF 0. Because $c_\mathcal{M}$ and $c_\mathcal{N}$ are simulation equivalent by Proposition 6.80, we know by Lemma 5.24 that $S/\zeta_\mathcal{M}$ is Polish space iff $S/\zeta_\mathcal{M}$ is Polish space. Proposition 6.69 implies that $(\mathbb{R}_+ \times s)^\infty/(\Delta_{\mathbb{R}_+} \times \zeta_\mathcal{M})$ is Borel isomorphic to $(\mathbb{R}_+ \times S/\zeta_\mathcal{M})^\infty$. We know from Theorem 5.46 that $M_\infty/c_\mathcal{M}$ and $N_\infty/c_\mathcal{N}$ are isomorphic stochastic relations.

1. We deal first with model $\mathcal{M}$. The stochastic relation

$$M_\infty/c_\mathcal{M} : S/\zeta_\mathcal{M} \rightsquigarrow (\mathbb{R}_+ \times S/\zeta_\mathcal{M})^\infty$$

yields a stochastic relation

$$\widetilde{M} : S/\zeta_\mathcal{M} \rightsquigarrow \mathbb{R}_+ \times S/\zeta_\mathcal{M}$$

with $\widetilde{M}_\infty = M_\infty/c_\mathcal{M}$ by the Markov property Lemma 6.27. Consequently,

$$\Phi_\mathcal{M} := (\eta_{\zeta_\mathcal{M}}, \Delta_{\mathbb{R}_+} \times \eta_{\zeta_\mathcal{M}}) : M \to \widetilde{M}$$

is a morphism for the stochastic relations. From Lemma 5.7 it is inferred that $\eta_{\zeta_\mathcal{M}}[\Sigma(S)]$ is a Borel set in $S/\zeta_\mathcal{M}$ with

$$\eta_{\zeta_\mathcal{M}}^{-1}[\eta_{\zeta_\mathcal{M}}[\Sigma(S)]] = \Sigma(S)$$

for each state variable S. Define for $S \in \mathsf{SV}, P \in \mathsf{PV}$ and for $a \in \mathsf{AP}$

$$\Sigma_\mathcal{M}^\sharp(S) := \eta_{\zeta_\mathcal{M}}[\Sigma(S)], \Pi_\mathcal{M}^\sharp((P)) := \eta_{w_\mathcal{M}}[\Sigma(S)], V_\mathcal{M}^\sharp((a)) := \eta_{\zeta_\mathcal{M}}[V(a)],$$

then

$$\Phi_\mathcal{M} : \mathcal{M} \to \mathcal{M}/\zeta_\mathcal{M} := (S/\zeta_\mathcal{M}, \widetilde{M}, (\Sigma_\mathcal{M}^\sharp, \Pi_\mathcal{M}^\sharp), V_\mathcal{M}^\sharp)$$

yields a morphism between these models.

2. In a similar way we construct together with a model a morphism

$$\Phi_\mathcal{N} : \mathcal{N} \to \mathcal{N}/\zeta_\mathcal{N}.$$

Since $\mathcal{M}_\infty/c_\mathcal{M}$ and $\mathcal{N}_\infty/c_\mathcal{N}$ are isomorphic, so are $\mathcal{M}/\zeta_\mathcal{M}$ and $\mathcal{N}/\zeta_\mathcal{N}$. This establishes the claim.   ▯

This, now, is the main result:

THEOREM 6.87

Let $\mathcal{M}$ and $\mathcal{N}$ be nontrivial models for μ**CSL**. Consider these statements:

a. $\mathcal{M}$ and $\mathcal{N}$ are behavioral equivalent.

b. $\mathcal{M}$ and $\mathcal{N}$ are logical equivalent.

c. $\mathcal{M}$ and $\mathcal{N}$ are bisimilar.

Then $a \Rightarrow b \Leftrightarrow c$, and if $S/\zeta_\mathcal{M}$ is a Polish space, then all three statements are equivalent.

PROOF   1. The implications $c \Rightarrow b$ and $a \Rightarrow b$ both follow from Proposition 6.77. If $S/\zeta_\mathcal{M}$ is Polish, then $b \Rightarrow a$ follows from Proposition 6.86, so that we have to take care of $b \Rightarrow c$.

2. Construct the Polish space A and the stochastic relation $L : A \rightsquigarrow \mathbb{R}_+ \times (A \times \mathbb{R}_+)^\infty$ together with the maps ℓ_∞ and r_∞ as in Proposition 6.85. Assume that the interpretation $\mathcal{J}$ for model $\mathcal{N}$ is $\mathcal{J} = (\Sigma', \Pi')$, and define

$$\mathcal{L} := (L, A, (\Sigma^*, \Pi^*), V^*)$$

with

1. $V^* := (V(a) \times W(a)) \cap A$ for the atomic propositions $a \in \mathsf{AP}$,

2. $\Sigma^*(Z) := (\Sigma(Z) \times \Sigma'(Z)) \cap A$ for the state variable $Z \in \mathsf{SV}$,

3. $\Pi^*(P) := \{\rho \in (A \times \mathbb{R}_+)^\infty \mid \ell_\infty(\rho) \in \Pi(P), r_\infty(\rho) \in \Pi'(P)\}$ for the path variable $P \in \mathsf{PV}$.

Then both $\ell_\infty : \mathcal{L} \to \mathcal{M}$ and $r_\infty : \mathcal{L} \to \mathcal{N}$ are morphisms.   ▯

Remark. Investigating probabilistic interpretations of the logics **CSL** and μ**CSL** with tools coming from stochastic relations displays this pattern: a congruence relation is defined through the satisfaction relation. In the case of logical equivalence this yields a pair of congruences that are bisimulation equivalent. From this both a span and a cospan of morphisms for the underlying stochastic relations are constructed, which are then manipulated into models for the corresponding logics. This proposes a more general treatment.

The context provided by coalgebraic logic may be of interest as well. Coalgebraic logic (Moss, 1999; Cîrstea and Pattinson, 2004) investigates behavioral properties of models in terms of coalgebras and predicate liftings. Assume that (S, γ) is a coalgebra for a functor in the category of sets with maps as morphisms. The general idea is that a modal formula $\langle\lambda\rangle\phi$ is valid for a state $s \in S$ iff the set $[\![\varphi]\!]$ of states for which formula ϕ is valid is transformed through predicate lifting λ into the set $\lambda_S([\![\varphi]\!])$ which contains $\gamma(s)$ as a member, so that $[\![\langle\lambda\rangle\phi]\!] = \{s \in s \mid \gamma(s) \in \lambda_S([\![\varphi]\!])\}$, or, equivalently, $[\![\langle\lambda\rangle\phi]\!] = \gamma^{-1} \circ \lambda_S([\![\varphi]\!])$. Here predicate liftings play the rôle of modal operators, a predicate lifting being a natural transformation for the contravariant power set functor and the functor governing the coalgebra. It can be shown that the usual semantic operations can be formulated in terms of suitably chosen predicate liftings. The coalgebraic approach permits a clearer view of the semantic mechanisms underlying the logic: it becomes clear which properties are attributed to the coalgebra, and which are due to the modal structure which in turn is modeled through predicate liftings.

These ideas may be translated into the realm of stochastic coalgebras for an investigation of bisimilarity and behavioral equivalence of stochastic coalgebras, using a logics for this that is based essentially on predicate liftings. The collection of predicate liftings requires then a certain selectivity similar to the separation properties proposed by Pattinson (Pattinson, 2004). It can be shown under that bisimilarity, behavioral and logical equivalence are equivalent under separation conditions, provided the underlying functor is compatible with the congruences involved; see (Doberkat, 2006b). It might be observed that in set based coalgebras the functor on is assumed to preserve at least weak pullbacks, and it is known that the probability functor proper does not have this property by Corollary 4.15.

6.6 Bibliographic Notes

Hennessy and Milner introduced in their 1980 paper (Hennessy and Milner, 1980) a very simple and negation free modal logic and related bisimilarity of image-finite Kripke models to the equivalence relation "accepting the same formulas" on states. Subsequently, the seminal paper by Larsen and

Skou (Larsen and Skou, 1991) introduced stochastic Kripke models, albeit over discrete state spaces, and established a Hennessy-Milner like Theorem for simple modal logics, among others a variant of the Hennessy-Milner logic (in which the diamond operator $\Diamond\varphi$ is replaced by a family of diamond operators $\Diamond_q\varphi$ with $0 \leq q \leq 1$). Changing the stage from discrete to analytic state spaces, Desharnais, Edalat and Panangaden investigated the problem of bisimilarity again. The research reported in (Desharnais et al., 2002) takes an analytic state space with universally measurable transition functions as a basic scenario. A Hennessy-Milner theorem is proved; the proof's idea is to produce a co-span of morphisms through injections into a suitably factored sum. This idea has left its traces in various parts of the present exposition. But the situation considered here is structurally subtly different: universal measurability, as assumed in (Desharnais et al., 2002), requires a somewhat elaborate completion process using all finite measures on that space. This is mainly due to the fact that the existence of semi-pullbacks could only be established under these circumstances. After the existence of semi-pullbacks could be established also for relations on the Borel sets of an analytic space (see Chapter 4 and (Doberkat, 2003)), the question of bisimulations became tractable also for the more general and natural case of analytic spaces with their Borel structure. These spaces are structurally much simpler and do not need additional considerations, since they are given through the morphisms of measurable spaces and nothing else (so one could work with them even if one would want to do without the real numbers).

In (Doberkat, 2003) a generalization of (Desharnais et al., 2002) is established for those labeled Markov transition systems which work over a Polish (rather than an analytic) state space and which have a certain smallness property. This technical condition is lifted in the present exposition. This is so since the technique of factoring stochastic relations is better understood now. Apart from a much wider class of modal logics which can be dealt with now (as witnessed in Section 3.5), the present discussion proposes a more general technical approach.

The logic **CSL** (Baier et al., 2003) is a stochastic version and variant of the popular logic **CTL** for model checking (Clarke et al., 1999). The logic has considerable expressive power, as is demonstrated convincingly in (Baier et al., 2003). Recently, Desharnais and Panangaden (Desharnais and Panangaden, 2003) have proposed an interpretation of a subset of **CSL** over a continuous domain, hereby providing a general framework for the treatment of bisimulations. The originally given interpretation in (Baier et al., 2003) is based on a finite state space in order to investigate the computational side of model checking using **CSL**. A comparison with (Desharnais and Panangaden, 2003) suggests that the wide and well-assorted toolkit provided by probabilities over analytic spaces is a welcome addition for investigating the properties of this logic. This is particularly true when it comes to investigating bisimulations.

Appendix A

Notations

A.1 Categories

$\mathbb{1}_{\mathfrak{C}}$	Identity functor on category $\mathfrak{C}$	p. 63
$\mathfrak{C}(a,b)$	Hom-set of a and b in category $\mathfrak{C}$	p. 64
$\mathfrak{C}(a,-), \mathfrak{C}(-,b)$	Co- and contravariant hom-set functors in category $\mathfrak{C}$	p. 64
$\mathfrak{C}^{\mathrm{op}}$	Category dual (or opposite) to category $\mathfrak{C}$	p. 63
$\mathfrak{Set}$	All sets with maps	p. 62
$\mathfrak{Meas}$	Measurable spaces with measurable maps	p. 63
$\mathfrak{cPol}$	Polish spaces with continuous maps	p. 66
$\mathfrak{BPol}$	Polish spaces with Borel maps	p. 66
$\mathfrak{Anl}$	Analytic spaces with Borel maps	p. 95
$\mathfrak{Stoch}$	Stochastic relations over measurable spaces	p. 95
$\mathfrak{PolStoch}$	Stochastic relations over Polish spaces	p. 95
$\mathfrak{anStoch}$	Stochastic relations over analytic spaces	p. 95
$\mathfrak{StrConv}$	Positive convex structures with continuous affine maps	p. 144
$\mathfrak{GPart}$	G-partitions with partition respecting continuous maps	p. 138
$\mathfrak{Alg}$	Algebras for the Giry monad with algebra morphisms	p. 138
$\mathfrak{pAlg}$	Subcategory of $\mathfrak{Alg}$ for probabilistic objects	p. 147
$\mathfrak{GTrip}$	G-triplets with G-triplet morphisms	p. 141
$\mathfrak{Prob}$	Measurable spaces with probability measures	p. 162
$\mathfrak{PolProb}$	Full subcategory of $\mathfrak{Prob}$ based on Polish spaces	p. 176
$\mathfrak{p.Kripke}$	Stochastic Kripke models	p. 265
$\overset{\bullet}{\to}$	Natural transformation between functors	p. 70
$\rightsquigarrow$	Kleisli morphism	p. 79
$\mathfrak{e}, \mathfrak{m}$	Unit and multiplication of a monad	p. 75

A.2 Spaces

$\mathcal{F}(N,\mathcal{N})$	$\mathcal{N}$-$\mathcal{B}(\mathbb{R})$-measurable and bounded functions $N \to \mathbb{R}$	p. 6
$\bigotimes_{i \in I}(X_i, \mathcal{A}_i)$	Product of the measurable spaces $(X_i, \mathcal{A}_i)_{i \in I}$	p. 5
$\bigoplus_{i \in I}(X_i, \mathcal{A}_i)$	Coproduct of the measurable spaces $(X_i, \mathcal{A}_i)_{i \in I}$	p. 6
$\prod_{i \in I}(X_i, \mathcal{T}_i)$	Product of the topological spaces $(X_i, \mathcal{T}_i)_{i \in I}$	p. 8
$\coprod_{i \in I}(X_i, \mathcal{T}_i)$	Coproduct of the topological spaces $(X_i, \mathcal{T}_i)_{i \in I}$	p. 8
$\mathfrak{S}(N,\mathcal{N})$	Subprobability measures on the measurable space $(N,\mathcal{N})$	p. 36
$\mathfrak{P}(N,\mathcal{N})$	Probability measures on the measurable space $(N,\mathcal{N})$	p. 36
$\mathcal{C}(X)$	All bounded continuous functions $X \to \mathbb{R}$	p. 41

A.3 Other

$\mathbb{N}$	Natural numbers $1, 2, 3, \ldots$	
$\mathbb{Q}$	Rational numbers	
$\mathbb{R}$	Real numbers	
$\mathbb{R}_+$	Nonnegative real numbers	
$\sigma(\mathcal{M})$	Smallest σ-algebra containing $\mathcal{M}$	p. 2
χ_A	Indicator function for set A	p. 6
$\mathcal{M} \cap A$	Trace of σ-algebra $\mathcal{M}$ on set A	p. 4
$[\cdot]_\rho$	Equivalence class for equivalence relation ρ	p. 5
η_ρ	Factor map for equivalence relation ρ	p. 5
X/ρ	Factor space for equivalence relation ρ	p. 5
$B_r(x), B_{r,d}(x)$	Ball with center x and radius r for metric d	p. 9
$\mathcal{B}(X)$	Borel sets of X, X is a topological or an analytic space	p. 8
$\mathrm{diam}(A)$	Diameter of set A	p. 14
$d(x, A)$	Distance of point x to set A	p. 10
X^∞	All infinite sequences over set X	p. 11
$\mathrm{graph}(f)$	Graph of map f	p. 22
$\ker(f)$	Kernel of map f	p. 28
$\mathrm{cl}(\cdot)$	Topological closure	p. 14
$\exists F(C), \forall F(C)$	Weak and strong inverse of set-valued map F	p. 32
$\mathcal{A}^\bullet$	Weak-*-σ-algebra on $\mathfrak{S}(X, \mathcal{A})$	p. 50
$\mathbf{d}_P$	Prohorov metric on $\mathfrak{S}(X)$ for the metric space X	p. 44

δ_a	Dirac measure on the point a	p. 43
$\rightharpoonup_w$	Weak convergence for probability measures	p. 41
D_x, D^y	Horizontal and vertical cuts of $D \subseteq X \times Y$	p. 13
$\bigotimes_{n \in \mathbb{N}} \mu_n$	Product of the measures $(\mu_n)_{n \in \mathbb{N}}$	pp. 38, 58
$\delta_{i,j}$	Kronecker's δ	p. 144
Ω	Set of positive convex coefficients	p. 133
Ω_c	Set of convex coefficients	p. 142
$\ell(R)$	Equivalence relation generated from R	p. 186
$\otimes [X, \alpha]$	Trace of the α-invariant Borel sets on α	p. 187
$\times_{n \in \mathbb{N}} \rho_n$	Infinite product of equivalence relations	p. 189
$\mathcal{INV}(\mathcal{B}(X), \rho)$	ρ-invariant Borel sets of an analytic space X	p. 190
$\rho + \sigma$	Sum of the equivalence relations ρ and σ	p. 189
$\rho \diamond \sigma$	Spawning sum for equivalence relations ρ and σ	p. 196
$\tau \bullet \rho$	Collects smooth equivalence relations τ and ρ	p. 150
Δ_X, U_X	Identity relation resp. universal relation on X	pp. 27, 182
$\mathsf{ker}(\mathsf{f})$	Kernel of morphism f	p. 199
$K_{\alpha,\beta}, \mathsf{K/c}$	Factor relation	p. 200
$\mathsf{c} \bullet \mathsf{d}$	Collects congruences c and d	p. 202
$(\alpha, \beta) \preceq (\alpha', \beta')$	Refinement of congruences	p. 204
$\mathsf{c} \propto \mathsf{c}'$	Simulation of congruences	p. 208
$\mathsf{K} \oplus \mathsf{K}'$	Direct sum of stochastic relations K and K'	p. 208
$K^{\bullet}$	Kleisli extension to K	p. 231
$\mu \otimes K$	Product of measure μ and relation K	p. 231
$K_{\mu}^{\smile}$	Converse of stochastic relation w.r.t. initial probability μ	p. 234
$\mathfrak{Mod}_b(\tau, P)$	Basic modal language	p. 254
$\mathfrak{Mod}_1(\tau, P)$	Negation free basic modal language	p. 254
$\mathfrak{Mod}_s(\tau, P)$	Extended modal language	p. 254
$\models$	Satisfaction relation	pp. 254, 255
$\rho(\triangle)$	Arity of modal operator $\triangle$	p. 254
$Th_{\mathcal{R}}(s)$	Theory for state s with Kripke model $\mathcal{R}$	p. 254
$\mathcal{K} \circledS \mathcal{R}$	$\mathcal{K}$ refines $\mathcal{R}$	p. 259
$\mathsf{supp}(\mu)$	Support of probability measure μ	p. 260
$\mathcal{K} \sim \mathcal{K}'$	Logical equivalence of Kripke models $\mathcal{K}$ and $\mathcal{K}'$	p. 264
$\mathfrak{L}_P$	All state formulas in **CSL**	p. 283
$\mathcal{S}_{\bowtie p}(\varphi)$	Steady-state operator in **CSL**, μ**CSL**	pp. 283, 299
$\mathcal{P}$	Path quantifier in **CSL**, μ**CSL**	pp. 283, 299
$\mathcal{X}^I \varphi$	Next operator in **CSL**, μ**CSL**	pp. 283, 299
$\varphi_1 \mathcal{U}^I \varphi_2$	Until operator in **CSL**, μ**CSL**	pp. 283, 299
$\sigma @ t$	State occupied by σ at time t	p. 283
$\mu P.\psi$	Fixed point of ψ in μ**CSL**	p. 299
$Th_{\mathcal{M}}(\sigma)$	Theory for path σ in μ**CSL**	p. 304
$\mathcal{M}[Z \backslash Q]$	Model with substitution	p. 300
$\mathcal{M} \approx \mathcal{M}'$	Logical equivalence of μ**CSL**-models $\mathcal{M}$ and $\mathcal{M}'$	p. 316

Bibliography

Abowd, G., Allen, R., and Garlan, D. (1993). Using style to understand descriptions of software architecture. In Notkin, D., editor, *SIGSOFT'93*, volume 18 of *Software Engineering Notes*, pages 9 – 20. ACM. ⟨p. 125, 126, 129⟩

Abramsky, S., Blute, R., and Panangaden, P. (1999). Nuclear and trace ideal in tensored *-categories. *J. Pure Appl. Algebra*, 143(1 – 3):3 – 47. ⟨p. 248, 249⟩

Arbab, F. and Rutten, J. J. M. M. (2002). A coinductive calculus of component connectors. Technical Report SEN-R0216, CWI, Amsterdam. ⟨p. xx, 129⟩

Arveson, W. (1976). *An Invitation to C*-Algebra*. Graduate Texts in Mathematics. Springer-Verlag, New York. ⟨p. xviii, 81, 248⟩

Aumann, G. (1952). *Relle Funktionen*. Number 68 in Grundlehren der mathematischen Wissenschaften. Springer-Verlag, Berlin, Heidelberg, New York. ⟨p. 222⟩

Baier, C., Haverkort, B., Hermanns, H., and Katoen, J.-P. (2003). Model-checking algorithms for continuous time Markov chains. *IEEE Trans. Softw. Eng.*, 29(6):524 – 541. ⟨p. 326⟩

Barbosa, L. M. (2001). *Components as Coalgebras*. PhD thesis, Universidade do Minho. ⟨p. 126, 129⟩

Barr, M. and Wells, C. (1985). *Toposes, Triples and Theories*. Number 278 in Grundlehren der mathematischen Wissenschaften. Springer-Verlag, New York, Berlin, Heidelberg, Tokyo. ⟨p. 82⟩

Barr, M. and Wells, C. (1999). *Category Theory for Computing Science*. Les Publications CRM, Montreal. ⟨p. 82, 86⟩

Billingsley, P. (1968). *Convergence of Probability Measures*. John Wiley & Sons, New York, 1st edition. ⟨p. 81⟩

Billingsley, P. (1995). *Probability and Measure*. John Wiley & Sons, New York. ⟨p. 81⟩

Billingsley, P. (1999). *Convergence of Probability Measures*. John Wiley & Sons, New York, 2nd edition. ⟨p. 82⟩

Blackburn, P., de Rjike, M., and Venema, Y. (2001). *Modal Logic*. Number 53 in Cambridge Tracts in Theoretical Computer Science. Cambridge University Press, Cambridge, UK. ⟨p. xi, 161, 254, 256, 316⟩

Borceux, F. (1994a). *Handbook of Categorical Algebra 1: Basic Category Theory*, volume 50 of *Encyclopedia of Mathematics and its Applications*. Cambridge University Press, Cambridge, UK. ⟨p. 82⟩

Borceux, F. (1994b). *Handbook of Categorical Algebra 2: Categories and Structures*, volume 51 of *Encyclopedia of Mathematics and its Applications*. Cambridge University Press, Cambridge, UK. ⟨p. 82⟩

Bourbaki, N. (1989). *General Topology, Chapters 5–10*. Elements of Mathematics. Springer-Verlag, Berlin, Heidelberg, New York. ⟨p. 11⟩

Broy, M. and Stølen, K. (2001). *Specification and Development of Interactive Systems*. Monographs in Computer Science. Springer-Verlag, New York. ⟨p. 127, 129⟩

Bruni, R., Lanese, I., and Montanari, U. (2005). Complete axioms for stateless connectors. In Fiadeiro, J. L. and Rutten, J., editors, *Proc. Algebra and Coalgebra in Computer Science*, number 3629 in Lect. Notes Comp. Sci., pages 98 – 113. Springer-Verlag. ⟨p. 129⟩

Bryans, J., Bowman, H., and Derrick, J. (2003). Model checking stochastic automata. *ACM Trans. Computational Logic*, 4(4):452 – 493. ⟨p. 127⟩

Carboni, A., Lack, S., and Walters, R. (1993). Introduction to extensive and distributive categories. *Journal of Pure and Applied Algebra*, 84:145 – 158. ⟨p. 127, 129⟩

Castaing, C. (1967). Sur les multi-applications mesurables. *Rev. Française d'Informatique et de Recherche Operationelle*, 1:91 – 126. ⟨p. 81⟩

Castaing, C. and Valadier, M. (1977). *Convex Analysis and Measurable Multifunctions*. Number 580 in Lect. Notes Math. Springer-Verlag, Berlin, Heidelberg, New York. ⟨p. 81⟩

Cîrstea, C. (2004). On logics for coalgebraic simulation. *Elctr. Notes in Theor. Comp. Sci.*, 106:63 – 90. ⟨p. 249⟩

Cîrstea, C. and Pattinson, D. (2004). Modular construction on modal logics. In Gardner, P. and Yoshida, N., editors, *Proc. CONCUR'04*, number 3170 in Lect. Notes Comp. Sci., pages 258 – 275. ⟨p. 325⟩

Clarke, E. M., Grumberg, O., and Peled, D. A. (1999). *Model Checking*. The MIT Press, Cambridge, MA. ⟨p. 326⟩

D'Argenio, P. (1999). *Algebras and Automata for Timed and Stochastic Systems*. PhD thesis, University of Twente. ⟨p. 127⟩

Desharnais, J., Edalat, A., and Panangaden, P. (2002). Bisimulation of labelled Markov-processes. *Information and Computation*, 179(2):163 – 193. ⟨p. 128, 178, 266, 270, 326⟩

Desharnais, J., Jagadeesan, R., Gupta, V., and Panangaden, P. (2000). Approximating labeled Markov processes. In *Proc. 15th Ann. IEEE Symp. on Logic in Computer Science*, pages 95 – 106, June. IEEE Computer Society. ⟨p. 128, 248⟩

Desharnais, J. and Panangaden, P. (2003). Continuous stochastic logic characterizes bisimulation of continuous-time Markov processes. *J. Log. Alg. Programming*, 56(1-2):99 – 115. ⟨p. 286, 326⟩

Doberkat, E.-E. (1981). Inserting an new element into a heap. *BIT*, 21(3):255 – 269. ⟨p. 245, 246⟩

Doberkat, E.-E. (2003). Semi-pullbacks and bisimulations in categories of stochastic relations. In *Proc. ICALP'03*, volume 2719 of *Lect. Notes Comp. Sci.*, pages 996 – 1007, Berlin. Springer-Verlag. ⟨p. 178, 270, 326⟩

Doberkat, E.-E. (2004). The converse of a probabilistic relation. *J. Logic and Algebraic Progr.*, 62(1):133 – 154. ⟨p. 238⟩

Doberkat, E.-E. (2006a). Hyperfinite approximations to labeled Markov transition systems. In Johnson, M. and Vene, V., editors, *Proc. AMAST 2006*, volume 4019 of *Lect. Notes Comp. Sci.*, pages 127 – 141, Berlin. Springer-Verlag. ⟨p. 128⟩

Doberkat, E.-E. (2006b). Stochastic coalgebraic logic: Bisimilarity and behavioral equivalence. Technical report, Chair for Software-Technology. ⟨p. 325⟩

Doberkat, E.-E., Engels, G., Hausmann, J. H., Lohmann, M., Pleumann, J., and Schröder, J. (2005). Software Engineering and eLearning: The MuSofT project. Technical report, Chair for Software Technology, University of Dortmund. ⟨p. 129⟩

Doberkat, E.-E., Schmidt, F., and Veltmann, C. (2000). Re-engineering the German integrated system for measuring and assessing environmental radioactivity. *Environmental Modelling & Software*, 15:267 – 278. ⟨p. 96⟩

Edalat, A. (1999). Semi-pullbacks and bisimulation in categories of Markov processes. *Math. Struct. in Comp. Science*, 9(5):523 – 543. ⟨p. 158, 170, 173, 177⟩

Elstrodt, J. (1999). *Maß- und Integrationstheorie*. Springer-Verlag, Berlin-Heidelberg-New York, 2 edition. ⟨p. xiv⟩

Fedorchuk, V. V. (1991). Probability measures in topology. *Russian Math. Surveys*, pages 45 – 93. (Uspekhi Mat. Nauk 46:1 (1991), 41 – 80). ⟨p. 155⟩

Fiadeiro, J. L. (2005). *Categories for Software Engineering*. Springer-Verlag, Berlin, Heidelberg. ⟨p. 82⟩

Fiadeiro, J. L. and Maibaum, T. (1996). A mathematical toolbox for the software architect. In Kramer, J. and Wolf, A., editors, *Proc. 8th Intl. Workshop on Software Specification and Design*. IEEE Computer Society Press. ⟨p. 97, 126, 129⟩

Fremlin, D. H. (2003). *Measure Theory — vol. 4: Topological Measure Spaces*. Torres Fremlin, Colchester. ⟨p. 151, 152⟩

Freyd, P. and Sčedrov, A. (1990). *Categories, Allegories*. North Holland. ⟨p. 249⟩

Fuchssteiner, B. and Lusky, W. (1981). *Convex Cones*, volume 56 of *North-Holland Mathematical Studies*. North-Holland Publishing Company, Amsterdam. ⟨p. 152⟩

Giry, M. (1981). A categorical approach to probability theory. In *Categorical Aspects of Topology and Analysis*, number 915 in Lect. Notes Math., pages 68 – 85, Berlin. Springer-Verlag. ⟨p. 89, 128⟩

Graham, R. E., Knuth, D. E., and Patashnik, O. (1989). *Concrete Mathematics: A Foundation for Computer Science*. Addison-Wesley, Reading, MA. ⟨p. 243⟩

Gumm, H. P. and Schröder, T. (2005). Types and coalgebraic structure. *Algebra univers.*, 53:229 – 252. ⟨p. 249⟩

Halmos, P. R. (1950). *Measure Theory*. Van Nostrand Reinhold, New York. ⟨p. 6, 36, 42, 53, 57, 67⟩

Heckmann, R. (1994). Probabilistic domains. In Tison, S., editor, *Proc. 19th Int. Colloquium on Trees in Algebra and Programming*, number 787 in Lect. Notes Comp. Sci., pages 142 – 156, Berlin. Springer-Verlag. ⟨p. 155⟩

Hennessy, M. and Milner, R. (1980). On observing nondeterminism and concurrency. In *Proc. ICALP'80*, number 85 in Lect. Notes Comp. Sci., pages 395 – 409, Berlin. Springer-Verlag. ⟨p. 248, 325⟩

Hewitt, E. and Stromberg, K. R. (1965). *Real and Abstract Analysis*. Springer-Verlag, Berlin, Heidelberg, New York. ⟨p. 66⟩

Himmelberg, C. J. (1975). Measurable relations. *Fund. Math.*, 87:53 – 72. ⟨p. 81⟩

Himmelberg, C. J. and van Vleck, F. S. (1974). Multifunctions on abstract measurable spaces and applications to stochastic decision theory. *Ann. Mat. Pura Appl.*, 101:229–236. ⟨p. 81⟩

Jacobs, K. (1978). *Measure and Integral.* Academic Press, New York. ⟨p. 164⟩

Joyal, A., Nielsen, M., and Winskel, G. (1996). Bisimulation from open maps. *Information and Computation*, 127(2):164 – 185. ⟨p. 248⟩

Kechris, A. S. (1994). *Classical Descriptive Set Theory.* Graduate Texts in Mathematics. Springer-Verlag, Berlin, Heidelberg, New York. ⟨p. xviii, 61, 68, 81, 82, 248⟩

Keisler, H. J. (1988). Infinitesimals in probability theory. In Cutland, N., editor, *Nonstandard analysis and its applications*, number 10 in London Mathematical Society Student Texts, pages 106 – 139. Cambridge University Press, Cambridge, UK. ⟨p. 93⟩

Kellerer, H. G. (Sommersemester 1972). Topologische Maßtheorie (in German). Lecture notes, Mathematisches Institut, Ruhr-Universität Bochum. ⟨p. 11, 82⟩

Knuth, D. E. (1973a). *The Art of Computer Programming. Vol. I, Fundamental Algorithms.* Addison-Wesley, Reading, Mass., 2 edition. ⟨p. 243⟩

Knuth, D. E. (1973b). *The Art of Computer Programming. Vol. III, Sorting and Searching.* Addison-Wesley, Reading, Mass. ⟨p. 244–246⟩

Kuratowski, K. (1966). *Topology*, volume I. PWN - Polish Scientific Publishers and Academic Press, Warsaw and New York. ⟨p. 16, 81⟩

Kurz, A. (2000). *Logics for Coalgebras and Applications to Computer Science.* PhD thesis, Institut für Informatik, Ludwigs-Maximilian-Universität München. ⟨p. 248⟩

Lajios, G. (2006). *Zur kategoriellen Beschreibung von Schichtenarchitekturen.* PhD thesis, Chair for Software Technology, University of Dortmund. ⟨p. 127, 129, 177⟩

Lang, S. (1965). *Algebra.* Addison-Wesley, Reading, Mass. ⟨p. 204, 223, 249⟩

Larsen, K. G. and Skou, A. (1991). Bisimulation through probabilistic testing. *Information and Computation*, 94:1 – 28. ⟨p. 128, 256, 326⟩

Lindstrøm, T. (1988). An invitation to nonstandard analysis. In Cutland, N., editor, *Nonstandard analysis and its applications*, number 10 in London Mathematical Society Student Texts, pages 1 – 105. Cambridge University Press, Cambridge, UK. ⟨p. 93, 261⟩

Loève, M. (1962). *Probability Theory.* Van Nostrand Company, Princeton, NJ, third edition. Later editions as Graduate Text in Mathematics, Springer. ⟨p. 81⟩

MacLane, S. (1997). *Categories for the Working Mathematician.* Graduate Texts in Mathematics. Springer-Verlag, Berlin. ⟨p. 74, 81, 82, 100, 128, 129, 132⟩

Medvidovics, N., Rosenblum, D. S., Redmiles, D. F., and Robbins, J. E. (2002). Modeling software architectures in the Unified Modeling Language. *ACM Trans. Softw. Eng. Method.*, 11(1):2 – 57. ⟨p. 129⟩

Moggi, E. (1989). An abstract view of programming languages. Lecture Notes, Stanford University. ⟨p. 129⟩

Moggi, E. (1991). Notions of computation and monads. *Information and Computation*, 93:55 – 92. ⟨p. 129⟩

Morgan, C., McIver, A., and Seidel, K. (1996). Probabilistic predicate transformers. *ACM Trans. Prog. Lang. Syst.*, 18(3):325 – 353. ⟨p. 94, 155⟩

Moss, L. S. (1999). Coalgebraic logic. *Annals of Pure and Applied Logic*, 96:277 – 317. ⟨p. 325⟩

Moss, L. S. and Viglizzo, I. D. (2004). Harsanyi type spaces and final coalgebras constructed from satisfied theories. *Electr. Notes Theor. Comp. Sci.*, pages 279 – 295. ⟨p. 249⟩

Panangaden, P. (1997). Stochastic techniques in concurrency. Technical report, BRICS, Dept. of Comp. Sci., Univ. of Aarhus. ⟨p. 128⟩

Panangaden, P. (1998). Probabilistic relations. In Baier, C., Huth, M., Kwiatkowska, M., and Ryan, M., editors, *Proc. PROBMIV*, pages 59 – 74. ⟨p. 128⟩

Parthasarathy, K. R. (1967). *Probability Measures on Metric Spaces.* Academic Press, New York. ⟨p. 48, 81, 82, 238⟩

Pattinson, D. (2004). Expressive logics for coalgebras via terminal sequence induction. *Notre Dame J. Formal Logic*, 45(1):19 – 33. ⟨p. 248, 325⟩

Pleumann, J. (2004). Erfahrungen mit dem multimedialen didaktischen Modellierungswerkzeug DAVE. In Engels, G. and Seehusen, S., editors, *Proc. 2nd German e-Learning Conference for Computer Science*, number 52 in LNI, pages 55 – 66. Gesellschaft für Informatik, Springer-Verlag. ⟨p. 129⟩

Pumplün, D. (2003). Positively convex modules and ordered normed linear spaces. *J. Convex Analysis*, 10(1):109 – 127. ⟨p. 143, 155⟩

Rutten, J. J. M. M. (2000). Universal coalgebra: a theory of systems. *Theoretical Computer Science*, 249(1):3 – 80. Special issue on modern algebra and its applications. ⟨p. xi, xviii, xx, 94, 177, 204, 222, 229, 248, 249⟩

Rutten, J. J. M. M. (2002). Bisimulation in enumerative combinatorics. *ENTCS*, 65(1):1 – 19. ⟨p. xx, 242⟩

Rutten, J. J. M. M. (2003). Behavioral differential equations: a coinductive calculus of streams, automata and power series. *Theor. Comp. Sci.*, 308:1 – 53. ⟨p. 242⟩

Schäl, M. (1974). A selection theorem for optimization problems. *Arch. Math.*, 25(219 – 224). ⟨p. 128⟩

Semadeni, Z. (1973). Monads and their Eilenberg-Moore algebras in functional analysis. *Queen's Papers Pure Appl. Math.*, 33. ⟨p. 128⟩

Shaw, M. (2001). The coming-of-age of software architecture research. In *Proc. 23rd International Conference on Software Engineering*, pages 656 – 664. ⟨p. 125⟩

Shaw, M., DeLine, R., Klein, D. V., Ross, T. L., Young, D. M., and Zelesnik, G. (1995). Abstractions for software architecture and tools to support them. *IEEE Trans. Softw. Eng.*, 21(4):314 – 335. ⟨p. 125⟩

Shaw, M. and Garlan, D. (1995). Formulations and formalisms in software architecture. In van Leeuwen, J., editor, *Computer Science Today: Recent Trends and Developments*, number 1000 in Lect. Notes Comp. Sci. Springer-Verlag. ⟨p. 125, 129⟩

Shaw, M. and Garlan, D. (1996). *Software Architecture — Perspectives on an Emerging Discipline*. Prentice-Hall. ⟨p. 96, 125, 129⟩

Shiryaev, A. N. (1996). *Probability*, volume 95 of *Graduate Texts in Mathematics*. Springer-Verlag, Berlin, Heidelberg, New York, second edition. ⟨p. 81⟩

Sokolova, A. (2005). *Coalgebraic Analysis of Probabilistic Systems*. PhD thesis, Department of Computer Science, University of Eindhoven. ⟨p. 128⟩

Spivey, J. M. (1989). *The Z Notation — A Reference Manual*. Prentice-Hall. ⟨p. 125⟩

Srivastava, S. M. (1998). *A Course on Borel Sets*. Graduate Texts in Mathematics. Springer-Verlag, Berlin. ⟨p. xviii, 24, 68, 81, 167, 178, 205, 222, 248⟩

Swirszcz, T. (1974). Monadic functors and convexity. *Bull. Acad. Polon. Sci. Ser. Sci. Math. Astronom. Phys.*, 22:39 – 42. ⟨p. 155⟩

Taylor, P. (1999). *Practical Foundations of Mathematics*, volume 59 of *Cambridge Studies in Advanced Mathematics*. Cambridge University Press, Cambridge. ⟨p. 127⟩

van Breugel, F., Mislove, M., Ouaknine, J., and Worrell, J. (2005). Domain theory, testing and simulation for labelled Markov processes. *Theoret. Comp. Sci.*, 333:171 – 197. ⟨p. 248⟩

van Breugel, F., Shalit, S., and Worrell, J. (2002). Testing labelled Markov processes. In *Proc. ICALP'2002*, number 2380 in Lect. Notes Comp. Sci., pages 537–548, Berlin. Springer-Verlag. ⟨p. 155⟩

Viglizzo, I. D. (2005). Final sequences and final coalgebras for measurable spaces. In Fiadeiro, J. L. and Rutten, J., editors, *Proc. Algebra and Coalgebra in Computer Science*, number 3629 in Lect. Notes Comp. Sci., pages 395 – 407. Springer-Verlag. ⟨p. 249⟩

Wadler, P. (1992). Comprehending monads. *Math. Struct. Comp. Sci.*, pages 461 – 493. ⟨p. 82⟩

Wagner, D. H. (1977). A survey of measurable selection theorems. *SIAM J. Control Optim.*, 15(5):859 – 903. ⟨p. 81⟩

Wagon, S. (1981). Circle-squaring in the twentieth century. *Math. Intell.*, 3(4):176 – 181. ⟨p. xiv⟩

Wermelinger, M. and Fiadeiro, J. L. (1998). Connectors for mobile programs. *IEEE Trans. Softw. Eng.*, 24(5):331 – 341. ⟨p. 97, 121, 126, 129⟩

Williams, J. W. J. (1964). Algorithm 232: Heapsort. *Comm. ACM*, 7:347 – 348. ⟨p. 245⟩

Index